# METHODS IN CELL BIOLOGY

VOLUME 29

*Fluorescence Microscopy of Living Cells in Culture*
*Part A. Fluorescent Analogs, Labeling Cells,*
*and Basic Microscopy*

*Series Editor*

## LESLIE WILSON

*Department of Biological Sciences*
*University of California, Santa Barbara*
*Santa Barbara, California*

# METHODS IN CELL BIOLOGY

# BIOLOGY

*Prepared under the Auspices of the American Society for Cell Biology*

## VOLUME 29

*Fluorescence Microscopy of Living Cells in Culture*
*Part A. Fluorescent Analogs, Labeling Cells,*
*and Basic Microscopy*

*Edited by*

### YU-LI WANG

CELL BIOLOGY GROUP
WORCESTER FOUNDATION FOR EXPERIMENTAL BIOLOGY
SHREWSBURY, MASSACHUSETTS

### D. LANSING TAYLOR

DEPARTMENT OF BIOLOGICAL SCIENCES
CENTER FOR FLUORESCENCE RESEARCH IN BIOMEDICAL SCIENCES
CARNEGIE-MELLON UNIVERSITY
PITTSBURGH, PENNSYLVANIA

ACADEMIC PRESS, INC.
**Harcourt Brace Jovanovich, Publishers**
Boston  San Diego  New York  Berkeley
London  Sydney  Tokyo  Toronto

ACADEMIC PRESS, INC.
San Diego, California 92101

*United Kingdom Edition published by*
ACADEMIC PRESS LIMITED
24-28 Oval Road, London NW1 7DX

LIBRARY OF CONGRESS CATALOG CARD NUMBER: 64-14220

ISBN   0-12-564129-X   (alk. paper)

PRINTED IN THE UNITED STATES OF AMERICA
89  90  91       9  8  7  6  5  4  3  2

# CONTENTS

8. *Fluorescent Labeling of Endoplasmic Reticulum*
   Mark Terasaki

9. *Fluorescent Labeling of Endocytic Compartments*
   Joel Swanson

10. *Incorporation of Macromolecules into Living Cells*
    Paul L. McNeil

11. *Hapten-Mediated Immunocytochemistry: The Use of Fluorescent and Nonfluorescent Haptens for the Study of Cytoskeletal Dynamics in Living Cells*
    Gary J. Gorbsky and Gary G. Borisy

CONTENTS

# CONTRIBUTORS

*Numbers in parentheses indicate the pages on which the authors' contributions begin.*

D. A. AGARD, Department of Biochemistry & Biophysics and Howard Hughes Medical Institute, San Francisco, California 94143 (291)

R. S. AIKENS, Photometrics Ltd., Tucson, Arizona 85745 (291)

KIMON J. ANGELIDES, Department of Physiology and Molecular Biophysics and Program in Neuroscience, Baylor College of Medicine, Houston, Texas 77030 (29)

GARY G. BORISY, Laboratory of Molecular Biology, University of Wisconsin, Madison, Wisconsin 53706 (175)

LAN BO CHEN, Dana-Farber Cancer Institute, Harvard Medical School, Boston, Massachusetts 02115 (103)

MICHAEL EDIDIN, Department of Biology, The Johns Hopkins University, Baltimore, Maryland 21218 (87)

GARY J. GORBSKY, Laboratory of Molecular Biology, University of Wisconsin, Madison, Wisconsin 53706 (175)

KARL KUTZ, Georgia Instruments, Inc., Atlanta, Georgia 30341 (239)

R. JOEL LOWY, Laboratory of Kidney and Electrolyte Metabolism, National Heart, Lung, and Blood Institute, National Institutes of Health, Bethesda, Maryland 20892 (269)

KATHERINE LUBY-PHELPS, Department of Chemistry, Center for Fluorescence Research in Biomedical Sciences, Carnegie-Mellon University, Pittsburgh, Pennsylvania 15213 (59)

FREDERICK R. MAXFIELD, Departments of Pathology and Physiology and Cellular Biophysics, College of Physicians & Surgeons, Columbia University, New York, New York 10032 (13)

NANCY M. MCKENNA, Worcester Foundation for Experimental Biology, Shrewsbury, Massachusetts 01545 (195)

PAUL L. MCNEIL, Department of Anatomy and Cellular Biology, Harvard Medical School, Boston, Massachusetts 02115 (153)

RICHARD E. PAGANO, Department of Embryology, Carnegie Institution of Washington, Baltimore, Maryland 21210 (75)

E. D. SALMON, Department of Biology, University of North Carolina, Chapel Hill, Chapel Hill, North Carolina 27514 (207)

J. W. SEDAT, Department of Biochemistry & Biophysics and Howard Hughes Medical Institute, San Francisco, California 94143 (291)

KENNETH R. SPRING, Laboratory of Kidney and Electrolyte Metabolism, National Heart, Lung, and Blood Institute, National Institutes of Health, Bethesda, Maryland 20892 (269)

JOEL SWANSON, Department of Anatomy and Cellular Biology, Harvard Medical School, Boston, Massachusetts 02115 (137)

D. LANSING TAYLOR, Department of Biological Sciences, Center for Fluorescence Research in Biomedical Sciences, Carnegie-Mellon University, Pittsburgh, Pennsylvania 15213 (207)

MARK TERASAKI, Laboratory of Neurobiology, IRP, N.I.N.C.D.S., National Institutes of Health at the Marine Biological Laboratory, Woods Hole, Massachusetts 02543 (125)

JOHN E. WAMPLER, Department of Biochemistry, University of Georgia, Athens, Georgia 30602 (239)

YU-LI WANG, Cell Biology Group, Worcester Foundation for Experimental Biology, Shrewsbury, Massachusetts 01545 (1, 195)

# PREFACE

Fluorescence techniques are uniquely suitable for probing living cells because of their sensitivity and specificity. Since fluorescence from a single cell can be detected with a microscope both as an image and as a photometric signal, fluorescence microscopy has great potential for qualitative and quantitative studies on the structure and function of cells. However, due to previous technical limitations, fluorescence has been used primarily for staining fixed cells for many years. It has not been until recently that the true power of the techniques has evolved for use with single living cells. The most important advances that have made this possible include the development of (1) probes for specific structures or environmental parameters; (2) methods for delivering fluorescent probes into living cells; (3) methods for detecting weak fluorescence signals from living cells; and (4) methods for acquiring, processing, and analyzing fluorescence signals with microscopes.

The primary purpose of this and the accompanying volume of *Methods in Cell Biology* is to provide readers with detailed descriptions of methods in these four areas. While techniques for fluorescence spectroscopy in solution are described in various sources, there has been no convenient source for the methods specifically applied to living cells. Even with an extensive literature search, one often finds crucial technical details, including instrumentation, sample handling, and precautions, left out in many research articles. It is our hope that these volumes will provide enough detail to make the new developments approachable by most investigators. Although some biological perspectives are provided in many chapters, the main emphasis of the volumes is practical laboratory methods; the job of biological reasoning and experimental design is left to individual investigators. The books are thus targeted primarily at experienced cell biologists who wish to apply modern fluorescence techniques. However, they should also be of great interest to biochemists and molecular biologists who attempt to correlate results in test tubes with activities in living cells. In addition, many chapters should be valuable to those specializing in instrumentation, including microscopy, electronic imaging, and digital image processing.

The two volumes represent a collective effort of many investigators. The chapters were assembled by specific areas which, in our view, were important or held great promise in the future. We then invited those researchers with extensive experience in the particular area to make contributions. There was a certain degree of subjectiveness in choosing the topics. On the one hand, we have included topics crucial to, but not specific for, fluorescence microscopy of living cells, including microscope cell culture, microinjection, microscope photometry, and low light level imaging. On the other hand, we decided to sacrifice several useful topics that were either not in a mature stage of development or where we were unable to obtain a commitment from an authority.

The first volume (Volume 29) deals with the preparation, delivery, and detection of fluorescent probes. The first half is focused on the preparation of specific structural probes, including fluorescent analogs that can be utilized by living cells in structural assembly, fluorescent molecules that bind to specific cellular components, and probes that can be used to label particular cellular compartments. There are special challenges in the preparation of each class of probes, including proteins, small peptides, heterocyclic compounds, lipids, and polysaccharides. Subsequent chapters discuss factors that determine the destination of probes and methods for delivering probes to specific sites in living cells. The second half of the first volume discusses the detection of fluorescent probes in living cells, including issues related to sample physiology (microscope cell culture), optics (basic fluorescence microscopy), and signal detection (electronic photometry and imaging, immunoelectron microscopic detection of fluorophores). The last few chapters introduce modern techniques in image detection and provide a continuity to quantitative, analytical methods covered in Volume 30.

The second volume (Volume 30) explores a combination of the theoretical and technical issues related to the quantitation of fluorescence signals in the living cell with a light microscope. The first section explores the engineering principles required in the characterization of the performance of an imaging system. The use of system validation procedures and quantitative fluorescent standards are explored in detail. The remainder of Volume 30 is devoted to specific applications and optical methods. A mix of theoretical and practical issues is discussed, including the measurement of membrane potential, ionic concentrations, tracer diffusion coefficients, total internal reflection, fluorescence polarization, and three-dimensional reconstruction. Thus, the two-volume set defines a technical continuum from organic chemistry, through biochemistry, cell biology, physics, and engineering, to computer science. The present status of the field reflects the occurrence of a revolution in cell biological research.

We would like to thank all contributing authors for providing us with their extensive experience in various areas. Most of them have worked closely with us in planning their chapters and minimizing overlaps, then submitting excellent manuscripts in a timely fashion and answering questions which arose during editing.

YU-LI WANG
D. LANSING TAYLOR

# Chapter 1

# Fluorescent Analog Cytochemistry: Tracing Functional Protein Components in Living Cells

## YU-LI WANG

*Cell Biology Group*
*Worcester Foundation for Experimental Biology*
*Shrewsbury, Massachusetts 01545*

## I. Introduction

The central task in modern cell biology is understanding how molecules interact in a cell to perform various functions. Unfortunately, few methods allow us to study directly the function of specific components inside living cells.

There is no reason why many techniques employed by biochemists to study molecules in test tubes or cuvettes cannot be adapted to the living cell. As long as the method has a high enough sensitivity and yields specific signals within a complicated environment, it should be equally useful within living cells. Fluorescence techniques readily satisfy these criteria (see,

1

for example, Taylor *et al.*, 1986; Weber, 1986). Fluorescent probes, covalently linked to specific molecules, have been used extensively in biochemistry to report molecular activities and interactions. By incorporating the fluorescently labeled molecules into a living cell, one should gain equally valuable information about the behavior of specific molecules inside living cells. This rationale led to the approach of fluorescent analog cytochemistry (Taylor and Wang, 1978, 1980).

Fluorescent analog cytochemistry involves preparation of fluorescently labeled cellular components (fluorescent analogs), followed by the introduction of the analogs into living cells. When properly prepared, the conjugates can maintain all or most of the original properties and functions. In addition, after entering the cell, many conjugates have been observed to associate with normal physiological structures.

So far, fluorescent analog cytochemistry has been applied primarily to cytoskeletal and surface components. In the simplest application, one can study the ability of the fluorescent analog to incorporate into native cellular structures. One example is the comparison of the incorporation of analogs prepared from different actin isoforms into stress fibers and myofibrils (McKenna *et al.*, 1985). By looking at very early time points after delivery, the site of incorporation can also be identified (Amato *et al.*, 1986). The study can also be combined with immunoelectron microscopy, using antibodies which recognize specifically microinjected components (Amato *et al.*, 1986; Gorbsky and Borisy, Chapter 11, this volume), to reveal the distribution of the analogs at a high resolution.

A second major application is to study the distribution of cellular components during certain processes, such as mitosis, or following treatment with agents which induce specific changes. For example, we have analyzed the distribution of $\alpha$-actinin in developing muscle cells to determine how myofibrils reach their high degree of organization (McKenna *et al.*, 1986). Finally, fluorescent analog cytochemistry has been combined with fluorescence photobleaching techniques to study the mobility of molecules associated with the cell surface and with the cytoskeletal structures (e.g., Wang *et al.*, 1982c; Wang, 1985). Both bulk translocation and random movement of molecules may be analyzed.

This chapter will focus on the practical methods involved in the application of fluorescent analog cytochemistry to intracellular protein components (see related chapters in this volume for other classes of fluorescent analogs). Readers are referred to several previous articles for general discussions of fluorescent analog cytochemistry (Taylor and Wang, 1980; Wang *et al.*, 1982a, b; Kreis and Birchmeier, 1982; Taylor *et al.*, 1984, 1986; Jockusch *et al.*, 1985; Simon and Taylor, 1986).

## II. Equipping the Laboratory for Fluorescent Analog Cytochemistry

The first step of fluorescent analog cytochemistry involves purification and fluorescent labeling of cellular components. Requirements for the methods involved, such as gel electrophoresis and column chromatography, are similar to those for a typical biochemistry laboratory. However, because of the sensitivity of many fluorophores to light, fluorescent labeling should be performed under reduced light; for example, in a room with dimmer control of the overhead illumination.

The cell culture facility is critical. Since cells will shuttle between the $CO_2$ incubator and the microscope during an experiment, the facility should be located close to the laboratory for fluorescence microscopy, especially if temperature-sensitive mutants are involved. Stocks of cells are maintained regularly. However, it is important to ensure that cells are in a highly healthy state. Poorly maintained cells often spread incompletely on the substrate, respond inconsistently to experimental manipulations, and sometimes emit a high level of autofluorescence. If necessary, culture conditions, such as the medium used or the level of glutamine, should be adjusted. We also routinely use antibiotics, such as penicillin and streptomycin, to prevent contamination during microinjection.

Special vessels are often required for plating cells during the fluorescence observation. Consideration should be given to the requirements for microinjection, cell culture, and fluorescence microscopy. For example, if cells will be injected directly with a needle, it is important to ensure that they are easily accessible for the needle and can be observed clearly at a relatively high (e.g., $\times 400$) magnification. Plastic Petri dishes, although quite accessible on an inverted microscope and ideal for cell culture, offer a very poor quality for both transmitted light and fluorescence optics. To maintain living mammalian cells on the microscope, it is also necessary to ensure proper controls of temperature, humidity, and pH. For extended observations, special devices, such as microscope stage incubators and/or perfusion chambers, should be used (McKenna and Wang, Chapter 12, this volume).

At least one high-quality fluorescence microscope is required (Taylor and Salmon, Chapter 13, this volume). We have chosen an inverted microscope for the convenience of microinjection and for the ease of the construction of the microscope cell culture system. For needle microinjection, it is important to ensure that the microscope has a sturdy base and is located on a table free of vibration. If necessary, 10-in. inner tube tires may be inserted under the legs of the table as an effective, economical means for the isolation of vibration. Additional equipment for direct microinjection are listed in the article by McNeil (Chapter 10, this volume). Fluorescence microscopy should be performed in a dark laboratory, to allow observations with dark-adapted eyes and low light level detectors. Air cleaning devices, such as HEPA filters, are also important for maintaining both the performance of the optical equipment and the sterility of the cell culture during microinjection.

In most laboratories, fluorescence microscopes are equipped with mercury arc lamps as the light source. However, unless ultraviolet light is required for excitation, a quartz–halogen lamp is a better and less expensive alternative. For observing living cells, the intensity of excitation light should be attenuated, either by adjusting the input voltage for the quartz–halogen lamp (see, however, Taylor and Salmon, Chapter 13, this volume, for precautions), or by inserting a diaphragm or a set of polarizers in the excitation light path (Wampler and Kutz, Chapter 14, this volume).

The signal level from injected cells is limited both by the very small number of fluorophores in the cell and by the use of very low levels of excitation light in order to minimize radiation damage to living cells (Spring and Lowy, Chapter 15, this volume). In order to collect multiple images from single microinjected cells, low light level detectors are required (Spring and Lowy, Chapter 15, this volume; Aikens *et al.*, Chapter 16, this volume). We have used the Intensified Silicon Intensified Target (ISIT) camera for our experiments. The less-sensitive Silicon Intensified Target (SIT) camera does not offer enough sensitivity for this purpose. Unfortunately, detectors for very low light levels, such as the ISIT camera, are usually quite noisy, and digital image processing is required to obtain acceptable images (Inoue, 1986).

Most investigators using image processing have the difficult experience of choosing an adequate system. For fluorescent analog cytochemistry, the two most crucial functions are frame averaging and background subtraction. They will be used repeatedly during an experiment and should be performed at video speed. A useful feature is the ability to store images on a hard disk. Because of the large number of values for storage, it requires not only a large capacity but also a high speed for the storage and retrieval of data (storing images on video tapes, although more convenient, generally results in a significant loss of resolution. In our experience, except for preliminary studies of short processes or for analyzing sequences stored in a computer, a VCR is not as useful as it might appear). The system should also contain several frame buffers to allow rapid comparison of images and to allow arithmetic manipulations of images. In addition, since images are often very low in contrast, the ability to stretch the contrast is useful. Pseudocolor, on the other hand, has been of little use to us so far. Finally, one should also consider the ability to perform quantitative analysis on fluorescence images. Even simple measurements of intensity and dimension can often provide invaluable insights. Issues related to quantitative measurement, such as corrections of shading and nonlinearity, are discussed by Aikens *et al.* (Chapter 16, this volume; see also Volume 30, this series).

Even if a computer disk storage is available, photographic equipment, such as a 35-mm camera body, macrolens, and tripod, is required for obtaining high-quality hard copies. It may also be useful to acquire a Polaroid camera for the CRT screen (such as model DS-34) or a thermoprinter for video images (such as Mitsubishi P70-U), to obtain rapid study prints. However, the quality of prints from these devices are not high enough for publication, and electronic or computer contrast enhancement may be required to obtain usable images with the thermoprinter.

## III. Preparation of Fluorescent Analogs

Not all proteins are suitable for the application of fluorescent analog cytochemistry. First, considerations should be given to the abundance of the protein inside cells (Wang *et al.*, 1982b). For minor components, it may be impossible to obtain detectable signals without inducing serious disruptions to the cell. Second, the application is limited to proteins which are soluble under conditions compatible with living cells. The presence of detergents, denaturing agents, and extremes of pH and osmolarity will result in cellular damage during delivery. Third, the destination of the analog should be considered. While proteins can be delivered readily into the cytoplasm and nucleus (McNeil, Chapter 10, this volume), incorporation of integral membrane proteins or intraorganelle components may be more difficult or impossible. Fourth, some criteria are required to determine whether the analogs have been utilized properly by the cell. This may be very difficult for proteins not involved in macromolecular assemblies. Finally, analogs in the order of milligrams are often required for each preparation to attain proper final concentration or volume for loading. Therefore it would be more efficient to choose proteins which can be isolated with high yields and stored conveniently.

Following purification of the protein under study, chemical reactions are used to couple a fluorescent probe to the protein. The optimal probe to use and the labeling condition vary with each protein, and must be determined empirically. In general, fluorophores with relatively long wavelengths of excitation and emission and high quantum yields are preferable (Simon and Taylor, 1986). However, other factors also need to be considered. For

example, phycobiliproteins, although ideal in terms of fluorescence properties (Oi *et al.*, 1982), suffer from their very large sizes, which may affect both
the efficiency of conjugation reaction and the properties of conjugates. In
addition, some fluorophores (notably, the rhodamine family) tend to associate noncovalently with proteins. In some cases such noncovalently associated probes are very difficult to remove, and a different probe has to be
used.

Consideration should also be given to the chemical reaction for the
conjugation (Simon and Taylor, 1986). Usually each fluorophore will be
available in several different reactive forms, e.g., isothiocyanate and maleimide. The optimal one to use is often difficult to predict and must be
determined empirically. Different reactions may have different effects on
the functional state of the conjugates, even if the reagents share similar
reactive properties. However, the reaction condition should always be mild
enough to preserve the native properties of the protein, and the bond should
be stable enough to ensure association after introduction into cells. It
should be noted that many amine-directed dyes (e.g., isothiocyanates) also
react readily with the sulfhydryl group via a relatively labile bond, which
may easily break in the cytoplasm. Such conjugates should be treated with
dithiothreitol to facilitate removal of reversibly associated probes.

Although many proteins (e.g., α-actinin) can be labeled directly, some
may require the protection of active sites. One approach is to label a
complex of the protein with ligands or accessory proteins, followed by
dissociation and purification of the analog(s). A variation of this approach is
to label partially purified preparations while the protein is still associated
with native ligands. Besides protecting active sites, these methods may yield
several useful analogs simultaneously. Alternatively, proteins may be labeled without protection and subsequently selected for functional conjugates. For proteins that form macromolecular assemblies, such as actin and
tubulin, several cycles of polymerization – depolymerization will provide
assembly-competent molecules. For other proteins, affinity chromatography with proper ligands may be applied. Such selections will ensure that the
conjugates maintain qualitatively the ability to bind, but quantitative characteristics, such as the rate of association, may still be altered.

The most important task after the labeling reaction is to remove unconjugated fluorophores (see also Chapter 2 by Maxfield and Chapter 6 by
Edidin, this volume). In our experience, dialysis is rarely adequate for this
purpose. Even if the dialysate appears nonfluorescent, the protein may still
be heavily contaminated with noncovalently associated fluorophores. The
most common method of separation, gel filtration in a column of Sephadex
G-25 or G-50, can also be seriously misleading. For example, even after an
apparent "clean" separation of the void volume (containing the conjugate)

from the included volume (containing the free dye), the conjugate may still contain a significant amount of noncovalently associated fluorophore. Conversely, commercial tetramethylrhodamine dyes often contain small aggregates which move in the void volume. An alternative, often more effective method for removing unconjugated fluorophores, is the Bio-Bead SM2 (Bio-Rad Laboratories, Richmond, California). These beads remove free fluorophores based on hydrophobic interactions and can be used with fluorescein or tetramethylrhodamine, but not lissamine rhodamine B. Ion-exchange chromatography may also be applied to separate adsorbed probes. In some cases, it has the additional advantage of fractionating conjugates according to the degree of labeling (Dandliker and Portmann, 1971). Finally, it should be noted that the effective method of separation varies not only with fluorophores, but also with reactive groups and proteins. The optimal method should be determined during the preparation of each new conjugate.

## IV. Assays of Fluorescent Analogs and Preparations for Microinjection

Serious effort should be made to ensure the absence of noncovalently associated fluorophores. For example, after electrophoresis in a SDS–polyacrylamide gel, noncovalently associated fluorophores will dissociate from the polypeptide and usually move in front of the Bromophenol blue tracking dye. Alternatively, a G-25 desalting column equilibrated in 1–2% SDS will also yield reliable information. Free fluorophores will appear in the included volume. After all fluorophores are proved to be covalently associated, measurements should be performed to determine the molar ratio of labeling (Wang et al., 1982b; Simon and Taylor, 1986). The value, although usually representing no more than an estimate, serves as a useful reference for adjusting the reaction condition and for ensuring consistency among experiments.

As emphasized in previous articles (Wang et al., 1982b; Taylor et al., 1984), functional assays should be performed to determine the effect of labeling on the protein. Changes or partial loss of activities is not uncommon and should not necessarily prevent one from using a particular fluorescent conjugate, as long as the effect is taken into account in the interpretation. Spectroscopic characterizations should also be performed if the experiment involves quantitative fluorometry of analogs in different structures or environments.

Sometimes extracted cell models are used to test the binding of the analog to residual structures (e.g., Sanger et al., 1984). However, while the information is valuable, the interpretation as a functional assay may be complicated by the possible disruption or creation of binding sites during the extraction.

For direct microinjection, it is usually much more desirable to introduce a small volume of concentrated solution ($\geq 1$ mg/ml), rather than a large volume of dilute solution. The total volume of the stock required for each experiment is very low (in the order of microliters). We regularly use the Amicon Centricon and Centriprep filters for concentrating protein analogs. However, some proteins (e.g., tropomyosin) appear to stick to the membrane. These proteins may be concentrated by vacuum dialysis through a Colloidin bag (Schleicher & Schuell, Keene, New Hampshire). Besides concentration, colloidin bags are also ideal for microdialysis, which is required to bring the analog into the solution for microinjection (Wang *et al.*, 1982b), and to remove any toxic materials which might be introduced during the concentration process. Finally, in order to perform direct microinjection, the solution should be clarified thoroughly. We have used a Beckman-type 42.2Ti rotor in a regular ultracentrifuge (20-minute spin at 25,000 rpm) and an Airfuge (23 psi, 20 minutes) for this purpose. Centrifugation in a microfuge or 0.22-$\mu$m filtration usually yields only marginal results. Clarified solutions should be handled very carefully: introduction of air bubbles readily induces aggregates which greatly aggravate the microinjection process.

## V.   Delivery of the Conjugates and Handling of Living Cells

Although the most common method for delivering the analog into living cells is needle microinjection, there is no reason to reject other, perhaps more convenient, methods (McNeil, Chapter 10, this volume). Some of these methods may actually be better tolerated by cells which are exceptionally sensitive to pressured flow from a microneedle (e.g., *Xenopus* myotomal muscle cells; N. M. McKenna, unpublished observations). However, bulk loading methods generally require a much larger amount of fluorescent analogs, and may not be compatible with some proteins. For example, both actin and myosin polymerize in balanced salt solutions used for bulk loading. Many bulk methods also have a serious limitation regarding the size of molecules to be delivered (McNeil, Chapter 10, this volume).

Fine adjustments are required for microinjecting different proteins. The amount of analogs a living cell can incorporate varies. For example, there is a clear limit for the incorporation of vinculin into adhesion plaques. Microinjecting a large amount of vinculin results in an increase in diffuse cytoplasmic fluorescence and obscures adhesion plaques. The shape of the needle should also be tailored according to the property of the solution. In general, a more rapid taper and larger tip opening should be used for more viscous solutions. For actin and myosin, it is also helpful to maintain a continuous flow between cells, otherwise the influx of salt in the medium will cause the protein to polymerize at the tip of the needle.

It is often necessary to identify a particular microinjected cell. This can be achieved by marking the surface of the cover slip with a diamond pen before microinjection. The position of a particular cell may then be noted relative to the scratch mark. Alternatively, a diamond marking device, which screws into the nosepiece of the microscope, can be purchased from Leitz, Inc. (Rockleigh, New Jersey). It makes neat, circular marks of different diameters around the center of field. A less expensive version of this approach uses rubber stamps. However, the ink dissolves in microscope immersion oil and is incompatible with oil-immersion objectives.

After microinjection, the medium should be replaced to remove any fluorescent analog located outside the cell. The dish is then returned to the incubator for 1–3 hours before observation. This allows both the recovery of cells from any damage inflicted by microinjection, and the incorporation of analogs into structures. Time required for the latter varies from 10 to 15 minutes for actin and from 2 to 3 hours for myosin.

Microinjected cells should first be examined for possible disruptive effects and intracellular distribution of the analog. Some fluorescent analogs may have toxic effects. For example, we found fluorescent caldesmon to cause rounding and detachment of fibroblasts (S. K. Stickel and Y.-L. Wang, unpublished observations). The analog may also form artifactual structures, and immunofluorescence should be performed to compare the distribution of the analog with that of the endogenous counterpart, although immunofluorescence itself may yield artifacts. Some proteins also exhibit variable behavior in different cells. For example, fluorescently labeled muscle actin incorporates consistently into native structures in chick embryonic fibroblasts and IMR33 gerbil fibroma cells, but forms paracrystalline structures in NRK cells and PtK cells (Wehland and Weber, 1980). In addition, different fluorophores also appear to affect the extent of actin incorporation into stress fibers (McKenna *et al.*, 1985).

## VI. Data Collection and Interpretation

If the purpose of the experiment is to study incorporation, then simple photography may be adequate. This often requires several minutes of exposure and results in photobleaching and cellular damage. In order to perform prolonged observations, low light level detectors should be employed (Spring and Lowy, Chapter 15, this volume; Aikens *et al.*, Chapter 16, this volume). However, even with a highly sensitive detector, cells should not be exposed to continuous illumination. Structural damage (e.g., breakdown of actin filament bundles) in cells injected with rhodamine-labeled actin may be noticed, for example, after 20 minutes of continuous recording with a video tape recorder. Efforts should also be made to minimize the time of visual observation of fluorescence. We usually use visual observation only to search for areas of microinjection, then switch immediately to a low light level video camera. The image detector should be used at the maximal acceptable sensitivity, while the excitation light should be at the minimal acceptable level. These adjustments are dependent on the purpose of the experiment (e.g., qualitative imaging or quantitative photometry) and on the availability of digital image-processing systems. Under ideal conditions, a single cell may be recorded off and on for at least 3 days.

If a digital image processor is unavailable, images may be recorded with a 35-mm camera and a motor winder mounted in front of the TV monitor. A simple electronic controller can be assembled to trigger the shutter for excitation light and the camera at the same time. This reduces unnecessary exposure of the cell to the excitation light. By adjusting the exposure time between 1/4 and 2 seconds, images may also be "integrated" on the film to improve the signal-to-noise ratio.

Ideally, images should be acquired with a digital image-processing system. In my laboratory, we first record an image of the "dark count" with no light entering the low light level video camera. Then the illumination shutter is activated briefly, and 64–128 frames of cellular fluorescence images are averaged. The image of dark count is then subtracted from the cellular image and the result stored in a disk file. This process requires typically 5–20 seconds, depending on the performance of the image-processing system. Although this method may not be suitable for very fast processes, it preserves the data with little distortion.

Many experiments require detailed comparison of fine differences among images. For images stored on computer disks, this may be done theoretically by loading one picture after another into the viewing frame buffer. However, the speed for loading is usually not high enough to allow perception of a continuous motion. If the number of images is small, the comparison may be done by loading different images into multiple frame buffers. A simple program can be written to continuously switch different frame buffers for display, and the differences will be much easier to perceive. If a large number of images are involved, such as in time-lapse studies, a time-lapse video recorder can be used. The images are retrieved one by one from the computer disk while recording at a low speed. Playing back at a high speed will then turn those images into motion pictures.

Most image-processing systems will also perform at least some quantitative analyses. This issue is covered in detail in the companion volume (Volume 30, this series). Moreover, fluorescence photobleaching recovery may be used, using equipment identical to that for membrane-bound probes (Wolf, Volume 30, this series), to study the mobility of incorporated analogs.

An important last step is obtaining hard copies of the images for presentation. High-quality prints should be prepared with a photographic camera aligned properly in front of the monitor screen. It should be noted that the large variation in the quality of published video images is due largely to the adjustment of monitors and the photography, not the quality of the monitors. The most important, often overlooked, adjustment is the vertical hold of the monitor. The method for adjusting vertical hold, as well as a general description of the photography of CRT screens, may be found in the book by Inoue (1986).

## VII. Prospectus

The discussion in this chapter represents our experience in using currently available equipment to trace protein molecules in living cells. The application so far has been limited primarily to cytoskeletal components, and the potential to study other classes of protein components as well as large complexes remains to be investigated. Similarly, little has been done to take advantage of various sophisticated analytical methods, as discussed in Volume 30 in this series. The basic application of fluorescent analog cytochemistry is also expected to undergo continuous changes with the advancement of related technologies. For example, the development of new detectors (Spring and Lowy, Chapter 15, this volume; Aikens *et al.*, Chapter 16, this volume), the application of confocal scanning microscopy (Brakenhoff *et al.*, Volume 30, this series), and the improvement of video recording equipment will all affect future applications of this approach. Advances in these areas should further enhance the power of fluorescent analog cytochemistry and facilitate an understanding of how different protein molecules work together in living cells.

### ACKNOWLEDGMENTS

The author wishes to thank Dr. Nancy McKenna for reading the manuscript. The work was supported by grants from National Institutes of Health (GM-32476), National Science Foundation (DCB-8796359), and the Muscular Dystrophy Association.

## References

Amato, P. A., and Taylor, D. L. (1986). *J. Cell Biol.* **102**, 1074–1084.

Dandliker, W. B., and Portmann, A. J. (1971). *In* "Excited States of Proteins and Nucleic Acids" (R. F. Steiner and I. Weinryb, eds.), pp. 199–275. Plenum, New York.

Inoue, S. (1986). "Video Microscopy." Plenum, New York.

Jockusch, B. M., Fuchtbauer, A., Wiegand, C., and Honer, B. (1985). *In* "Cell and Molecular Biology of the Cytoskeleton" (J. W. Shay, ed.), Vol. 7, pp. 1–40. Plenum, New York.

Kreis, T., and Birchmeier, W. (1982). *Int. Rev. Cytol.* **75**, 209–228.

McKenna, N. M., Meigs, J. B., and Wang, Y.-L. (1985). *J. Cell Biol.* **100**, 292–296.

McKenna, N. M., Johnson, C. S., and Wang, Y.-L. (1986). *J. Cell Biol.* **103**, 2163–2171.

Oi, V. T., Glazer, A. N., and Stryer, L. (1982). *J. Cell Biol.* **93**, 981–986.

Sanger, J. W., Mittal, B., and Sanger, J. M. (1984). *J. Cell Biol.* **99**, 918–928.

Simon, J. R., and Taylor, D. L. (1986). *In* "Methods in Enzymology" (R. B. Vallee, ed.), Vol. 134, 487–507. Academic Press, Orlando, Florida.

Taylor, D. L., and Wang, Y.-L. (1978). *Proc. Natl. Acad. Sci. U.S.A.* **75**, 857–861.

Taylor, D. L., and Wang, Y.-L. (1980). *Nature (London)* **284**, 405–410.

Taylor, D. L., Amato, P. A., Luby-Phelps, K., and McNeil, P. L. (1984). *Trends Biochem. Sci.* **9**, 88–91.

Taylor, D. L., Amato, P. A., McNeil, P. L., Luby-Phelps, K., and Tanasugarn, T. (1986). *In* "Applications of Fluorescence in the Biomedical Sciences" (D. L. Taylor, A. S. Waggoner, R. F. Murphy, F. Lanni, and R. R. Birge, eds.), pp. 347–376. Liss, New York.

Wang, Y.-L. (1985). *J. Cell Biol.* **101**, 597–602.

Wang, K., Feramisco, J., and Ash, J. (1982a). *In* "Methods in Enzymology" (D. W. Frederiksen and L. W. Cunningham, eds.), Vol. 85, pp. 514–562. Academic Press, New York.

Wang, Y.-L., Heiple, J. M., and Taylor, D. L. (1982b). *Methods Cell Biol.* **25** (Pt. B), 1–11.

Wang, Y.-L., Lanni, F., McNeil, P., Ware, B., and Taylor, D. L. (1982c). *Proc. Natl. Acad. Sci. U.S.A.* **79**, 4660–4664.

Weber, G. (1986). *In* "Applications of Fluorescence in the Biomedical Sciences" (D. L. Taylor, A. S. Waggoner, R. F. Murphy, F. Lanni, and R. R. Birge, eds.), pp. 601–615. Liss, New York.

Wehland, J., and Weber, K. (1980). *Exp. Cell Res.* **127**, 397–408.

# Chapter 2

## *Fluorescent Analogs of Peptides and Hormones*

### FREDERICK R. MAXFIELD

*Departments of Pathology and Physiology and Cellular Biophysics*
*College of Physicians & Surgeons*
*Columbia University*
*New York, New York 10032*

## I.  Introduction

Fluorescent analogs of peptides and hormones have proved to be invaluable tools for studying the interaction of these molecules with living cells. Several fluorescence methods can provide information about the behavior of hormones and other peptide ligands in living cells which would be very difficult to obtain by alternative techniques. Many of the specific methods are described in detail in other chapters of these volumes (Volumes 29 and 30, this series). These include measurement of the mobility of receptor-bound molecules on the cell surface, measurement of rates of endocytosis, direct observation of the movement of endocytosed molecules through the endocytic compartments, measurement of pH of endosomes and lyso-

13

somes, and quantification of receptor binding properties on individual cells. One of the exciting new developments in the field is the ability to measure a response in a target cell (e.g., a change in intracellular free calcium or pH) using one fluorescent dye as a cell is stimulated with a fluorescent analog of a hormone. This allows one to take advantage of the high time resolution and sensitivity of fluorescence methods to study the detailed time course of hormone binding and action.

There are many specific examples of the use of fluorescent analogs of hormones and other peptides which bind to cellular receptors. These include derivatives of $\beta$-melanotropin (Varga *et al.*, 1976), propanolol (Melamed *et al.*, 1976), insulin (Shechter *et al.*, 1978), epidermal growth factor (Shechter *et al.*, 1978), enkephalins (Hazum *et al.*, 1979a), chemotactic peptides (Niedel *et al.*, 1979), thyroid hormones (Cheng *et al.*, 1979), $\alpha_2$-macroglobulin (Maxfield *et al.*, 1978), low-density lipoproteins (Barak and Webb, 1981), transferrin (van Renswoude *et al.*, 1982; Enns *et al.*, 1983), and many others. In addition, other molecules can be used in similar ways as tracers of processes in living cells. For example, FITC-labeled dextrans (see Luby-Phelps, Chapter 4, this volume), which enter cells by nonspecific fluid pinocytosis (see Swanson, Chapter 9, this volume), are often used as tracers of endocytic processes for comparison with other molecules which enter by specific receptor-mediated endocytosis. Similarly, fluorescent lectins are often used to label many membrane proteins (see Edidin, Chapter 6, this volume).

In this chapter, the practical aspects of using fluorescent analogs of peptides and hormones will be described. The strategy used to choose a labeling procedure will be discussed. Various procedures must be used to validate the fluorescent analog to show that it retains the properties of the parent molecule when interacting with the living cell (e.g., specificity of binding). In order to obtain a complete description of such interactions, it is usually necessary to use fluorescence techniques in combination with other methods such as radioligand binding and electron microscopic localizations. The methods for characterizing the analogs and their interaction with cells will be described. Once a valid fluorescent analog has been obtained, it can be used for many types of experimental procedures. Examples of the successful use of fluorescent probes will be provided along with a discussion of some of the experimental difficulties which are frequently encountered.

## II.  Designing a Fluorescent Analog

The methods for the synthesis of fluorescent conjugates of proteins and peptides are described in chapters by Haugland and by Wang in Volume 30 (this series). In addition to the general considerations which pertain to the labeling of any peptide, labeling of hormones and other peptides that bind to specific receptors places special constraints on the labeling procedure. These include retaining the specificity of the binding to the appropriate receptor, preserving the physiological response to the hormone, and obtaining a sufficiently high level of fluorescence for the signal to be visible above interference from intrinsic cellular autofluorescence.

One of the first steps in a study using fluorescent analogs should be an estimate of the overall signal likely to be obtained when the fluorescent analogs are incubated with cells. The intensity of the signal will depend on the number of binding sites per cell, the number of fluorophores per peptide, and the properties of the fluorophore (e.g., quantum yield). For imaging studies, the detectability of the signal will also depend in part on the distribution of the fluorescent analogs in the cell. When fluorescent analogs are diffusely distributed throughout the cell, the fluorescence intensity is more difficult to observe than when the fluorescence is concentrated into a few discrete *foci*.

The fluorescent signal intensity must then be compared with the sensitivity of the fluorescence detection methods available to determine whether the proposed studies are feasible. In most cases the detectability of the fluorescence is not limited primarily by the sensitivity of the detector (see Taylor and Salmon, Chapter 13, this volume; Spring and Lowy, Chapter 15, this volume; Aikens et al., Chapter 16, this volume). The major problem is most frequently detecting the fluorescence above a background of cellular autofluorescence. The level of this autofluorescence, its spatial distribution, and its wavelength dependence vary among different cell types and within a cell type depending on culture conditions. Thus, it is not possible to provide completely general rules for the amount of fluorescence required to obtain a usable signal from living cells. However, some general guidelines can be used to predict the feasibility of a study.

In most cases where fluorescent hormones have been applied to cells, it has been necessary to have about $10^5$ rhodamine or fluorescein molecules associated with the cell in order to obtain a satisfactory signal above the background noise and fluorescence. This number can be reduced if the conditions are favorable. In the most favorable cases, where very brightly fluorescent molecules can be prepared, it has even been reported that single fluorescent molecules can be observed with image intensification fluorescence microscopy in living cells (Gross and Webb, 1986). This is possible with low-density lipoproteins when the nonpolar core is filled with highly fluorescent molecules (Barak and Webb, 1981), and it may also be feasible when proteins are covalently coupled to highly fluorescent molecules such as phycobiliproteins (Oi et al., 1982; see also chapter by Haugland, Volume 30, this series). In this case, the fluorescence is observable because a high amount of fluorescence comes from a very small area. Even if imaging of single molecules is not required, the coupling of a hormone or small peptide to a highly fluorescent compound can increase the detectability of the signal. For example, conjugation of insulin or epidermal growth factor (EGF) to highly fluorescent rhodamine–lactalbumin allowed visualization of these hormones in living cells (Schlessinger et al., 1978b). An important trade-off in this strategy is that coupling to a large protein often alters the specific interaction with the appropriate receptor, and this must be assessed carefully (see below).

If the study appears feasible, then a fluorescent labeling procedure must be chosen. A variety of methods are available as described by Haugland in Volume 30 (this series). There are some special considerations in labeling hormones and receptor-binding peptides. First, the derivatization procedure should have a minimal effect on the binding properties of the ligand. In the case of hormones, the biological effects should also be minimally affected. If data are available on the effects of chemical modification on a particular hormone, these can be used to guide the selection of a labeling procedure. For example, it was known that the N-terminal amino group of EGF was not required for biological activity, so this group could be derivatized to prepare fluorescent analogs (Shechter et al., 1978; Haigler et al., 1978). In the case of enkephalins (Hazum et al., 1979a) and chemotactic peptides (Niedel et al., 1979), lysine residues were added to provide a site for derivatization which would not interfere with the biological activity of the peptides. If chemical modification data or polypeptide sequence requirements are not available, it may be necessary to try several labeling protocols in order to obtain highly fluorescent derivatives which retain the specific biological properties of the parent molecule.

Several properties of the fluorophore must be considered. Usually, cellular autofluorescence decreases at longer wavelengths (Aubin, 1979; Benson et al., 1979). This is a significant advantage of rhodamine and other dyes that excite above 500 nm (see Waggoner, Volume 30, this series). If a cell is to be observed for longer than a few seconds, then photobleaching of the fluorophore may be a significant factor, and dyes that are less sensitive to photobleaching should be considered. In this case, rhodamine labels are preferable to fluorescein. Methods are available for reducing photobleaching in fixed cells, but these are not applicable to studies of living cells.

Most fluorophores are affected significantly by their environment. Especially if the fluorescent analogs are to be used for quantitative studies, it is essential to consider the factors that could alter the fluorescence in the cell. When fluorescein is exposed to low pH, the excitation profile changes, with an overall decrease in fluorescence intensity. This can be very useful for measuring pH in lysosomes and endosomes (Ohkuma and Poole, 1978; Tycko and Maxfield, 1982; Murphy et al., 1984; van Renswoude et al.,

1982), but it would create potential difficulties in other studies of the passage of fluorescent analogs through endocytic compartments. Molecules in low pH compartments would have their fluorescence suppressed, and this would create an incorrect view of the distribution of the fluorescent molecules in the cell.

In addition to pH, many other factors alter fluorescence properties. These include the polarity of the medium, proximity to other dye molecules, and stacking interactions with aromatic groups. In many cases, the change in fluorescence caused by changes in the polarity of the microenvironment has been a useful property of fluorescent analogs. For example, lipid probes that only fluoresce brightly in a nonpolar environment can be used without significant interference from molecules dissolved in the aqueous buffer (Prendergast et al., 1981). Polymerization of some proteins can be observed by changes in the fluorescence intensity of attached fluorophores as the molecules associate (Kawasaki et al., 1976). Especially with small hormones and peptides, it is possible that binding to a receptor could bury the attached fluorophore in a hydrophobic binding pocket with a resulting change in the fluorescence.

The presence of other molecules with absorption spectra that overlap the emission profile of the fluorophore can result in quenching of fluorescence by resonance energy transfer. Since the efficiency of quenching strongly depends on the distance between the molecules, this property can be used to measure distance between the molecules (Stryer, 1978; see also Herman, Volume 30, this series). By the same mechanism, if the concentration of a fluorophore becomes too high, self-quenching may be observed. This could potentially occur as molecules are concentrated on the cell surface (e.g., in coated pits) or within endocytic organelles. However, in most cases this has not proven to be a significant factor in practice. Resonance energy transfer has been observed between molecules present at very high density on the cell surface. These include lipid analogs (Fung and Stryer, 1978) and fluorescent lectins (Fernandez and Berlin, 1976). However, several attempts to quantify resonance energy transfer between receptor-bound proteins such as $\alpha_2$-macroglobulin which are present at approximately $10^5$ molecules/cell have indicated that the efficiency of transfer is quite low (F. R. Maxfield, unpublished). This would indicate that self-quenching is not likely to be a significant artifact for most studies of fluorescent hormone analogs.

## III.   Characterization of Fluorescent Analogs

Once an analog has been prepared, it must be characterized with regard to its molecular properties and its interaction with cells. First, it must be verified that the dye molecules are covalently attached to the hormone or peptide. This can be checked in part by column chromatography, by thin-layer chromatography, by high-pressure liquid chromatography (HPLC), or by dialysis, depending on the size of the hormone. Especially with tetramethylrhodamine isothiocyanate derivatives, slow release of the dye is frequently noted. Thus, covalent attachment must be verified at intervals when the analog is stored, and, if necessary, free dye must be removed by dialysis or chromatography. Even at extremely low concentrations, free rhodamine can create significant artifacts. In many cells the rhodamine becomes highly concentrated in the nucleus. This can lead to artificially high fluorescence levels and to misleading localizations.

The dye-to-protein ratio can be estimated using the spectroscopic properties of the dye and the protein (The and Feltkamp, 1970a,b). Frequently, the ratio of absorbance at 280 nm and at the absorption maximum of the dye can be used to estimate the number of fluorophores per protein or peptide (see Haugland, Volume 30, this series; Simon and Taylor, 1986). If necessary, the dye-to-protein ratio can also be estimated by amino acid analysis, and the sites of derivatization can be analyzed by sequencing the peptide.

The extent of derivatization can alter the properties of the hormone, so it is important, with each synthesis, to test the properties of the analog. If the dye-to-peptide ratio is too low, the fluorescence may be too weak to be useful. On the other hand, overderivatization will often alter the specific interactions of the hormone or peptide with its receptor.

In many respects, the characterization of a fluorescent hormone analog is the same as the more common characterization of radiolabeled analogs. It is essential to know that the analog behaves like the parent molecule in its interactions with the cell. Since the analog will be followed by its fluorescence properties, it is also important to know that the fluorophore remains covalently associated with the peptide within the cell and that the cell does not alter the fluorescence properties in unexpected ways.

The first step is to characterize the specificity and affinity of the fluorescent analog for its receptor. The binding of the analog to cells should be competed by the unlabeled hormone, and this competition should be observed at concentrations comparable to the $K_d$ for the hormone binding to its receptor. Similarly, the cell-associated fluorescence should increase in a saturable pattern as the concentration of the fluorescent analog is increased. If appropriate instrumentation is available, the competition and binding curves should be checked quantitatively. With a microscope spectrofluorometer, the fluorescence intensity of individual cells can be measured following incubation with the fluorescent analog in the presence or absence of excess unlabeled peptide (Maxfield et al., 1981a). Binding can also be measured in a flow cytometer (Murphy *et al.*, 1984; Sklar *et al.*, 1984) or in detergent extracts of cells (Maxfield *et al.*, 1981a). The concentration dependence of the binding can be measured directly, and the percentage of nonspecific binding at each concentration can be determined. Ideally, the binding curve and percentage of nonspecific binding should be comparable for fluorescent analogs and for radiolabeled analogs of hormones. However, since the addition of fluorophores is generally a larger chemical modification than radiolabeling, it is not unusual to find some reduction in affinity and specificity. If the perturbation is unacceptably large, then the ratio of fluorophore to peptide should be reduced or an alternative labeling procedure should be considered.

In many laboratories, equipment for measuring binding of radiolabeled hormones is more readily available than equipment for quantifying fluorescence in cells. The ability of the fluorescent-labeled hormone or peptide analog to compete for binding sites with a radiolabeled analog can be measured using standard techniques for determining the $K_d$ of a radiolabeled hormone (Levitzki, 1985). Although this procedure is simpler than measurements of the fluorescence intensity, it has several limitations. First, the assay will show that the fluorescent analog can compete for the normal binding site, but it will not indicate whether there is additional nonspecific binding of the fluorescent analog. Also, most preparations of fluorescent analogs will contain a mixture of molecules labeled at different positions or to different extents at each position. Some of these forms may bind to the normal receptor with good affinity, but others may bind elsewhere on the cell. This would not be detected by a competition assay with radiolabeled hormone.

Depending on the analog and the experiments to be carried out with it, other preliminary characterizations should also be performed. First, with peptide hormones the biological activity of the analog should be measured and compared to the native hormone. As with binding, the chemical modification may reduce the potency of the hormone, and alternative labeling procedures may be necessary. If endocytosis of the analog will be observed, then the kinetics of internalization and degradation or release of the analog should be measured. Internalization kinetics can be measured by techniques similar to those used for radiolabeled analogs (Haigler *et al.*, 1980). In this case, a biochemical difference between internal and external probe (e.g., protease sensitivity or resistance to acid release) is used to distinguish internalized ligand from surface bound ligand. Alternatively, one can take advantage of special properties of the fluorescence to measure internal versus external fluorescence. Addition of a cell impermeant quenching agent to the cells can rapidly distinguish external fluorescence. In the case of fluorescein-labeled analogs, antifluorescein antibodies or rapid pH changes can be used to distinguish extracellular versus intracellular fluorescence (Sklar *et al.*, 1984).

Once the fluorescent analog has been internalized, it is important to monitor the stability of the fluorophore–peptide complex. Unlike [125]I, fluorophores can remain cell associated long after the peptide is degraded. For example, when tetramethylrhodamine isothiocyanate-labeled $\alpha_2$-macroglobulin is incubated with NRK cells, a rat kidney fibroblastic cell line, the $\alpha_2$-macroglobulin is rapidly degraded ($t_{1/2} \sim 15$ minutes), but the fluorescence remains cell associated (and in a punctate distribution) for several hours (Maxfield *et al.*, 1981a). Degradation of the peptide or release of the fluorophore from the peptide can be monitored by polyacrylamide gel electrophoresis of cell extracts (Maxfield *et al.*, 1981a). The degradation can be quantified by scanning the gels using a microscope spectrofluorometer.

In some cases, cells may modify the covalent structure of the fluorophore. When fluorescein-labeled molecules are subjected to the oxidative burst of neutrophils, the fluorescein is covalently modified with a resulting change in its fluorescence properties (Hurst *et al.*, 1984).

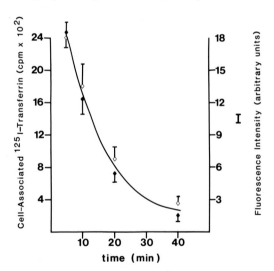

FIG. 1.   Release of [125]I-labeled transferrin and FITC–transferrin from Chinese hamster ovary cells. Cells were incubated with [125]I-labeled transferrin (2 $\mu$g/ml) or FITC–transferrin for 10 minutes at 23°C. The cells were rinsed and reincubated at 23°C for various times. For fluorescence measurements, the cells were fixed with 2% formaldehyde, and the intracellular pH gradients were collapsed by addition of methylamine. The remaining FITC–transferrin (◊) was determined by measuring the intensity of individual cells using a microscope spectrofluorometer. Approximately 50 cells were used for each measurement. Cell-associated radioactivity (◆) was determined in parallel experiments.

The release of fluorescein transferrin and $^{125}$I-labeled transferrin from cells is shown in Fig. 1. Transferrin is internalized by receptor-mediated endocytosis. After release of iron within the cell, apotransferrin is returned to the cell surface and released from the cell (Hanover and Dickson, 1985). This is in contrast to many other internalized peptides which are degraded in lysosomes. As shown in Fig. 1, fluorescein transferrin is released from cells with kinetics which are indistinguishable from the release of $^{125}$I-labeled transferrin. Transferrin is a good illustration of how subtle differences in labeling techniques can influence the results obtained. The derivative used for the studies shown in Fig. 1 was prepared by reaction of transferrin with fluorescein isothiocyanate. This procedure does not involve the cross-linking of proteins. One of the ways to make highly fluorescent analogs is to cross-link a highly fluorescent protein with the peptide being studied (Shechter et al., 1978). If the reaction conditions and purification steps are not carefully checked, more than one transferrin could be attached to a fluorescent protein. Rather than being returned to the cell surface, a significant percentage of cross-linked transferrin receptor oligomers are delivered to lysosomes and degraded (Weissman et al., 1986). This artifact would not have been detected in an analysis of binding data or by measuring the rate of internalization.

## IV.  Examples of the Use of Fluorescent Analogs of Hormones and Peptides

Specialized techniques for the use of fluorescent analogs are described throughout this book. In this section some specialized applications of fluorescence techniques used to study hormones will be discussed.

## A.  Tracing Endocytic Pathways

The most frequent use for fluorescent hormone analogs has been to visualize their interaction with living cells. Among the fluorescent hormone analogs which have been prepared are derivatives of $\beta$-melanocyte stimulating hormone (Varga et al., 1976), epidermal growth factor (Schlessinger et al., 1978a,b; Haigler et al., 1978), insulin (Schlessinger et al., 1978a,b), chemotactic peptides (Niedel et al., 1979), enkephalins (Hazum et al., 1979b), gonadotropin-releasing hormone (Hazum et al., 1980), and thyroid hormone (Cheng et al., 1980). Protein analogs which have been used for studies of endocytosis include $\alpha_2$-macroglobulin (Willingham and Pastan, 1978; Maxfield et al., 1978), low-density lipoproteins (Barak and Webb,

1981), diphtheria toxin (Keen *et al.*, 1982), transferrin (Enns *et al.*, 1983; Yamashiro *et al.*, 1984), and asialoglycoproteins (Tycko *et al.*, 1983).

The synthesis of fluorescent analogs of EGF illustrates many of the principles discussed earlier in this chapter. Working independently, two groups prepared fluorescent analogs of EGF and visualized their interaction with living cells. A difficulty in visualizing EGF in fibroblastic cell lines which respond mitogenically to the hormone is the relatively low number of surface receptors (approximately 10,000/cell). In order to overcome this limitation in one set of studies, EGF was covalently linked through the $\alpha$ amino group to a protein (lactalbumin) that was heavily labeled with rhodamine (Shechter *et al.*, 1978). The coupling procedure was selected to take advantage of the fact that derivatization at the amino terminus has little effect on the biological potency of EGF. It was reported that rhodamine $_7$–lactalbumin–EGF retained the same ability to displace $^{125}$I-labeled EGF binding to Swiss 3T3 fibroblasts as native EGF, and it retained 40% of the potency in stimulating DNA synthesis (Shechter *et al.*, 1978). The rhodamine $_7$–lactalbumin derivative provided enough signal to be observed by image intensification fluorescence microscopy even when the receptors were diffusely distributed on the cell surface (Schlessinger *et al.*, 1978b). A direct conjugate of rhodamine with the amino terminus of EGF could only be observed in the mouse fibroblasts after the fluorescent analog had been concentrated in endocytic vesicles.

Haigler *et al.* (1978) took advantage of a cell line, the A-431 human carcinoma cells, which possess approximately $2-3 \times 10^6$ EGF receptors/cell to visualize a direct conjugate of fluorescein at the amino terminus of EGF. In these cells, the large number of receptors allowed visualization of the fluorescein–EGF when bound diffusely on the surface of the cells. These workers found that fluorescein–EGF was about 45% as effective as EGF in displacing $^{125}$I-labeled EGF binding to cells. In a separate study, Haigler *et al.* (1979) prepared ferritin–EGF by coupling ferritin to the amino terminus of EGF. This conjugate was used for electron microscopic localizations, but the methods for initial characterization are essentially the same as for fluorescent protein conjugates. It was reported that the conjugate was only 8% as effective as EGF in displacing $^{125}$I-labeled EGF binding, but it was 75% as effective as EGF in an assay for stimulation of DNA synthesis. In retrospect, this apparent discrepancy may be explained in part by the presence of low- and high-affinity EGF receptors on the surface of cells (Schlessinger, 1986). It is evident that there is not necessarily a simple relationship between one property of an analog (e.g., binding affinity) and other properties (e.g., biological potency).

In other cases, one is not as fortunate as with EGF in having an easily derivatized site which has relatively small effects on the interaction with cells. In the case of insulin, a more elaborate derivatization scheme was required to protect the $\alpha$ amino group of the $\beta$ chain while allowing derivatization at the side chain of lysine B29 (Shechter et al., 1978). The $\alpha$ amino groups of insulin were protected by reaction with citraconic anhydride at pH 6.9, and then Lys B29 was reacted with m-maleimidobenzoyl-N-hydroxysuccimide ester, producing a thiol-reactive center at that site. The derivatized insulin was reacted with a large excess of partially reduced rhodamine $_7$-lactalbumin, producing a covalent conjugate of rhodamine $_7$-lactalbumin-insulin. The remaining free sulfhydryl groups were blocked, and the amino blocking was removed by treatment at pH 2.0. The resulting conjugate was sufficiently fluorescent to be easily observed bound to surface receptors on cultured fibroblasts by image intensification fluorescence microscopy (Schlessinger et al., 1978b). Unfortunately, the binding affinity and biological potency of this analog were not as well retained as in the case of EGF. The analog was approximately 8% as effective as insulin in displacing $^{125}$I-labeled insulin binding and retained about 1% of the biological potency in an assay for stimulation of hexose transport. Nevertheless, the fluorescence could be competed with excess unlabeled insulin, and the analogs were useful markers of the binding, surface mobility, and endocytosis of insulin.

These fluorescent analogs and many others have been used to follow endocytosis in many cells. In many cases, (e.g., insulin, EGF, or $\alpha_2$-macroglobulin) diffuse binding to surface receptors is seen initially (Schlessinger et al., 1978b; Willingham et al., 1979). Within a few minutes, the fluorescence can be observed in patches on the cell surface and in endocytic vesicles. The motion of these vesicles can be observed and recorded by image intensification fluorescence microscopy (Willingham et al., 1979). Using two different fluorescent labels, the colocalization of various analogs in the same endosome can be demonstrated. This has been done for several ligands, including $\alpha_2$-macroglobulin, insulin, EGF, and diphtheria toxin, among others (Maxfield et al., 1978; Keen et al., 1982). This type of experiment helped to establish the similarity of endocytic pathways for many different ligands. Transferrin remains bound to its receptor throughout the receptor recycling pathway (Hanover and Dickson, 1985). Using fluorescein-, transferrin-, and rhodamine-labeled $\alpha_2$-macroglobulin or EGF, it has been possible to observe the divergence of the receptor recycling pathway from the pathway leading to degradation in lysosomes (Fig. 2).

A biologically active analog of enkephalin, Tyr-D-Ala-Gly-Phe-Leu-Lys-rhodamine, was prepared and used for localization of opiate-binding sites on neuroblastoma cells (Hazum et al., 1979a,b). In contrast to the observa-

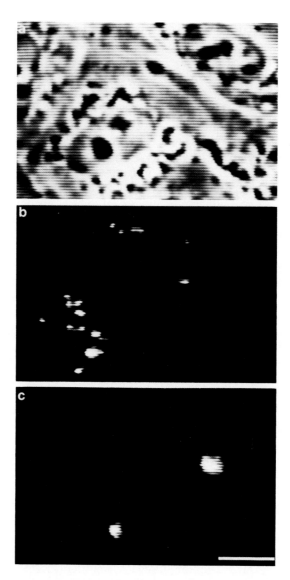

Fig. 2. Localization of both FITC–transferrin and rhodamine–$\alpha_2$-macroglobulin in Chinese hamster ovary cells. Cells were incubated with FITC–transferrin (100 $\mu$g/ml) and rhodamine–$\alpha_2$-macroglobulin (100 $\mu$g/ml) for 15 minutes at 37° C. (a) A phase contrast image of two cells. (b) The rhodamine–$\alpha_2$-macroglobulin in endocytic vesicles distributed throughout the cytoplasm. (c) FITC–transferrin concentrated in recycling endosomes near the nucleus. Electron microscopic studies revealed that these juxtanuclear structures are collections of small vesicles or tubules near the Golgi complex. Bar, 8 $\mu$m. Reprinted from Yamashiro *et al.* (1984).

tions with insulin and EGF, the clustering of enkephalin on the surface of cells appeared to be very slow and did not lead to internalization and degradation in lysosomes.

## B. Measurement of Receptor Mobility

One of the earliest uses of fluorescent analogs of hormones was to demonstrate and quantify the lateral mobility of hormone–receptor complexes. The methods used to quantify receptor mobility and the molecular basis for the reduced mobility of membrane proteins are described in several chapters in this series (see Webb and Ghosh, and Wolf, Volume 30). The fluorescent derivatives of insulin and EGF described above were used in fluorescence photobleaching recovery experiments to show that the occupied receptors were mobile and had a diffusion coefficient of approximately $4 \times 10^{-10}$ cm$^2$ second$^{-1}$. Although this is considerably lower than the diffusion coefficient for lipid analogs, the lateral mobility is sufficiently rapid to allow interaction between occupied receptors within fractions of a second (Schlessinger et al., 1978a). It is thought that interaction of occupied receptors may be part of the signaling mechanism for both insulin (Kahn et al., 1978) and EGF (Schlessinger, 1986). Similar diffusion coefficients were obtained for other receptors including $\alpha_2$-macroglobulin (Maxfield et al., 1981a) and for a surface binding site for thyroid hormone (Maxfield et al., 1981b). The diffusion coefficient obtained for low-density lipoprotein receptors was significantly lower ($D = 1-5 \times 10^{-11}$ cm$^2$ second$^{-1}$; Barak and Webb, 1982). This may be related to the observation that low-density lipoprotein receptors are clustered even prior to occupancy (Goldstein et al., 1985), whereas receptors for insulin, EGF, and $\alpha_2$-macroglobulin are mainly diffusely distributed prior to occupancy (Schlessinger et al., 1978b; Willingham et al., 1979).

## C. Measuring the pH of Endosomal Compartments

Since the excitation profile of fluorescein is pH dependent, quantitative measurements of fluorescein fluorescence in living cells can be used to estimate the pH of endocytic compartments (Ohkuma and Poole, 1978). This technique was first used to measure the pH of lysosomes after endocytosis of FITC–dextran (Ohkuma and Poole, 1978). Subsequently, a number of analogs which bind to specific surface receptors have been used, including $\alpha_2$-macroglobulin, asialoglycoproteins, transferrin, and insulin (Tycko and Maxfield, 1982; van Renswoude et al., 1982; Tycko et al., 1983; Murphy et al., 1984; Yamashiro et al., 1984; Sipe and Murphy, 1987). Measurements have been made on cells in a cuvette (Ohkuma and Poole,

1978; van Renswoude *et al.*, 1982), by flow cytometry (Murphy *et al.*, 1984; Sipe and Murphy, 1987), and by microscope spectrofluorometry (Tycko *et al.*, 1982, 1983; Heiple and Taylor, 1982). Measurements in a cuvette are relatively straightforward, provided that sufficient signal above the light scattering, autofluorescence, and extracellular dye can be obtained. An advantage of flow cytometry is that kinetic data can be obtained on rates of acidification from large numbers of cells (Murphy *et al.*, 1984; Sipe and Murphy, 1987). Microscope spectrofluorometry, particularly when combined with digital image analysis, has the advantage of being able to provide pH measurements on morphologically distinct types of endosomes (Tycko *et al.*, 1983). For example, separate measurements can be made on recycling endosomes containing fluorescein transferrin and prelysosomal endosomes containing $\alpha_2$-macroglobulin (Yamashiro *et al.*, 1984, 1987). In CHO cells, it was found that the recycling endosomes had a pH of about 6.5, while the prelysosomal endosomes maintained a pH of about 5.4. Distributions of pH in individual endosomes or lysosomes can be measured to provide an indication of the heterogeneity of pH in these compartments (Tycko *et al.*, 1983; Yamashiro and Maxfield, 1987). The methods for measuring pH by fluorescence ratio imaging are described elsewhere in this series (see chapters by Bright *et al.* and Tsien and Poenie, Volume 30).

## D.   Correlation of Binding with Signaling

When optical methods can be used to measure the effect of a hormone, fluorescence analogs allow the simultaneous observation of receptor occupancy and response. This has been shown elegantly in the case of chemotactic peptides binding to neutrophils (Sklar *et al.*, 1984). Both fluorescein and rhodamine derivatives of a chemotactic peptide, *N*-formyl-Nle-Leu-Phe-Nle-Tyr-Lys, have been prepared and shown to bind specifically to receptors on neutrophils (Niedel *et al.*, 1979; Sklar *et al.*, 1984). The fluorescent analogs are internalized within minutes at 37°C. Internalization of the rhodamine analog was shown by direct observation (Niedel *et al.*, 1979). For the fluorescein analog, the pH dependence of fluorescein fluorescence was exploited. Rapid lowering of the extracellular pH quenched the fluorescence of surface-bound fluorescein without affecting the fluorescence from the internalized analog. Thus, the internalization kinetics could be quantified (Sklar *et al.*, 1984). The binding of the fluorescein analog could also be quantified with high time resolution by fluorescence methods. An antifluorescein antibody was prepared which quenched fluorescence from the free fluorescein-labeled peptide but did not quench when the peptide was bound to receptors. Addition of the antibody at different times allowed the quantification of free versus receptor-bound peptide (Sklar *et al.*, 1984). These quantitative fluorescence techniques for measuring binding and internalization have been combined with fluorescence assays for elastase secretion (Sklar *et al.*, 1982), membrane potential, and cytoplasmic free calcium (Sklar *et al.*, 1984) to provide a high time-resolution depiction of the response to chemotactic peptides.

With the development of many new fluorescent methods as described in these volumes (Volumes 29 and 30, this series), it should be possible to extend this type of analysis to many hormone responses in other cell types.

## REFERENCES

Aubin, J. (1979). *J. Histochem. Cytochem.* **27**, 36–43.

Barak, L. S., and Webb, W. W. (1981). *J. Cell Biol.* **90**, 595–602.

Barak, L. S., and Webb, W. W. (1982). *J. Cell Biol.* **95**, 846–852.

Benson, R., Meyer, R. A., Zaruba, M., and McKhann, G. M. (1979). *J. Histochem. Cytochem.* **27**, 44–48.

Cheng, S.-Y., Eberhardt, N. L., Robbins, J., Baxter, J. D., and Pastan I. (1979). *FEBS Lett.* **100**, 113–116.

Cheng, S.-Y., Maxfield, F. R., Robbins, J., Willingham, M. C., and Pastan, I. H. (1980). *Proc. Natl. Acad. Sci. U.S.A.* **77**, 3425–3429.

Enns, C. A., Larrick, J. W., Suomalainen, H., Schroder, J., and Sussman, H. H. (1983). *J. Cell Biol.* **97**, 579–585.

Fernandez, S. M., and Berlin R. D. (1976). *Nature (London)* **264**, 411–415.

Fung, B. K., and Stryer, L. (1978). *Biochemistry* **17**, 5241–5248.

Goldstein, J. L., Brown, M. S., Anderson, R. G. W., Russell, D. W., and Schneider, W. J. (1985). *Annu. Rev. Cell Biol.* **1**, 1–39.

Gross, D., and Webb, W. W. (1986). *Biophys. J.* **49**, 901–911.

Haigler, H., Ash, J. F., Singer, S. J., and Cohen S. (1978). *Proc. Natl. Acad. Sci. U.S.A.* **75**, 3317–3321.

Haigler, H. T., McKanna, J. A., and Cohen, S. (1979). *J. Cell Biol.* **81**, 382–395.

Haigler, H. T., Maxfield, F. R., Willingham, M. C., and Pastan, I. (1980). *J. Biol. Chem.* **255**, 1239–1241.

Hanover, J. A., and Dickson, R. B. (1985). *In* "Endocytosis" (I. Pastan and M. C. Willingham, eds.), pp. 131–161. Plenum, New York.

Hazum, E., Chang, K.-J., Shechter, Y., Wilkinson, S., and Cuatrecasas, P. (1979a). *Biochem. Biophys. Res. Commun.* **88**, 841–846.

Hazum, E., Chang, K.-J., and Cuatrecasas, P. (1979b). *Science* **206**, 1077–1079.

Hazum, E., Cuatrecasas, P., Marian, J., and Conn, P. M. (1980). *Proc. Natl. Acad. Sci. U.S.A.* **77**, 6692–6695.

Heiple, J. M., and Taylor, D. L. (1982). *J. Cell Biol.* **94**, 143–149.

Hurst, J. K., Albrich, J. M., Green, T. R., Rosen, H., and Klebanoff, S. (1984). *J. Biol. Chem.* **259**, 4812–4821.

Kahn, C. R., Baird, K. L., Jarrett, D. B., and Flier, J. S. (1978). *Proc. Natl. Acad. Sci. U.S.A.* **75**, 4209–4213.

Kawasaki, Y., Mihashi, K., Tanaka, H., and Ohnuma, H. (1976). *Biochim. Biophys. Acta* **351**, 205–213.

Keen, J. H., Maxfield, F. R., Hardegree, M. C., and Habig, W. H. (1982). *Proc. Natl. Acad. Sci. U.S.A.* **79**, 2912–2916.

Levitzki, A. (1985). *In* "Endocytosis" (I. Pastan and M. C. Willingham, eds.), pp. 45–68. Plenum, New York.

Maxfield, F. R. (1985). *In* "Endocytosis" (I. Pastan and M. C. Willingham, eds.), pp. 235–257. Plenum, New York.

Maxfield, F. R., Schlessinger, J., Schechter, Y., Pastan, I., and Willingham, M. C. (1978). *Cell* **14**, 805–810.

Maxfield, F. R., Willingham, M. C., Haigler, H. T., Dragsten P., and Pastan, I. (1981a). *Biochemistry* **20**, 5353–5358.

Maxfield, F. R., Willingham, M. C., Pastan, I., Dragsten, P., and Cheng, S.-Y. (1981b). *Science* **211**, 63–65.

Melamed, E., Lahav, M., and Atlas, D. (1976). *Nature (London)* **261**, 420–422.

Murphy, R. F., Powers, S., and Cantor, C. R. (1984). *J. Cell Biol.* **98**, 1757–1762.

Niedel, J. E., Kahane, I., and Cuatrecasas, P. (1979). *Science* **205**, 1412–1414.

Ohkuma, S., and Poole, B. (1978). *Proc. Natl. Acad. Sci. U.S.A.* **75**, 3327–3331.

Oi, V. T., Glazer, A. N., and Stryer, L. (1982). *J. Cell Biol.* **93**, 981–986.

Prendergast, F. G., Haugland, R. P., and Callahan, P. J. (1981). *Biochemistry* **20**, 7333–7338.

Schlessinger, J. (1986). *J. Cell Biol.* **103**, 2067–2072.

Schlessinger, J., Shechter, Y., Cuatrecasas, P., Willingham, M. C., and Pastan, I. (1978a). *Proc. Natl. Acad. Sci. U.S.A.* **75**, 5353–5357.

Schlessinger, J., Shechter, Y., Willingham, M. C., and Pastan, I. (1978b). *Proc. Natl. Acad. Sci. U.S.A.* **75**, 2659–2663.

Schechter, Y., Schlessinger, J., Jacobs, S., Chang, K.-J., and Cuatrecasas, P. (1978). *Proc. Natl. Acad. Sci. U.S.A.* **75**, 2135–2139.

Simon, J. R., and Taylor, D. L. (1986). *In* "Methods in Enzymology" (R. B. Vallee, ed.), Vol. 134, p. 407, Acadamic Press, Orlando, Florida.

Sipe, D. M., and Murphy, R. F. (1987). *Proc. Natl. Acad. Sci. U.S.A.* **84**, 7119–7123.

Sklar, L. A., McNeil, V. M., Jesaitis, A. J., Painter, R. G., and Cochrane, C. G. (1982). *J. Biol. Chem.* **257**, 5471–5475.

Sklar, L. A., Finney, D. A., Oades, Z. G., Jesaitis, A. J., Painter, R. G., and Cochrane, C. G. (1984). *J. Biol. Chem.* **259**, 5661–5669.

Stryer, L. (1978). *Annu. Rev. Biochem.* **47**, 819–846.

The, T. H., and Feltkamp, T. E. W. (1970a). *Immunology* **18**, 865–873.

The, T. H., and Feltkamp, T. E. W. (1970b). *Immunology* **18**, 875–881.

Tycko, B., and Maxfield, F. R. (1982). *Cell* **28**, 643–651.

Tycko, B., Keith, C. H., and Maxfield, F. R. (1983). *J. Cell Biol.* **97**, 1762–1776.

van Renswoude, J. K., Bridges, K. R., Harford, J. B., and Klausner, R. D. (1982). *Proc. Natl. Acad. Sci. U.S.A.* **79**, 6186–6190.

Varga, J. M., Moellman, G., Fritsch, P., Godawska, E., and Lerner, A. B. (1976). *Proc. Natl. Acad. Sci. U.S.A.* **73**, 559–562.

Weissman, A. M., Klausner, R. D., Rao, K., and Harford, J. B. (1986). *J. Cell Biol.* **102**, 951–958.

Willingham, M. C., and Pastan, I. (1978). *Cell* **13**, 501–507.

Willingham, M. C., Maxfield, F. R., and Pastan, I. (1979). *J. Cell Biol.* **82**, 614–625.

Yamashiro, D. J., and Maxfield, F. R. (1987). *J. Cell Biol.* **105**, 2723–2733.

Yamashiro, D. J., Tycko, B., Fluss, S. R., and Maxfield, F. R. (1984). *Cell* **37**, 789–800.

# Chapter 3

## *Fluorescent Analogs of Toxins*

### KIMON J. ANGELIDES

*Department of Physiology and Molecular Biophysics and Program in Neuroscience*
*Baylor College of Medicine*
*Houston, Texas 77030*

## I. Introduction

In the past decade the isolation, chemical characterization, and use of toxins in cell biology has increased significantly. The increased interest in toxins has been largely motivated by the fact that they provide some of the most powerful and specific tools to dissect the molecular and cellular details of vital physiological processes (1). For example, several important receptors and ion channels found in the nervous system are inhibited by toxins. Because of the low density of these receptors in the nervous system, the isolation, molecular characterization, and description of their cellular dy-

29

namics would have been virtually impossible without the use of these specific and high-affinity toxins.

A particularly good example of the use of toxins is illustrated by the rapid progress that has been made with the molecular and cellular characterization of the nicotinic acetylcholine receptor. The discovery of the specific snake toxin, $\alpha$-bungarotoxin, was shown to bind with very high specificity and irreversibly to the acetylcholine receptor and for the past 20 years the toxin has provided an essential tool for labeling, identifying, and characterizing acetylcholine receptors in several preparations (2). Fluorescent analogs of this toxin have provided important insights into the cellular dynamics of acetylcholine receptor ontogeny (3), distribution (4), and lateral mobility (5). Recently, toxins as useful as $\alpha$-bungarotoxin have been applied to the molecular and cellular biology of the voltage-dependent sodium channel (6–12). As nature's arsenal of new toxins specific for other ion channels or receptors are uncovered, it is with certainty that new fluorescent tools will be constructed and used to determine the cellular topography of their receptors.

The use of toxins in cell biology, however, has not been restricted to the nervous system or to receptors and ion channels. Toxins have provided important tools as probes of regulatory proteins such as pertussis toxin on the modulatory G proteins (13), as enzyme inhibitors, such as amanitine of the DNA polymerase reactions (14), and as probes for cytoskeletal structures (21).

Toxins can be conveniently grouped into three categories based on their chemical structures.

*Group 1.* Proteins and polypeptides such as $\alpha$-bungarotoxin, the polypeptide toxins that act on voltage-dependent sodium channels (15), and cholera (16), pertussis (19), and diphtheria toxins (18) that exert their effects on regulatory proteins or at the cell membrane surface.

*Group 2.* Heterocyclic natural products, such as histrochiotoxin, that act on acetylcholine receptors (19), tetrodotoxin and batrachotoxin that act on voltage-dependent sodium channels (15), strychnine that inhibits glycine receptors (20), and taxol, colchicine, or phalloidine (21) that act on cellular cytoskeletal components.

*Group 3.* Lipopolysaccharides and gangliosides that act on membrane lipids (22).

Because of the high affinity of toxins for their receptors, toxins provide ideal probes for examining the cellular distribution and dynamics of their receptors on the cell surface. In many instances they provide more superior ligands than fluorescent antibodies since they do not induce cross-linking of the receptors, are not membrane permeable, and in some cases are not internalized (see Edidin, Chapter 6, this volume; Maxfield, Chapter 2, this volume).

The scope of this chapter is (1) to describe those fluorescent toxin analogs that have been used to label specific receptors; (2) to describe general synthetic methods to prepare toxin analogs; and (3) to outline general approaches to prepare biologically active fluorescent toxin conjugates with a step-by-step procedure. Finally, the practical uses and the potential of these probes to examine living cells is discussed.

## II. Examples of Fluorescent Toxins

Because of the availability of reactive probes and the well-described chemistry of fluorophore conjugation, the most common fluorescent toxin analogs are polypeptide derivatives. Selected examples of polypeptide toxins that have been exploited for the molecular characterization of several ion receptors and channels are discussed below.

### Receptor Toxins

#### 1. FLUORESCENT α-BUNGAROTOXIN: A PROBE FOR THE NICOTINIC ACETYLCHOLINE RECEPTOR

Some of the most important fluorescent toxin probes have developed from the discovery of isolated polypeptide toxins from venomous snakes (23, 24). Of these, the best characterized and most widely used toxin for cellular neurobiology has been α-bungarotoxin. This small (MW ~7,500) polypeptide binds selectively, specifically, and irreversibly to the nicotinic acetylcholine receptor. Before the development of a fluorescent conjugate of α-bungarotoxin, rather cumbersome, time-consuming, and limited electrophysiological (25) or autoradiographic methods (26) were used to examine the distribution of acetylcholine receptors in skeletal muscle. For cellular neurobiologists, fluorescently labeled α-bungarotoxin provided a new and versatile tool to examine the cellular dynamics of acetylcholine receptors on living cells and to follow changes in receptor topography during myogenesis (27).

The first fluorescent derivative of α-bungarotoxin was reported by Anderson and Cohen who modified the protein with fluorescein isothiocyanate (4). The reaction led to several modified derivatives and partial fractionation of the reaction mixture led to a derivative that was biologically active and that intensely labeled endplates of *Xenopus* skeletal muscle fibers. Improvements in the fluorescent signal were made by Ravdin and Axelrod who prepared tetramethylrhodamine derivatives of $\alpha$-bungarotoxin and chromatographically separated the derivatized products into mono- and disubstituted toxin that had different quantum efficiencies (28). As with the fluorescein $\alpha$-bungarotoxin, the fluorescent toxin irreversibly labeled acetylcholine receptors on mouse diaphragm muscle. The derivatives have also been used in fluorescence recovery after photobleaching experiments where it was shown that acetylcholine receptors located at the neuromuscular junction were immobile ($D < 3 \times 10^{-12}$ cm$^2$/second) (5) and that extrajunctional receptors that were also organized as clusters were partly mobile ($10^{-11}$ cm$^2$/second) (29). The latter receptors were observed to participate in receptor cluster formation and reorganize to the neuromuscular junction as a consequence of innervation (30). The tetramethylrhodamine derivative was superior to the fluorescein derivative in that experiments employing fluorescein $\alpha$-bungarotoxin were often limited by the background autofluorescence of the cells.

Several of these fluorescent derivatives of $\alpha$-bungarotoxin have recently become commercially available through Molecular Probes (Eugene, Oregon). These are the fluorescein, tetramethylrhodamine, and phycoerythrin derivatives that can be used in single- or double-labeling studies.

## 2. FLUORESCENT TOXIN ANALOGS FOR THE VOLTAGE-DEPENDENT SODIUM CHANNEL

Pharmacological dissection of the ionic currents responsible for the action potential showed several classes of toxins that act specifically on sodium channels (15). Fluorescent derivatives of these sodium channel toxins have been prepared to map the molecular structure, the cellular topography, and lateral mobility of sodium channels on nerve and muscle cell surfaces (6–12). There are four different receptors for toxins on the sodium channel. One receptor binds tetrodotoxin, a heterocyclic guanidine molecule, a second receptor binds the alkaloid toxin, batrachotoxin, a secretion from the skin of the poisonous frog, *Phylobates;* and the third and fourth receptors bind the polypeptide scorpion toxins, referred to as $\alpha$- or $\beta$-scorpion toxins according to their distinct pharmacology. These polypeptides are small (MW 7000) high-basic molecules with four disulfide bridges.

The labeling of sodium channels on cultured neurons by several fluorescent toxin analogs has shown that, on cultured neurons, voltage-dependent sodium channels are localized to morphologically distinct regions of these cells (11). Fluorescent neurotoxin probes specific for the voltage-dependent sodium channel stain the hillock 5 – 10 times more intensely than the cell body and demonstrate punctate fluorescence confined to the hillock region which can be compared to the more diffuse labeling in the cell body. Using fluorescence recovery after photobleaching, the lateral diffusion coefficient of the mobile fraction of sodium channels was measured at specific regions of the neuron. Nearly all sodium channels stained with specific neurotoxins are free to diffuse within the cell body with diffusion coefficients on the order of $10^{-9}$ cm²/second, while >80% of the labeled channels located at the hillock and presynaptic terminal are immobile on the time scale of the measurement $(D \leq 10^{-12}$ cm²/second). The small, mobile fraction had diffusion coefficients in the range $10^{-10} - 10^{-11}$ cm²/second. In contrast, however, no regionalization or differential mobilities were observed for either tetramethylrhodamine – phosphatidylethanolamine as a probe of lipid diffusion, or fluorescein isothiocyanate – succinylconcanavalin A as a probe for glycoproteins. The results suggested that, in distinct regions of the neuron, sodium channels experience stronger retarding interactions with the underlying cytoskeleton than most membrane proteins.

When sodium channels were labeled with fluorescent toxin analogs on muscle cells it was shown that on uninnervated muscle cells sodium channels are diffusely distributed and very mobile ($10^{-9}$ cm²/second). After innervation, however, fluorescently labeled sodium channels reorganize, colocalize with acetylcholine receptors, and are immobilized to the sites of neuronal contact (6).

In addition to these fluorescent scorpion toxin conjugates, one report on the preparation and pharmacological characterization of a commercially available (Calbiochem, San Diego, California) polypeptide toxin from isolated sea anemone has appeared. However, because of the relatively low affinity of the derivative (1 $\mu M$) no subsequent study on the labeling of excitable cells with this derivative has been reported (31).

### 3. FLUORESCENT TOXINS FOR $Ca^{2+}$ AND $K^+$ CHANNELS AND THE GLYCINE RECEPTOR

Recently, interest in voltage-dependent calcium channels has been motivated by its central role in a variety of cellular processes including exocytosis, neurotransmitter release, and metabolic processes. The discovery of specific toxins that inhibit these channels has aided in this progress. Short polypeptide toxins (33 amino acids) from *Conus geographus* appear to exert their effects specifically on the neuronal $Ca^{2+}$ channel (33). Unfortunately, the availability of the venom to purify the active component or the active peptide toxin itself has not allowed investigators, at least until recently, to develop fluorescent derivatives. However, recently the chemical synthesis of the peptide and the successful reoxidation of the disulfide bonds have yielded biologically and chemically indistinguishable synthetic products of this toxin (34). The peptide is now available through Pennisula Laboratories (San Carlos, California). We have now prepared fluorescent derivatives of the synthetic toxin, characterized the products biologically, and have used these analogs to label neuronal cells in culture. The preparation of the fluorescent toxin is similar to that described in detail below for fluorescent toxin analogs used as probes of the voltage-dependent sodium channel.

Glycine is a major inhibitory transmitter in the mammalian spinal cord. The excitatory action of strychnine on the central nervous system has been attributed to its ability to interact with glycine-sensitive postsynaptic receptors, thereby blocking the inhibitory effects of this amino acid transmitter (2). Strychnine binds specifically to synaptic membranes and spinal cord cells and the binding is inhibited by glycine. The preparation of a fluorescein derivative of strychnine has been reported and the labeling of neuronal cells has been demonstrated. This derivative maintains the high affinity for binding to the glycine receptors (35).

The next section describes our general strategy to select, design, and prepare fluorescent toxin conjugates using as a general model the polypeptide and heterocyclic toxins active on voltage-dependent sodium channels.

# III. Selection, Preparation, and Utilization of Fluorescent Toxin Analogs

Our first efforts to prepare fluorescent toxin conjugates focused on the voltage-dependent $Na^+$ channel, a channel which is crucial to the propagation of nerve impulses. This was due in part to the number of high affinity and specific $Na^+$ channel ligands available and the well-described pharmacology. The approaches described in the subsequent section to prepare fluorescent derivatives are general and can be applied to other toxins that are rapidly becoming available. Current work with the preparation of fluorescent $Ca^{2+}$, $K^+$, and $Cl^-$ channel toxin probes and strychnine, using the same general approach, illustrates the applicability of these synthetic methods.

## A. Chemical Requirements

What features are required and desired for fluorescent toxins? First, the specificity and pharmacology should be well described and characterized as to their mode of action. Second, sufficient chemical reactivity, for example, amino, sulfhydryl, tyrosine, and arginine, on proteins or reactive groups, such as ketones, aldehydes, alkylhalides, carboxylic acids, or hydroxyls, should be present. Ideally, the chemistry of the modification should be such that the introduction of a single chemical functionality would produce a convenient intermediate. Third, since toxins are typically active at very low concentrations, a powerful method must be available to separate the modified from the unmodified toxin, so that the unmodified ligand does not mask the biological properties of the derivatized molecule. Fourth, the derivatives must be characterized biologically as to their physiological actions and to retention of their specificity.

## B.   Spectroscopic Requirements

In choosing a fluorescent probe for receptor or channel work the comparatively low density of ion channels in most tissues requires the selection of probes that have both high extinction coefficients ($\epsilon \geq 50,000 \ M^{-1} \ cm^{-1}$) and quantum yields ($\Phi \geq 0.4$) since these spectral attributes will facilitate microscopic visualization of the toxin receptors. In addition, both the absorption and fluorescence should be at long wavelengths to circumvent cellular autofluorescence, and selected to correspond to the available lines of gas ion or pulse dye lasers, and the emission directed to regions where most photomultipliers and cameras are maximally sensitive.

Of course, no single fluorescent probe satisfies all these requirements and the synthesis of a variety of toxin derivatives is usually necessary. However, it is easier if a modified toxin is prepared with a single reactive group to serve as an intermediate in the synthesis of a fluorescent ligand. Our general strategy has been to introduce a unique chemical functional group onto a toxin that can then be reacted with fluorescent reagents with different spectral properties depending on the needs for a particular study. We have found, because most peptide toxins lack free sulfhydryl residues, that the introduction of a reactive thiol group often allows selective and nonperturbing modifications of the toxin.

The next section details a representative procedure to prepare fluorescent conjugates of polypeptide toxins, the characterization of the derivative chemically, biologically, and spectrally, and its use in cell biological studies.

## C.   Preparation and Characterization of Fluorescent Toxins as Cellular Probes for the Voltage-Dependent Sodium Channel

Pharmacological dissection of the $Na^+$ channel has revealed at least four separate receptor sites for neurotoxins (15). Among these the heterocyclic guanidinium compounds, tetrodotoxin and saxitoxin, reversibly block ion conductance presumably by occluding the external mouth of the channel and preventing $Na^+$ ions from entering the channel. Both these reagents bind with high selectivity and affinity to the channel. Batrachotoxin (BTX) and veratridine, two alkaloid toxins, interact at a distinct $Na^+$ channel locus, shift the voltage dependence of activation and inactivation, and elicit long open times in single channel recordings. Upon repetitive firing the actions of BTX and veratridine are enhanced, suggesting that they bind to a receptor site that undergoes a voltage-dependent conformational change. A third receptor of the $Na^+$ channel binds the scorpion toxins and modifies the gating properties of the channel. These toxin reagents are highly positively charged polypeptides of molecular weight 7000 with four disulfide bonds. The scorpion toxins can be further divided into two general classes. The $\alpha$-scorpion toxins, which include the general class of North African toxins, modulate the inactivation kinetics of the channel by prolonging the action potential. The $\beta$-scorpion toxins, or toxins from North American scorpions, elicit their action by modifying the activation kinetics of channel opening, inducing repetitive firing due to abnormal activation but have no effects on the activation kinetics. Each class of scorpion toxins binds to the channel at different sites. These reagents that elicit distinct pharmacological modifications of sodium channels have been extremely useful in the elucidation of the molecular structure, biochemistry, development, and cellular dynamics of the voltage-dependent $Na^+$ channel. Their conserved mode of action in almost all excitable tissues has placed these reagents as the most specific and versatile tools for the study of this ion channel.

Because of their high affinity and specificity, these neurotoxins provide ideal reagents to label and explore the distribution and lateral mobility of $Na^+$ channels on excitable cell surfaces. Therefore, we have found it important to create a large and diverse set of reagents that can be applied to a variety of experimental systems.

## 1.  POLYPEPTIDE TOXINS

Chemical modification of these polypeptide toxins reveal that toxin activity is lost after acylation, alkylation, and citraconylation of amino groups (36). Limited iodination of tyrosine and guandidation preserve the toxic activities. These results suggested that the charge of the toxin is critical for toxic and biological activity. Therefore, strategies have been devised to retain the positive charge on the modified toxin and to introduce a functional group that serves as an intermediate to allow the preparation of a number of fluorescent analogs having different spectral properties. Four different modification routes have been taken where acylated, amindinylated, thioamindinylated, and reductively aminated polypeptide toxins were prepared (7).

*a. Acylation.* Modification by acylation results in the loss of positive charge. However, if appropriately directed, we find that, under mild conditions, acylation results in the modification of only one lysine. The conditions are such that the fluorescent modifying reagent is limiting (i.e., 0.5 : 1.0 mol ratio of reagent to toxin). With the net loss of one positive charge, the derivatives can be purified by ion-exchange chromatography, isoelectric focusing, or preparative high-performance liquid chromatography. At least with the sodium channel, toxins confirmation by sequence analysis indicates that only one lysine has been modified. The derivatives that were biologically active were modified at position lysines 60 and 13 of the $\alpha$- and $\beta$-scorpion toxins, respectively (7, 10). However, modification of the adjacent lysine at position 58 in toxin V from *Leuirus quiquestriatus* (LqqV) when higher concentrations of the reagent (10 : 1) are used lead to an inactive derivative. The results have indicated that, although a small loss in positive charge can be tolerated by these toxins, a specific residue is usually critical for toxin activity.

*b. Amidinylation and Reductive Alkylation.* Amidinylation and reductive alkylation on the other hand do not result in the loss of positive charge on the toxin and, in the cases that we have examined, even with multiple modifications the biological activity is retained. However, because the charge is not significantly altered in these derivatives, separation of modified from unmodified by conventional chromatography is very difficult. Alternative methods to handle small amounts of modified toxin using immunoaffinity chromatography and/or in certain cases by high-performance liquid chromatography on $C_{18}$ using a gradient of 5–60% acetonitrile buffered with 0.1% trifluoroacetic acid will resolve the modified and unmodified toxins. In some cases, antibodies raised against the fluorophore can be coupled to Sepharose 4B and used to immunoaffinity purify the fluorescently modified toxin, separating any unmodified toxin which may otherwise mask the biological properties of the modified toxin. This method may be required when very limited quantities of the toxin are available (Fig. 1).

*c. Thioamidation.* Although the above methods yield fluorescent toxins via acylation or amidinylation, the fluorescent reagents with the reactive groups must be synthesized. The lability of imidoesters often makes these reactions difficult and of low yield. We searched for a more versatile method to prepare fluorescent toxins that would retain the positive charge of the toxin yet would serve as an intermediate to allow the coupling of standard available fluorescent reagents with the precise spectral properties desired for individual experiments. Since most polypeptide toxins do not contain cysteine, a thiol group was introduced onto the polypeptide, through amidinylation, by reaction with the cyclic imidoester, 2-iminothiolane. The method is shown in Fig. 1. Not only does this reaction preserve the toxin's net positive charge but the sulfhydryl that is generated can be used to affinity purify the modified toxin on a thiol–disulfide exchange column prior to modification of the toxin with sulfhydryl specific reagents. The sulfhydryl bearing toxin that is separated can then be modified with any number of fluorescent maleimides covering the whole visible spectrum. For example, fluorescent derivatives bearing coumarin, fluorescein, or tetramethylrhodamine can be easily prepared using the same intermediate.

## 2. PREPARATION OF FLUORESCENT TOXIN CONJUGATES: EXPERIMENTAL DETAILS

*a. Acylation of Toxins.* To 116 $\mu$g of LqqV (16 nmol) in 1.0 ml of 0.1 $M$ Na$_2$CO$_3$ buffer, pH 9.0, is added 16 nmol of TmRhd–succinimidyl ester (Molecular Probes, Eugene, Oregon). The concentration of reagent was selected to give a 1:1 mol ratio of reagent to toxin but a 7 mol excess of total reactive lysine groups to reagent. The reaction is allowed to proceed for 60 minutes at 25°C. Unreacted and hydrolyzed probes are removed by gel filtration on P-2 (0.5 × 15 cm) in 300 m$M$ ammonium acetate, pH 6.38. Fractions of 1.0 ml are collected and scanned for the absorbance at 277 nm (toxin) and by fluorescence ($\lambda_{excit} = 565$; $\lambda_{emiss} = 600$ nm). The intensely pink fluorescent protein fractions eluting in the void volume were collected and lyophilized. The material collected herein represents both fluorescently modified toxin and unmodified toxin. It is useful at this point to determine the extent of the modification by measuring the ratio of the absorbances of dye and peptide. Since the acylation of lysine results in a loss of net positive charge, the two (or more) possible species can be characterized, isolated, and purified by either (1) preparative isoelectric focusing on 7.5% polyacrylamide gels with 7–9 and 9–11 Immobilines or (2) by cation-exchange chromatography on CM-cellulose. Samples of 30–50 $\mu$g of modified toxin are applied in 0.1% 7–9 Immobiline and run on gels containing 7.5% acrylamide, 0.25% bisacrylamide, 0.1% 7–9 Immobilines, and 0.2% 9–11 Immobilines. Slab gels are run at 400 V (~2 mA/well) for 2 hours, after which current decreases to 1 mA. The modified bands can be immediately identified by their intense red fluorescence while the migratory position of the native toxin can be determined by staining with Coomassie blue. The fluorescent bands are excised and the modified toxins eluted overnight with 20% acetic acid. On the basis of a molar extinction coefficient at 565 nm of 65,000 $M^{-1}$ cm$^{-1}$ of the TmRhd chromophore and the extinction coefficient of the toxin at 277 nm, the preparative method typically yields pure

## I. PREPARATION OF THIOAMIDINYLATED TOXIN

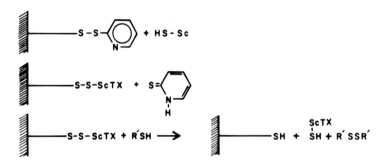

SCORPION TOXIN        2-IMINOTHIOLANE              ScTX-SH
   (SCTX)

## 2. AFFINITY COLUMN PURIFICATION

—S—S—⟨◯⟩  + HS—Sc
        N

—S—S—ScTX  +  S=⟨◯⟩
                  N
                  H

—S—S—ScTX + R'SH  ⟶      |—SH  +  ScTX + R'SSR'
                                   SH

## 3. DERIVATIZATION OF FLUOROPHORES

ScTX—SH + R—N⟨ ⟩  ⟶  ScTX—S—⟨ ⟩N—R

WHERE R = FLUORESCENT CHROMOPHORES

## 4. IMMUNOPRECIPITATION WITH RABBIT IgG

ANTIFLUOROPHORE IgG + FLUORESCENT TOXIN / NATIVE TOXIN  ⟶

IgG ⟨ FLUORESCENT TOXIN PRECIPITATE  $\xrightarrow{HOAC}$

IgG − IgG + PURIFIED FLUORESCENT TOXIN

FIG. 1.   Preparation and purification of fluorescent scorpion toxin conjugates by affinity chromatography.

disubstituted (1.8 mol of TmRhd/mol of toxin) and monosubstituted toxins (0.98 mol of TmRhd/mol of toxin), and represents 15 and 35%, respectively, of the total toxin applied to the gel. Typical yields from 16 nmol of starting toxin are 1.4 nmol disubstituted and 3.4 nmol monosubstituted toxin.

Alternatively, if larger amounts of toxin are available, the species of TmRhd-modified toxin can be separated by ion-exchange chromatography on either CM-cellulose or Bio-Rex 70 (1 × 200 cm) equilibrated and eluted with 200 m$M$ ammonium acetate, pH 6.42, at a flow rate of 20 ml/hour. The column should be sialyzed to avoid toxin absorption to the glass. Fractions can be analyzed by the absorbances at 277 and 565 nm and by the fluorescence intensity at 600 nm with excitation at 565 nm. Typically, three peaks are observed. The first two are somewhat closely spaced and on occasion overlap, whereas the third is separated by at least eight column volumes. Only the first two fractions should show fluorescence corresponding to the TmRhd moiety. The first time the fractions can be analyzed by isoelectric focusing, usually the leading edges of peaks I and II show homogeneous fluorescent bands with fractions in peak II corresponding to the more basic of the two. Determination of the mole ratio of dye to protein by absorbance spectrophotometry indicates that peak I is usually dimodified (2.1 mol of TmRhd/mol of toxin) and peak II corresponds to the monomodified derivative (1.08 mol of TmRhd mol of toxin). The third peak typically shows no fluorescence at the appropriate TmRhd excitation or emission wavelengths and is unlabeled toxin. Under conditions in which fluorescent reagent is limiting, the third peak is usually the majority. This unmodified toxin can be recovered and used once again in modifying reactions.

When toxin is reacted with a 1 : 1 mole ratio (toxin : reagent), the amount of toxin modified is usually very low (10%) and requires rather extensive workup to eliminate unmodified toxin. The resultant product, however, is mostly a monomodified derivative that is biologically active. The unmodified toxin is not lost and can be recycled for further modifications.

*b. Reductive Alkylation.* An aldehyde can be added to an amino to form a Schiff base that can be reduced. For example, NBD–methylaminoacetaldehyde (0.90 μmol) is added to 108 μg of LqqV (15.6 nmol) dissolved in 1.0 ml of 0.1 $M$ sodium phosphate buffer, pH 6.97, at 4°C. Sodium cyanoborohydride (1.9 μmol) is added, followed by two further additions (0.2 μmol) of reagent at 30-minute intervals. The reaction is terminated by filtration on P-2 (0.7 × 20 cm) eluting with 300 m$M$ ammonium acetate, pH 6.37. Fractions of 1.0 ml are collected and scanned for absorbance at 277 nm for the toxin and for the chromophore.

Purification and isolation of imidoester and reductively alkylated modified toxins from unmodified toxin can be accomplished by analytical and preparative isoelectric focusing as described for acylated toxins (7). Yields of 10–20% modified toxin based upon starting toxin in reductive alkylations are about 20–45%.

For difficult separations with small quantities of modified toxin the reductively alkylated-modified toxin can be separated from native toxin by immunoprecipitation with a rabbit antifluorophore IgG. For the NBD chromophore, it is a structural and functional analog of the 2,4-dinitrophenyl group and has an association constant of $2.5 \times 10^7 M$ (21) with the antibody (DNP). Affinity-purified rabbit anti-DNP IgG linked to Sepharose 4B can be added to equivalence in 50 m$M$ sodium phosphate buffer, pH 7.4. The mixture is incubated for 90 minutes at 37°C, cooled to 2°C for 10 minutes, and then centrifuged at 12,000 $g$ for 2 minutes. The precipitate formed is washed three times with 0.15 $M$ NaCl, and then 30% acetic acid is added. Immunoprecipitated modified toxin is released from IgG by the addition of 30% acetic acid after incubation at 37°C for 30 minutes. The acetic acid is evaporated by a gentle stream of nitrogen gas, after which the precipitate is taken up in a small volume of 0.1 $M$ sodium phosphate buffer, pH 7.4, containing 0.15 $M$ NaCl. The mixture is then centrifuged at 12,000 $g$ for 2 minutes to remove insoluble IgG, and the supernatant carefully removed. Yields of modified toxin using immunoprecipitation usually exceed those of preparative high-performance liquid chromatography or isoelectric focusing.

   c. *Thioamidination: Preparation of LqqV-SH, Purification by Affinity Chromatography, and Chemical Modification with Fluorescent Maleimides (7).*   To 136 $\mu$g (18.8 nmol) of LqqV in 50 m$M$ sodium borate buffer, pH 8.5, is added 2.5 $\mu$mol of 2-iminothiolane. This concentration of reagent is selected to give a 17-fold mole excess of reagent to total reactive lysine groups. The reaction mixture is incubated at room temperature for 70 minutes. Aliquots of the reaction mixture are periodically removed and titrated with 5,5′-dithiobis(2-nitrobenzoic acid) to monitor the appearance of sulfhydryl groups and the kinetics of the modification. Controls performed in the absence of toxin show that the rate of spontaneous hydrolysis of 2-iminothiolane at 8.5 is low (e.g., 0.018 sulfhydryl group/mol of reagent after 2 hours). The modification of the toxin is terminated by gel filtration on a P-2 column (0.07 $\times$ 20 cm) with 100 m$M$ sodium phosphate, pH 7.0, 1 m$M$ EDTA (this buffer should be deaerated and preflushed with nitrogen). The eluant is monitored at 280 nm and fractions which elute at the void volume are collected, purged with nitrogen, and tightly sealed. The sulfhydryl-to-protein ratio of the peak fractions should be determined by titration with 5,5-dithiobis(2-nitrobenzoic acid), and by measurement of the absorbance at 280 nm. Typically, 60–70% of the applied protein is eluted from the P-2 column.

Material from the gel filtration step, usually consisting of 30–40 $\mu$g of toxin (5 nmol), is added to 200 $\mu$mol of activated thiol-Sepharose 4B (for optical monitoring of the affinity coupling) or thiopropyl-Sepharose 4B, which is equilibrated with 100 m$M$ Na$^+$ phosphate, pH 8.1, 1 m$M$ EDTA buffer. A 50-fold excess of total sulfhydryl capacity of the gel should be added to the fraction and allowed to equilibrate with the gel-filtered fraction for 6 hours at room temperature with agitation. The total volume is about 1.0 ml. The material is washed with five volumes of 100 m$M$ Na phosphate, 1 m$M$ EDTA buffer, and centrifuged in an Eppendorf minifuge at 12,000 $g$ for 2 minutes. A 100-$\mu l$ aliquot of the supernatant from each wash is removed and examined for the concentration of nonbound unmodified toxin at 277 nm. Alternatively, when spectral monitoring of the coupling is not necessary, a small column of thiopropyl-Sepharose 4B is poured after equilibration of the modification mixture with the gel, and the nonbound toxin washed with five volumes of 100 n$M$ Na phosphate, pH 7.0, and 1 m$M$ EDTA buffer. Typically, 30% of the applied gel filtration fraction, representing unmodified toxin, is removed at these washing steps.

The sulfhydryl-containing toxin can be released from the matrix by washing the gel with five volumes of 100 m$M$ Na phosphate, pH 8, 1 m$M$ EDTA, containing 10 m$M$ L-cysteine. L-Cysteine at this concentration and pH is more suitable than more powerful reducing agents like dithiothreitol in order to preserve the toxin's disulfide bonds. The supernatants at each washing step are carefully removed and assayed for their absorbance at 277 nm. All the toxin applied to the affinity column can usually be recovered with yields of approximately 25 $\mu$g of modified, affinity purified, sulfhydryl-containing toxin. The sulfhydryl-containing toxin is passed over Sephadex G-10, equilibrated with a degassed and nitrogen-flushed 100 m$M$ Na phosphate, 1 m$M$ EDTA, pH 7.0 buffer after which modification with fluorescent maleimides can take place immediately.

Modification with fluorescein–5-maleimide, tetramethylrhodamine–maleimide or coumarin–phenylmaleimide is performed at 4°C for 2 hours using a 50-fold excess of maleimide dissolved in dimethyl sulfoxide. Excess fluorescent reagent is removed by gel filtration on Sephadex G-10 or P-2 (0.7 × 20 cm) eluting with 300 m$M$ ammonium acetate buffer, pH 6.6, and the fluorescent toxin samples collected and lyophilized.

## 3. HETEROCYCLIC NATURAL PRODUCTS

*a. Tetrodotoxin.* TTX has been the most widely used probe of the voltage-dependent Na$^+$ channel. For a long time the insolubility and chemical liability of TTX thwarted chemical derivatization. An early work reported that the hydroxyls could be esterified with carbonyldiimadazole and a carboxylic acid (37). Although this material was reported to be biologically active, significant amounts of unreacted TTX were responsible for the activity. However, further reports showed that TTX could be oxidized at either C-6 or C-11 to form a C-6 ketone or C-11 aldehyde generating a functional group that is unique to the molecule (38). Ketones and aldehydes can either be coupled with hydrazides to form hydrazones or reductively aminated with amines in the presence of $NaCNBH_3$ to form primary amines at C-6 (amino-TTX) (Fig. 2). This latter intermediate is particularly useful since commercially available and commonly used protein reagents specific for lysine can be coupled via acylation or alkylation.

　　*b.   Synthesis of Fluorescent TTX.*   Citrate-free TTX (2.0 mg, 6.28 $\mu$mol) (Calbiochem) is dissolved in 0.2 ml of distilled water with 2.0 $\mu$l of trifluoroacetic acid added (pH in test solutions 2.3). The solution is lyophilized and redissolved in 0.2 ml of distilled water (pH final in test solutions 5.5) in a polyethylene Eppendorf minifuge tube. Periodic acid (6.3 $\mu$mol) is added, and the reaction mixture is allowed to react at 25°C for 30 minutes. Lead (II) acetate (6.4 $\mu$mol) is added to the reaction mixture at 4°C to precipitate excess ions, and the precipitate of lead diiodide is removed by centrifugation in an Eppendorf minifuge. The supernatant is carefully removed and lyophilized. The oxidized product can be identified on a silica gel G plate by spraying with a 0.5% solution of 2,4-DNPH in 1 $N$ HCl or pyrenebutyryl hydrazide in methanol: $R_f$ (chloroform – methanol, 2 : 1, v/v)

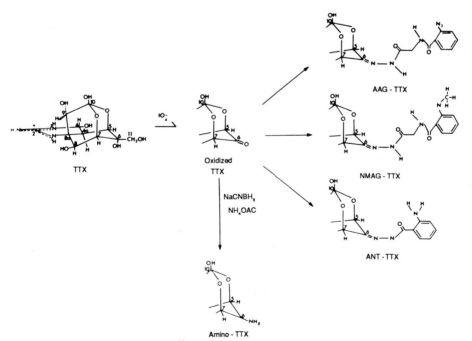

FIG. 2.   Synthesis of fluorescent TTX derivatives.

0.08. In addition, migration of TTX can be followed by spraying with alcoholic KOH and heating: $R_f$ (ethanol – acetic acid, 96 : 4, v/v) 0.38.

The preparation of 6-amino-TTX and 6-(methylamino)-TTX follows the pattern detailed by Varlet *et al.* (39) for 2-aminoalkanephosphoric acids. The reactions are performed by dissolving a mixture of oxidized TTX (6.0 $\mu$mol), NH$_4$OAc or methylammonium acetate (32 $\mu$mol) with [$^{14}$C]methylammonium acetate as tracer, and 4.2 $\mu$mol of NaCNBH$_3$ in 3.0 ml of absolute methanol in a screw-capped conical vial. The kinetics of the reaction are followed by removing aliquots of the reaction mixture (10 $\mu$l) at selected times and reacted with 2 – 5 $\mu$l of fluorescamine (0.4 mg ml$^{-1}$ in acetonitrile). Excitation is at 390 nm and emission at 480 nm. The reaction is complete usually after 48 hours when no further increase in the fluorescamine fluorescence appears.

    *c.  Coupling of Hydrazide Derivatives to Oxidized TTX.* Oxidized TTX (1 mg, 3.14 $\mu$mol) is dissolved in 300 $\mu$l of acidified methanol (pH$_{app}$ 5.3), and 9.6 $\mu$mol of fluorescent hydrazide derivative is added. The reaction is carried out over molecular sieves (3 Å) to remove the water which is produced as a result of the coupling. The reaction should be allowed to proceed at 37°C, and the kinetics should be followed by thin-layer chromatography in acetonitrile/0.1 $M$ ammonium formate (pH$_{app}$ 5.2). After 5 hours, all the insoluble TTX solubilized and the kinetics of the reaction indicated that equilibrium had been reached. Alternatively, 1 mg of oxidized TTX is dissolved in 300 $\mu$l of water acidified with 1 $\mu$l of trifluoroacetic acid, lyophilized, and then redissolved in 200 $\mu$l of anhydrous dimethylsulfoxide. Dimethyl sulfoxide is then used as the solvent and can be removed by lyophilization later. Reduction by NaCNBH$_3$ is performed by adding 5.0 $\mu$mol of NaCNBH$_3$ in absolute methanol to the coupled fluorescent TTX in 300 $\mu$l of methanol (pH$_{app}$ 6.8), and reduction proceeds for 52 hours at 37°C. The reaction mixture is redissolved in a minimum (200 $\mu$l) of acidified methanol (pH$_{app}$ 5.3) and applied to a Sephadex LH-20 column preequilibrated with acidified methanol. The column is eluted with acidified methanol (pH$_{app}$ 5.3) and a linear salt gradient consisting of 0.01 – 0.15 $M$ ammonium formate (final pH 5.5). Fractions of 200 $\mu$l are collected. The fractions are then subjected to analysis for unreacted tetrodotoxin by the fluorescence method under alkaline conditions (excitation 390 nm, emission 500 nm) and for the product formation (excitation 325 or 350 nm, emission 405 or 425 nm) as well as for determination of radioactivity when [$^{14}$C]glycine ethyl ester was used as reagent. The fractions which showed evidence for coupled tetrodotoxin product are then analyzed and repurified by two-dimensional thin-layer chromatography on silica gel G plates with choloroform – methanol (3 : 1) in the first dimension followed by acetonitrile/0.1 $M$ ammonium formate (pH$_{app}$ 5.2) (3 : 1) in the second. The product remained at the origin in the first dimension while the uncoupled hydrazides migrate with an $R_f$ of ~0.85. In the second dimension, these are the $R_f$ values: 2-azidoanthraniloylglycy-TTX, 0.33; N-methylanthraniloylglycl, 0.63; anthraniloyl-TTX, 0.46. (see Fig. 2 for structures).

## D. Characterization of Fluorescent Toxins

### 1. PHARMACOLOGY AND BIOLOGICAL ACTIVITY

The assay for toxin activity must be established prior to the conjugation with a fluorophore. Although, for toxins, one of the most commonly used methods is *in vivo* toxicity, the assays involving animals are neither reliable nor quantitative. For receptor and ion channel studies, sensitive and quantitative tests using electrophysiological assays are the preferred methods. Using these procedures and multiple sampling techniques in conjunction with digital recording methods, one should be capable of resolving significant differences as small as 5% and thus be able to assay concentrations of toxin at 1 n$M$ or less with excellent reproducibility (7). Similarly, for toxins that exert their actions on regulatory processes, e.g., pertussis and cholera toxin, the biological assays are ADP-ribosylation of G protein (40). In some instances, the "biological activity" can also be determined by labeling of cells and competition and blocking with unlabeled toxin.

### 2. EQUILIBRIUM BINDING OF TOXIN ANALOGS

If a radioactive derivative is available, then the activity can be assessed by equilibrium binding methods and competition. Displacement of [125]I-labeled toxin by competition binding to isolated cell membranes with unlabeled fluorescent analogs can be done to test the activity of the analogs. In some cases of the sodium channel toxin analogs, dimodified toxin shows no competition even at concentrations 500 times higher than that of native toxin. On the other hand, derivatives where the positive charge has been retained (e.g., the dimodified amidinated thio-LqqV derivatives) show excellent displacement with $K_d$ values which are only two to three times higher than that of the native toxin.

### 3. CHEMICAL STRUCTURE

In some instances where the reaction yields multiple derivatives it is useful to determine the modification site (10). This can usually be done by conventional amino acid analyses and by direct protein sequencing.

# E.  Application of Fluorescent Toxin Analogs to Cells

The application of fluorescent toxins to cells is similar to those procedures discussed for fluorescent peptide analogs (see Maxfield, Chapter 2, this volume). Usually, in the case of a fluorescent analog, it is essential to know the dissociation constant and association constant of the toxin binding to cells since (1) the toxin should be applied to saturate the channels if the dissociation rate is very slow or (2) in the case of ligands where the dissociation rate is sufficiently rapid (e.g., 3 minutes or less) the concentration of the fluorescent toxin should be considerably below the $K_d$ to ensure that almost all the fluorescent toxin is bound to the receptor on the cell surface. For most fluorescent toxin analogs the dissociation rates are in fact slow ($t_{1/2} \simeq 30$ minutes) and in some cases are essentially irreversible (e.g., $\alpha$-bungarotoxin).

Generally, cells are incubated with the fluorescent toxin until equilibrium is reached (30–40 minutes). Cells are washed five times with ice-cold phosphate-buffered saline and then used for microscopy. If the cells are to be examined later the cells can be fixed with 2.5% paraformaldehyde to cross-link the fluorescent polypeptide toxin followed by several washes of PBS containing ammonium chloride to block reactive aldehyde groups. For examination of the distribution and lateral mobility on living cells we routinely cultured cells on 0.13-mm-thick glass cover slips. These are then either mounted inversely onto a serological plate for use with a $100\times$ oil-immersion objective, or we now routinely immerse a $63\times$ high numerical aperature water objective directly into the culture dish.

### 1.  LABELING OF VOLTAGE-DEPENDENT SODIUM CHANNELS: GENERAL PROCEDURE (11)

The $\alpha$- and $\beta$-scorpion toxins bind reversibly to a single class of receptor sites on the $Na^+$ channel of excitable cells with $K_d$s between 2 and 8 n$M$. The half time for dissociation is about 65 minutes at 22°C for NBD–LqqV, TmRhd–LqqV, 7-diethylaminocoumarin-4-phenyl-maleimido toxin (CPM)–Css II, 5 minutes for NBD–TTX, and ~3 hours for TmRhd–Tityus $\gamma$. Prior to microscopy or photobleaching experiments, confluent cultures, grown under standard conditions, are incubated with 10 n$M$ fluorescent LqqV, Css II, Tityus $\gamma$, or TTX for 30 minutes at room temperature in standard binding medium consisting of 130 m$M$ choline, 5.4 m$M$ KCl, 0.8 m$M$ MgSO$_4$, 5.5 m$M$ glucose, and 50 m$M$ HEPES adjusted to pH 7.4 with Tris base. The cultures are then rapidly washed three times with ice-cold standard binding medium or ice-cold phosphate-buffered saline. In some experiments the fluorescent scorpion toxins can be incubated in phosphate-buffered saline that includes 2 $\mu M$ TTX which blocks ion movement through the channels that lead to both membrane depolarization and dissociation of the toxin. Since the binding of ligands such as NBD–LqqV and NBD–TTX to the $Na^+$ channel results in a 35-fold enhancement of the fluorescence quantum yield, it is not necessary to remove the unbound reagent as the contribution of the free signal to the cellular bound fluorescence emission is negligible. In the case of fluorescent derivatives of LqqV which bind to the channel in a voltage-dependent manner, the nonspecific labeling can be reduced to 10–15% by depolarization of the cells with 150 m$M$ external $K^+$ or by dissipation of the membrane potential by the addition of 10 $\mu$g/ml gramicidin A. With CPM–Css II, TmRhd–Tityus $\gamma$, or NBD–TTX, the specific binding is blocked by excess unlabeled toxin (2 $\mu M$).

## 2. NONSPECIFIC LABELING

For some toxins the nonspecific binding can be problematic. The contribution of the nonspecific binding to the fluorescence signal for such ligands can be measured as follows. The total binding is measured after equilibration of the fluorescent toxin and either the image is taken or the photons counted for dwell times between 40 msec and 1 second. No fewer than 20 cells in the culture should be sampled in each region. The nonspecific binding is measured after competitive displacement by unlabeled toxin to these same cultures to which fluorescent toxin had already been added (e.g., 200 n$M$ LqqV, 2 $\mu M$ Css II, 2 $\mu M$ Tityus $\gamma$, or 2 $\mu M$ TTX), allowed to remain with the cultures for 15 minutes at room temperature, and the fluorescence intensities remeasured under identical counting conditions. In addition, nonspecific fluorescence can be evaluated by comparing the intensities from the cell membrane of interest with the intensities from control cells such as fibroblasts.

For some toxin receptors the densities are extremely low and conventional microscopic methods are not sufficiently sensitive to map the distribution. There are several ways in which to produce a more highly amplified signal from the toxin or the amplification of small signals with sensitive detectors (see Spring and Lowy, Chapter 15, this volume; Aikens et al., Chapter 16, this volume). If viewing the steady-state distribution is satisfactory, the signal from toxin analogs can be amplified immunocytochemically. Usually the fluorophore or the chemical group introduced into the toxin serves as a hapten for an antibody (see Gorbsky and Borisy, Chapter 11, this volume). Typically we use DNP and anti-DNP IgG coupled with a goat anti-rabbit IgG conjugated with phycoerythrin. The signal in these cases is amplified about 20 times. However, they cannot be used to study the lateral mobilities.

### 3. Biological Effects on Cells of Fluorescent Toxins

One issue that should be addressed concerns the biological effects when using fluorescent conjugates of toxins as "nonpertubing probes." At least in the instances that have been reported on the use of fluorescent toxins, there appear to be no major deleterious effects on cell survival. For example, application of $\alpha$-bungarotoxin inhibits muscle contraction, but the other "essential" properties of the muscle, such as generation of the action potential, are unaffected. For fluorescent sodium channel probes, each probe exhibits different pharmacological profiles and affects the channel in different fashions. In all cases examined, however, cell death has not been accelerated. These observations stem in part from the selectivity of the toxins for specific receptors such that they are not usually lethal.

# IV.   Use of Fluorescent Toxin Analogs

## Distribution of Toxin Receptors

Fluorescent toxins, in addition to probes of molecular structure, also provide tools to map the cellular distribution and mobility of their cellular receptors. In neurobiology the distribution and segregation of ion channels to specific cellular domains is a very important and as yet relatively unexplored area since the precise and regulated distribution of these ion channels likely confers the specialized excitability properties expressed by each cell. With the preparation of ligands specific for cell receptors, the distribution and ontogeny of the receptor can be examined on a single cell and correlated with the physiological properties that these cells express.

### 1. Localization of Na⁺ Channels by Fluorescence Microscopy

The $Na^+$ channel specific fluorescent neurotoxins that we have prepared have allowed us to microscopically localize $Na^+$ channels in nerve and muscle fibers and to determine channel distribution under some pathological conditions. Our first studies were aimed at visualizing $Na^+$ channels on myelinated and unmyelinated nerve fibers. In studies of myelinated nerve, NBD–Lqq, coumarin–TTX, and DNP–Lqq have been very useful. NBD–Lqq has very low fluorescence in solution, but is substantially enhanced when bound to the channel and so the fluorescence signal comes almost entirely from the channel-bound ligand. Furthermore, the fluorescence signal could be amplified when the $Na^+$ channel density was low by an antibody against the fluorophore (anti-NBD IgG) that is conjugated with phycobiliproteins. The latter was also applied as a collodial gold suspension so that, subsequent to fluorescence microscopy, electron microscopy was performed. Figure 3 shows a video-enhanced photograph of a single fiber from mouse sciatic nerve stained specifically for $Na^+$ channels with toxin analogs. Staining is restricted to the nodal region (41).

Even though the histochemical studies provide a much needed view of the distribution of ion channels, an essential question in neurobiology centers on the mechanisms that segregate channels to morphologically distinct regions of nerve and muscle during development and the maintenance of this heterogenous distribution. It is known that segregation of voltage-dependent $Na^+$ channels to the hillock of motoneurones and nodes of Ranvier in myelinated axons is crucial for conduction of the nerve impulse, while other ion channels are restricted to other domains of the same cell.

## 2. REGIONALIZATION AND LATERAL MOBILITY OF VOLTAGE-DEPENDENT $Na^+$ CHANNELS

*a. $Na^+$ Channel Distribution and Mobility on Mature Nerve.* Using microfluorimetry and fluorescence recovery after photobleaching (FPR) and channel specific fluorescent toxin analogs, we have taken a direct approach to measuring $Na^+$ channel mobility and localization on living cells. This general approach has been extended to other ion channels (see below). On cultured neurons, voltage-dependent sodium channels are segregated to morphologically distinct regions. Fluorescent neurotoxin probes specific for the voltage-dependent sodium channel stain the hillock five to ten times more intensely than the cell body and demonstrate punctate fluorescence confined to the hillock region which can be compared to the more diffuse labeling in the cell body (Fig. 4). Using fluorescence recovery after photobleaching we demonstrated that $Na^+$ channels on the cell body are diffusely distributed and freely mobile with diffusion coefficients on the order of $10^{-9}$ cm/second, while >80% of the labeled channels located at the hillock and presynaptic terminal are immobile on the time scale of the measurement (11). The small mobile fraction had diffusion coefficients in the range $10^{-10} - 10^{11}$ cm$^2$/second. In comparison, however, no regionalization of differential mobilities were observed for either TmRhd-phosphatidylethanolamine which moves at rates $10^{-9}$ cm$^2$/second over all parts of the cell or FITC-succinyl concanavalin A as a probe for glycoproteins which move at $10^{-10}$ cm$^2$/second (Fig. 5). The rate of lateral mobility of $Na^+$ channels located on the cell body is at the high end for cell surface membrane protein mobility, and their movement is limited only by the viscosity of the membrane lipid.

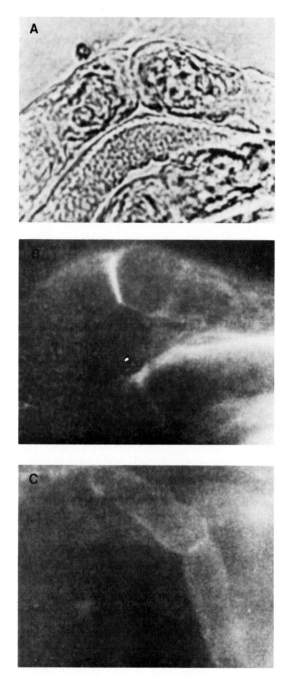

Fig. 3. Staining of voltage-dependent sodium channels by TmRhd–Lqq in mouse sciatic nerve. (A) Phase micrograph, (B) specific labeling, and (C) nonspecific labeling.

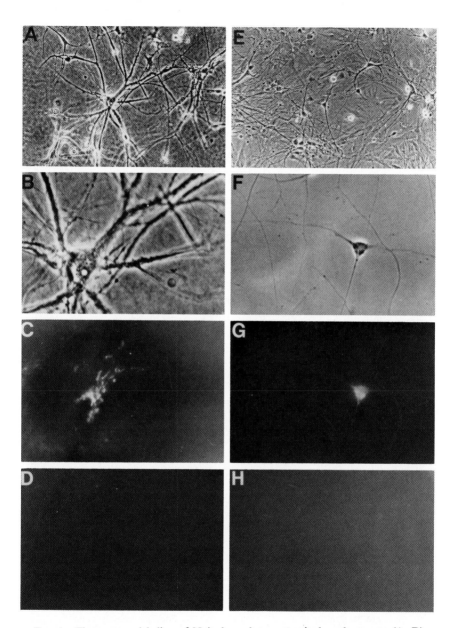

FIG. 4. Fluorescence labeling of Na$^+$ channels on rat spinal cord neurons (A–D) or cortical neurons (E–H) maintained *in vitro* for 6 weeks by TmRhd–LqqV. TmRhd–LqqV binds reversibly to these cells with a $K_d$ of 6.5 n$M$, and a dissociation half-time of 65 minutes at 22°C. The nonspecific staining for LqqV is shown in D and H. The photographs were obtained with a Zeiss Photomicroscope III employed in an FPR system where the laser served as the illumination source. The laser beam was dispersed by both a diffusion lens and an opaque rotating disk placed in the exciting light path. The micrographs were recorded on Tri-X film pushed to ASA 3200 through a Zeiss 63× water immersion 1.2 numerical aperture objective. Note the absence of staining of the underlying nonneuronal cells forming the substratum. (From 11.)

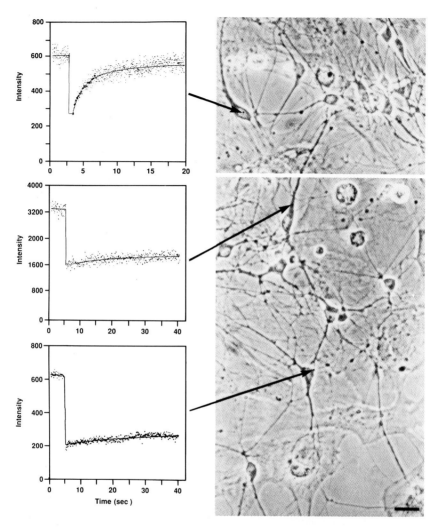

FIG. 5. Representative FPR curves of TmRhd–LqqV on spinal cord neurons. Cells were labeled with 10 n$M$ TmRhd–LqqV and FPR measurements were performed at 22°C. The points represent the photons counted/50 msec dwell time. Solid lines are the computer-generated nonlinear regression best fit fluorescence recovery curves obtained for a lateral diffusion process with a single diffusion coefficient, $D_L$. Photobleaching with the 0.85-$\mu$m beam radius of the l00× objective. The right-hand panels show the location of the photobleaching spot superimposed on a phase-contrast image of a spinal cord neuron preparation. The diffusion coefficients are $2 \times 10^{-9}$ cm$^2$/second at the cell body, $1.6 \times 10^{-10}$ cm$^2$/second at the hillock, and $2.1 \times 10^{-10}$ cm$^2$/second at the synapse with fractional recoveries of 0.9, 0.35, and 0.17, respectively. Bar, 50 $\mu$m.

*b. Organization of Na+ Channels in Muscle.* Using several of these Na$^+$ channel specific toxins, we examined Na$^+$ channel distribution in muscle. Although the segregation of voltage-dependent sodium channels to the hillock of motoneurones and nodes of Ranvier in myelinated axons has received considerable attention, much less is known about the distribution of voltage-dependent Na$^+$ channels on muscle fibers. From the excitability properties of skeletal muscle, it was assumed that Na$^+$ channels are uniformly distributed along the muscle surface. Recently, however, Beam *et al.* showed that the neuromuscular junction is a region of high Na$^+$ conductance and that Na$^+$ channels might be concentrated there.

We explored this possibility by specifically labeling Na$^+$ channels with fluorescent toxins and examining those mechanisms that govern the distribution of voltage-dependent Na$^+$ channels on muscle using microfluorimetry and fluorescence photobleach recovery (FPR). To our surprise, we found that on uninnervated myotubes, Na$^+$ channels are diffusely distributed and freely mobile, whereas innervation, Na$^+$ channels concentrate at neuronal contact sites (Fig. 6). These clustered Na$^+$ channels are immobile and colocalize with acetylcholine receptors (AChRs) (12). At extrajunctional regions, the Na$^+$ channel density is lower and Na$^+$ channels are mobile. The sequelae of Na$^+$ channel accumulation at newly formed synapses is reminiscent of AChR cluster formation, where diffusely distributed AChRs cluster and become immobilized to the neuromuscular junction. However, in contract to AChRs, there do not appear to be regions of preexisting, dense, and immobile Na$^+$ channel patches on uninnervated myotubes. Rather, all Na$^+$ channels are diffusely distributed and very mobile with lateral motion rates of $10^{-9}$ cm$^2$/second before innervation. The factors that specifically control the reorganization and subsequent immobilization of Na$^+$ channels are unknown. Apparently, AChR clustering per se does not induce Na$^+$ channel clustering. Further, the specific recruitment of Na$^+$ channels and the conspicuous absence of Na$^+$ channels colocalizing with all AChR hot spots suggest that the factors governing Na$^+$ channel cluster formation show some degree of temporal and molecular specificity. These results suggest that the nerve induces Na$^+$ channels to redistribute, immobilize, and colocalize with AChRs at sites of neuronal contact. A major question arises whether neuronal contact per se is required or some factor contributed by the incoming nerve is sufficient to induce Na$^+$ channel reorganization. It is known that a factor from neurons increases the number of AChR aggregates on muscle cells as well causing their redistribution. It was found that a similar factor isolated from *Torpedo* electric organ, whose antibodies cross-react with components at the basal lamina of frog muscle, reduced the reorganization of AChRs. Furthermore, AChRs on primary rat muscle cells redistribute to reach junctional site

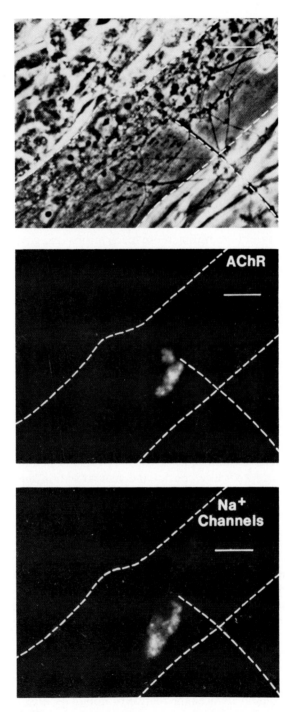

Fɪɢ. 6.  Fluorescence labeling of Na$^+$ channels by TmRhd–LqqV in 11-day-old chick myotubes innervated by spinal cord neurons. AChRs were visualized on the same fiber after labeling with $10^{-7}$ $M$ FITC–$\alpha$-bungarotoxin. Bar, 10 $\mu$m.

densities after exposure to soluble neuronal extracts. At present it is not known whether $Na^+$ channels redistribute with AChRs and whether these factors will induce a reorganization of muscle $Na^+$ channels. We have tested the latter possibility by measuring $Na^+$ channel distribution and mobility of preparations of extracellular matrix. These preparations induce AChR clustering and the formation of hot spots. However, the mechanisms of $Na^+$ channel and AChR segregation appear to be different since diffusely distributed $Na^+$ channels redistribute with AChRs.

### 3. LOCALIZATION AND MOBILITY OF OTHER TOXIN RECEPTORS

*a. Voltage-Dependent $Ca^{2+}$ Channels.* As for the $Na^+$ channels, we have applied the basic techniques described in this review to visualize the distribution and lateral mobility of $Ca^{2+}$ channels in nerve. A new toxin specific for the $Ca^{2+}$ channel, $\omega$-conotoxin has been derivatized with a fluorescent label via acylation and amindinylation and shown to be biologically active by blocking $Ca^{2+}$ currents. The fluorescent derivative has been used to visualize the distribution of $Ca^{2+}$ channels on chick dorsal ganglion cells maintained in tissue culture while the distribution of $Na^+$ channels on these same cells has been examined with fluorescein sea anemone toxin II.

*i. Distribution and mobility.* Examination of the fluorescent micrographs reveals that $Ca^{2+}$ channel distribution/$\omega$-conotoxin receptors on these cells is not homogeneous and that patches or clusters of channels are observed. These dense clusters, which appear to be randomly distributed on the neuronal cell surface, do not seem to colocalize with $Na^+$ channels. These clusters are mobile (30–50% recovery) with lateral mobilities of $10^{-11}$ cm²/second. This is about 10-fold slower than general classes of glycoproteins, which on neurons, move with rates of $10^{-10}$ cm²/second. The dense immobile clusters do not appear to be the result of the segregation of general glycoproteins or lipids on these cells since these proteins appear to be homogeneously distributed and mobile with lipids diffusing at rates of $10^{-9}$ cm²/second.

*ii. Development.* It is known that both $Na^+$ and $Ca^{2+}$ channels contribute to the propagation of the action potential and that the action potential goes from one of $Na^+$ spike to predominantly $Ca^{2+}$ current. In development neuron, binding and electrophysiological studies have examined the developmental appearance of $Na^+$ and $Ca^{2+}$ channels. The interesting question that is raised by these studies is that, in the absence of colocalization of these channels, what are the mechanisms that segregate the channels at different developmental stages. Use of these channel-specific fluorescent neurotoxins will provide insights into those mechanisms that segregate $Na^+$ and $Ca^{2+}$

channels to parts of the cell or the induction of $Ca^{2+}$ channel clusters. At the moment very little is known about the distribution of these channels and the developmental mechanisms that segregate $Na^+$ and $Ca^{2+}$ channels to specific regions of the neuronal cells.

b. *Distribution and Mobility of Chloride Channels and the Glycine Receptor.* One of the most abundant inhibitory neurotransmitters in the central nervous system, glycine, exerts its main effects via a receptor that gates a chloride channel in the subsynaptic membrane. These receptors can contain a modulatory unit, the strychnine receptor, through which ligands of different chemical classes can increase or decrease glycine receptor function. We have prepared a fluorescent congener of strychnine which has $K_d$ of 10 n$M$ on binding to cortical neurons. No information, however, has appeared at the individual cell level or on the mechanisms that regulate this distribution. We have performed some preliminary experiments with fluorescent strychnine and have shown on these receptors that (1) not all neurons in the tissue culture system have these receptors and (2) the highest receptor density appears to be on the neuronal processes that correspond to dendrites and the cell body. No measurable fluorescence is seen on the axon or on the presynaptic terminal. Measurement of the lateral mobility of these receptor at these locations show that dendritic receptors have some mobility with diffusion coefficients of $10^{-10}$ cm$^2$/second, similar to most glycoproteins on these domains, while those on the cell body are immobile. The behavior of these channels – receptors is in marked contrast to voltage-dependent $Na^+$ channels which are diffusely distributed and freely mobile ($10^{-9}$ cm$^2$/second) on the cell body. This is a good example of the differential segregation and molecular specificity displayed by the neuron in the maintenance of its ion channels. This is obviously related to the diverse physiology of the cells, and the mechanisms of signaling via these cells.

# V.  Requirements for Successful Application of Toxin Analogs

First and foremost is that the fluorescent toxin conjugate should behave in an identical manner to the unlabeled molecule. The high affinity must be retained for their receptor sites and the nonspecific binding should be low, on the order of 15–20%. The dissociation rate of the toxin from the cell surface should be sufficiently slow to examine the distribution of the receptors on a time scale of the photography as well as during measurements of the lateral mobility by FPR.

The major problem with the use of fluorescent toxin conjugates is that often their receptors are of low density on cell surfaces. This is true for the voltage-dependent sodium channel and other ion channels in nerve and muscle. Visualization of the AChR is facilitated because it is sufficiently concentrated at the neuromuscular junction. In most cases the toxin conjugate should be prepared with high extinction coefficient and high quantum efficiency probes that absorb beyond the 480-nm region so as not to compete with cellular autofluorescence that can be several times more intense than the specific signal. Thus tetramethylrhodamine, Bo-dippy, fluorescein, and phycoerythrin are fluorophores of choice. In most cases, because of the small size of the toxin and the limited sites that can be modified without loss of biological activity, only one fluorophore per toxin molecule can be accommodated. However, in extreme cases, the signal may be amplified by the preparation of a hapten-bearing toxin (e.g., DNP), followed by anti-DNP antibodies that can be used with a second antibody or in biotin–avidin–biotin–phycoerthryin sandwiches.

With the development of more sophisticated and sensitive cameras and imaging systems, it will be possible to obtain information on the distribution of as few as 100 copies of a cell surface or internal receptor that are labeled with a toxin.

## ACKNOWLEDGMENTS

This work has been supported by grants from the National Institutes of Health and the Muscular Dystrophy Association. I would also like to acknowledge the many contributions of my students and colleagues at Baylor to this work.

## REFERENCES

1. Ceccarelli, B., and Clementi, F. (1979). "Neurotoxins: Tools in Neurobiology." Raven, New York.
2. Schmidt, J., and Raftery, M. A. (1973). *Biochemistry* **12**, 852.
3. Fambrough, D. M. (1979). *Physiol. Rev.* **59**, 165–216.
4. Anderson, M. J., and Cohen, M. W. (1974). *J. Physiol. (London)* **237**, 385–396.
5. Axelrod, D. A., Kuppel, D. E., Schlessinger, J., Elson, E. L., and Webb, W. W. (1976). *Proc. Natl. Acad. Sci. U.S.A.* **77**, 4815–4827.
6. Angelides, K. J. (1981). *Biochemistry* **20**, 4107–4118.
7. Angelides, K. J., and Nutter, T. J. (1983). *J. Biol. Chem.* **258**, 11948–11957.
8. Angelides, K. J., and Nutter, T. J. (1983). *J. Biol. Chem.* **258**, 11958–11967.
9. Angelides, K. J., and Brown, G. B. (1984). *J. Biol. Chem.* **259**, 6117–6126.
10. Darbon, H., and Angelides, K. J. (1984). *J. Biol. Chem.* **259**, 6074–6084.
11. Angelides, K. J., Elmer, L. W., Loftus, D., and Elson, E. L. (1986). *J. Cell Biol.,* in press.
12. Angelides, K. J. (1986). *Nature (London)* **321**, 63–66.
13. Wreggett, K. A. (1986). *J. Receptor Res.* **6**, 95–126.
14. Weiland, M., and Faulstich, M. (1979). *CRC Crit. Rev. Biochem.* **5**, 185–759.

15. Catterall, W. A. (1980). *Annu. Rev. Pharmacol. Toxicol.* **20**, 15–41.
16. Ludwig, D. S., Holmes, R. K., and Schoolnick, G. K. (1985). *J. Biol. Chem.* **260**, 12528–12534.
17. Koch, B. D., Dorflinger, L. J., and Schonbrunn, A. (1985). *J. Biol. Chem.* **260**, 13138–13145.
18. Olsnes, S., Carvagal, E., Sundan, E., and Sanduig, K. (1985). *Biochim. Biophys. Acta* **846**, 334–341.
19. Eldetrawi, A. T., Eldefrawi, M. E., Albuquerque, E. X., Oliveira, A. C., Mansour, N., Adler, M., Daly, J., Brown, G. B., Burgermeister, W., and Witkop, B. (1977). *Proc. Natl. Acad. Sci. U.S.A.* **74**, 2172–2176.
20. Young, A. B., and Synder, S. H. (1973). *Proc. Natl. Acad. Sci. U.S.A.* **70**, 2832–2836.
21. Barak, L. S., Yocum, R. R., Nothnagel, E. A., and Webb, W. (1980). *Proc. Natl. Acad. Sci. U.S.A.* **77**, 980–984.
22. Spiegel, S. (1985). *Biochemistry* **74**, 594–597.
23. Tu, A. T. (1973). *Annu. Rev. Biochem.* **42**, 235–258.
24. Tu, A. T. (1977). *In* "Venoms: Chemistry and Molecular Biology" (Tu, A. T., ed.), Wiley, New York.
25. Katz, B., and Miledi, R. (1964). *J. Physiol. (London)* **170**, 389–396.
26. Fertuck, H. C. and Salpeter, M. M. (1974). *Proc. Natl. Acad. Sci. U.S.A.* **71**, 1376–1381.
27. Podleski, T., Axelrod, D., Ravdin, P., Greenberg, I., Johnson, M. M., and Salpeter, M. (1978). *Proc. Natl. Acad. Sci. U.S.A.*, **75**, 2035–2039.
28. Ravdin, P., and Axelrod, D. A. (1977). *Anal. Biochem.* **58**, 585–592.
29. Stya, M., and Axelrod, D. A. (1983). *Proc. Natl. Acad. Sci. U.S.A.* **80**, 449–453.
30. Role, L. W., Matossian, V. R., O'Brien, R. J., and Fischback, G. D. (1986). *J. Neurosci.* **5**, 2197–2204.
31. Rack, M., Meures, H., Beress, L., and Grunhagen, H. H. (1983). *Toxicon* **21**, 231–237.
32. Tsien, R. W. (1983). *Annu. Rev. Physiol.* **45**, 341.
33. Olivera, B. M., Gray, W. R., Zeijus, R., McIntosh, J. M., Varga, J., Rivier, J., deSantos, V., and Cruz, L. (1985). *Science* **230**, 1338–1343.
34. Rivier, J., Galyean, R., Gray, W. R., Zonooz-Azimi, A., McIntosh, J. M., Cruz, L. J., and Olivera, B. M. (1987). *J. Biol. Chem.* **262**, 1194–1198.
35. Bhattacharyya, P. K., and Bhattacharyya, A. (1981). *Biochem. Biophys. Res. Commun.* **101**, 273–280.
36. Habersetzer-Rochat, C., and Sanpieri, F. (1976). *Biochemistry* **15**, 2254–2261.
37. Guillary, R. T., Rayner, M. D., and O'Arrigo, J. S. (1977). *Science* **196**, 883–885.
38. Tsien, R. Y., Green, D. P. L., Levinson, S. R., Rudy, B. R., and Sanders, J. K. M. (1975). *Proc. R. Soc. London Ser B* **191**, 555–559.
39. Varlet, T.-M., Collignon, N., and Savignac, P. (1978). *Synth. Commun.* **8**, 335–343.
40. Kahn, R. A., and Gilman, A. G. (1986). *J. Biol. Chem.* **261**, 7906–7911.
41. Angelides, K. J., and Nutter, T. J. (1984). *Biophys. J.* **45**, 31–34.

# Chapter 4

# Preparation of Fluorescently Labeled Dextrans and Ficolls

## KATHERINE LUBY-PHELPS

*Department of Chemistry*
*Center for Fluorescence Research in Biomedical Sciences*
*Carnegie-Mellon University*
*Pittsburgh, Pennsylvania 15213*

## I. Introduction

Electroneutral, hydrophilic polysaccharides, such as dextran and Ficoll, are valuable and versatile tools in the biomedical sciences. Traditionally used as inert colloids in perfusion studies and in the isolation of cells and organelles by phase separation or density gradient centrifugation, dextran and Ficoll are also utilized in a variety of experimental strategies that require inert macromolecules. Two examples of this are the use of size-fractionated dextran for studying the permeability of capillaries and basement membranes (e.g., Grotte, 1956; Simionescu and Palade, 1971; Farquhar, 1982), and the use of Ficoll derivatives as carriers for eliciting antibodies to small haptens (Mosier *et al.,* 1974; Inman, 1975).

Dextran is a poly-D-glucose of very high molecular weight, with sparse, short branches that are produced by several strains of bacteria (Granath, 1958; Larm *et al.,* 1971). Commercially available dextran fractions of

59

METHODS IN CELL BIOLOGY, VOL. 29

different average molecular weight ($\overline{M}_w$) are obtained from the natural product of *Leuconostoc mesenteroides* by partial acid hydrolysis, followed by size fractionation. The polydispersity of the commercial product prepared by Pharmacia (Pharmacia Fine Chemicals, Piscataway, New Jersey), defined as the ratio of $\overline{M}_w$ to the number average molecular weight ($\overline{M}_n$), is about 1.6–2.0 (Basedow and Ebert, 1979; Laurent and Granath, 1967). In solution, dextan molecules of $\overline{M}_w$ greater than 2000 are almost perfect statistical coils, and show no detectable self-association (Basedow and Ebert, 1979). Ficoll is a synthetic polymer of sucrose and epichlorohydrin (Holter and Moller, 1956). Its high degree of branching and internal cross-linking make it a more compact molecule than dextran on a molecular-weight basis (Laurent and Granath, 1967; Bohrer *et al.*, 1984). The polydispersity of Ficoll from Pharmacia is about 1.6 (Laurent and Granath, 1967; Bohrer *et al.*, 1984).

The addition of a fluorophore to dextran or Ficoll extends their usefulness to the realms of fluorescence spectroscopy and fluorescence microscopy. The focus may be on either the polysaccharide or the fluorophore. Thus, an environment-insensitive fluorophore may be used as a marker for detecting the polysaccharide, or the polysaccharide may be used as in a carrier for an environment-sensitive fluorophore in order to probe parameters of the chemical environment, such as pH, calcium ion concentration, and solvent polarity.

Fluorescein dextran was originally used by Arfors and Hint (1971) to visualize the microcirculation. The use of fluorescent dextrans in microcirculation studies has been reviewed recently by Thorball (1981).

Like the underivatized polysaccharides, size-fractionated, fluorescent derivatives of dextran and Ficoll can be used as "molecular rulers" in studies of permeability. For example, fluorescein dextran has been used to monitor changes in the permeability of the blood–brain barrier (Mayhan and Heisted, 1985) and to study the diffusion of macromolecules in the extracellular matrix of tumors (Nugent and Jain, 1984). Fluorescein dextran has also been used to estimate the size of nuclear pores in studies of the exchange of macromolecules between cytoplasm and nucleus (Lang *et al.*, 1986; Jiang and Schindler, 1986). By a similar rationale, narrowly size-fractionated fluorescein and rhodamine derivatives of dextran and Ficoll have been used as inert tracer particles to probe the submicroscopic structure of the cytoplasmic ground substance in living cells by fluorescence recovery after photobleaching (FRAP) and digital fluorescence microscopy (Luby-Phelps *et al.*,1986, 1987, 1988; Luby-Phelps and Taylor, 1988).

Using a technique originally developed by Okhuma and Poole (1978), several investigators have used fluorescein derivatives of dextran to measure intracellular pH (e.g., Heiple and Taylor, 1982; Rothenberg *et al.*,

1983; Tanasugarn *et al.*, 1984; Bright *et al.*, 1987; Paradiso *et al.*, 1987) and to follow the acidification of endosomes (e.g., Tycko and Maxfield, 1982; McNeil *et al.*, 1983; Murphy *et al.*, 1984). A modification of this approach has been used to monitor proton uptake by vesicles reconstituted with a $Na^+/H^+$-ATPase (Hara and Nakao, 1986).

The fact that dextran and Ficoll cannot cross biological membranes makes them ideal as markers and probes of specific subcellular compartments. Fluorescein dextran was first reported as a marker for fluid phase pinocytosis by Berlin and Oliver (1980). More recently, a red-fluorescent, cyanine derivative of dextran has been used to mark the pinosome compartment in a multiparameter study of living Swiss 3T3 cells (DeBiasio *et al.*, 1987). Because they do not appear to bind to intracellular components, fluorescent dextran derivatives have been microinjected into living cells as volume markers and as a control for fluorescent analogs of functional macromolecules (Lanni *et al.*, 1985; Gingell *et al.*, 1985; Luby-Phelps *et al.*, 1985; DeBiasio *et al.*, 1987). Dextrans small enough to penetrate the nuclear envelope can be used to mark the total accessible volume of the cell, exclusive of membrane-bounded organelles or vesicles, while dextrans larger than the size of nuclear pores can be microinjected into either the nucleus or the cytoplasm to mark those compartments separately. Internalized fluorescein dextran has recently been utilized in the isolation of endosomes from cell lysates by flow cytometry and fluorescence-activated sorting (Murphy, 1985). Fluorescent dextran or Ficoll trapped within membrane-bounded vesicles can be used to assay vesicle fusion, both *in vivo* (Goren *et al.*, 1984) and *in vitro* (Sowers, 1986; Stutzin, 1986). Fluorescein dextran has been used to assay the efficiency of methods for bulk-loading macromolecules into living cells (McNeil *et al.*, 1984; Fechheimer *et al.*, 1986) and has also been microinjected into embryos as a marker for single cells to obtain fate maps and cell lineages during development (e.g., Gimlich and Braun, 1985). Additional references on the properties and uses of dextran, Ficoll, and fluorescent derivatives of dextran are available in bibliographies furnished upon request from Pharmacia and from Molecular Probes, Inc. (Eugene, Oregon).

Fluorescein isothiocyanate derivatives of dextran (FTC-dextran)[1] are available from Pharmacia or Sigma (St. Louis, Missouri,), and Molecular Probes now offers dextrans labeled with several different fluorophores. Molecular Probes dextrans generally have high substitution ratios and are intended for studies where sensitive detection of small amounts of material

---

[1]Abbreviations: FITC, fluorescein isothiocyanate; FTC-, fluorescein thiocarbamoyl-; TRITC, tetramethylrhodamine isothiocyanate; TRTC-, tetramethylrhodamine thiocarbamoyl; BSA, bovine serum albumin; AECM, aminoethylcarboxymethyl.

is desired. However, at present, no fluorescent derivatives of dextran of $\overline{M}_w$ > 150,000 are available commercially, nor are any fluorescent derivatives of Ficoll. In addition, many useful fluorescent probes have been developed recently, but are not yet available on dextrans (for review, see Waggoner, 1986). Fortunately, custom-tailored fluorescent derivatives can be prepared from commercially available dextran (e.g., T500, T1000, T2000 from Pharmacia or Sigma) and Ficoll (Ficoll 70, Ficoll 400 from Pharmacia) by the methods described below.

TABLE I

DeBelder and Granath Method

*Materials*
| | |
|---|---|
| Anhydrous DMSO (10 ml) | Dextran or Ficoll (1 g) |
| Pyridine (0.3 ml) | FITC or TRITC (100 mg) |
| Dibutyl tin dilaurate (20 μl) | 95% Ethanol (1 liter) |

*Procedure*

1. Mix 10 ml anhydrous dimethyl sulfoxide (DMSO), 0.3 ml pyridine, and 20 μl dibutyl tin dilaurate in a 50 ml KIMAX screw top tube. After opening, DMSO should be stored with molecular sieve in a closed container with desiccant.
2. Add 1 g dextran. Place tube in 95°C water bath and allow dextran to dissolve. This may take several hours.
3. Add 100 mg FITC or TRITC, mix well, and incubate at 95°C for 2 hours.
4. Divide reaction mixture evenly into two or three 30-ml Corex tubes and add 15 ml ethanol to each tube while vortexing. This precipitates the dextran, leaving free dye in the supernatant.
5. Add an additional 10 ml ethanol to each tube.
6. Centrifuge at room temperature for 10 minutes at 10,000 g to collect labeled dextran.
7. Remove the supernatant and test it for unprecipitated dextran by adding ethanol. If none, discard supernatant. Otherwise, add more ethanol and centrifuge as before to collect precipitate.
8. Add 10 ml ethanol to each pellet. Resuspend the pellets as well as possible. They will be very gooey.
9. Centrifuge as before to collect precipitate.
10. Repeat steps 7–9 twice more. The pellets will become less gooey and more brittle each time.
11. After final spin, discard the supernatant and resuspend the pellets in 4 ml glass distilled water. Let sit at room temperature for 1 hour.
12. Transfer resuspended pellets to dialysis tubing and dialyze versus 2 × 4 liters distilled water overnight to redissolve labeled dextran. For dextrans of MW ≤ 20,000, low-molecular-weight dialysis tubing should be used to avoid loss of material to the dialysis medium.
13. Clarify dextran solution by centrifugation at 20,000 rpm in a Sorvall SS-34 rotor, or equivalent, for 30 minutes at room temperature.
14. Shell freeze and lyophilize the supernatant. Lyophilized dextran should be stored as a powder, desiccated, at −20°C for best results.

# II.  Methods

## A.  Preparation of Fluorescent Derivatives of Polysaccharides

Detailed procedures for preparing fluorescent derivatives of polysaccharides, such as dextran and Ficoll, are given in Tables I–III. Comments on the procedures are presented below.

TABLE II

INMAN METHOD: A. PREPARATION OF AECM-DERIVATIVES

*Materials*

| | |
|---|---|
| Chloroacetic acid (6.44 g) | 2 $M$ NaH$_2$PO$_4$ (1 ml) |
| 5 $N$ NaOH (20 ml) | 6 $N$ HCl, for pH titration |
| 10 $N$ NaOH (5 ml) | 1 $N$ NaOH, for pH titration |
| Ficoll or dextran | Ethylenediamine dihydrochloride (1 g) |
| 1-ethyl-3-(3-dimethylaminopropyl) carbodiimide (1 g) | |

*Procedures*

1. Prepare a stock solution of 1.35 $M$ sodium chloroacetate by dissolving 6.44 g of chloroacetic acid in a mixture of 30 ml of distilled water and 13.5 ml of 5.0 $N$ NaOH. Cool to 25°C, adjust pH to 6.8–7.2 with either 5 $N$ NaOH or 10% (w/v) chloroacetic acid, and dilute to 50 ml with distilled water. This should be made fresh each time.
2. Dissolve 1.33 g of Ficoll or dextran in 18.5 ml of 1.35 $M$ sodium chloroacetate, and place in 25°C water bath.
3. Add 5 ml of 10 $N$ NaOH, followed by 1.5 ml distilled water, to bring the total volume to 25 ml.
4. Incubate for 30 minutes.
5. Stop the reaction with 0.2 ml of 2 $M$ NaH$_2$PO$_4$.
6. Titrate the mixture to pH 7.0 with 6 $N$ HCl.
7. Dialyze extensively (4 × 2 liters) at 4°C against distilled water (saturated with toluene to retard microbial growth. Do not use azide). The dialyzed product may either be used directly, or lyophilized for storage.
8. Measure the volume of the dialyzed solution and add ethylenediamine dihydrochloride to 1.0 $M$ (14.3 g/100 ml). If using lyophilized sample, dissolve 25 mg/ml in distilled water and add 5.7 mg ethylenediamine dihydrochloride per mg of sample.
9. Titrate to pH 4.7 with 1 $N$ NaOH (approximately one drop).
10. Add 0.5 mg of 1-ethyl-3-(3-dimethylaminopropyl)carbodiimide hydrochloride per mg of Ficoll or dextran (665 mg for 1.33 g), with stirring, over the course of 10 minutes, while maintaining the pH near 4.7 with 1 $N$ NaOH.
11. React for 3.5 hours at room temperature with stirring, while maintaining the pH near 4.7.
12. Dialyze at 4°C versus 3 × 4 liters of toluene-saturated distilled water, followed by one change of 4 liters of distilled water without toluene.
13. Lyophilize and store desiccated.

TABLE III

INMAN METHOD: B. LABELING OF AECM-DERIVATIVES WITH FITC OR TRITC

*Materials*

AECM-Ficoll or -dextran (100 mg)

10 m$M$ Carbonate–bicarbonate buffer, pH 9.2 (3 ml)

FITC or TRITC (15 mg)

0.1 $N$ NaOH for pH titration

Sephadex G-25

*Procedures*

1. Dissolve 100 mg AECM-Ficoll or -dextran in 2 ml of carbonate–bicarbonate buffer at pH 9.2.
2. Dissolve dye in 1 ml of same buffer:
    a. Dissolve 15 mg of TRITC in 400 $\mu$l dimethyl formamide (DMF) and add dropwise, with vortexing, to 1 ml of carbonate–bicarbonate buffer. Adjust pH to $\geq$ 90 with 0.1 $N$ NaOH.
    b. Dissolve 15 mg of FITC in 1 ml of buffer. Adjust pH to $\geq$ 9.0 with 0.1 $N$ NaOH. FITC may not go into solution completely until pH has been adjusted.
3. Add dye solution dropwise with stirring to AECM-Ficoll or -dextran solution.
4. Incubate at 40°C for 30 minutes.
5. Desalt on 1 × 30 cm column of Sephadex G-25 to remove unreacted dye.
6. Dialyze versus 2 × 1 liter of distilled water and lyophilize for storage.

## 1. DIBUTYL TIN DILAURATE METHOD (DeBelder and Granath, 1973)

This procedure has the advantage that it is a one-step process, and is therefore quick and easy. It is the method by which Pharmacia labels the fluoresceinyl-dextrans that are currently available through them and through Sigma. The same method should be directly applicable to labeling Ficoll, although we have not used it for this purpose.

The product of the procedure, as described here, generally has a substitution ratio of $\geq$ 0.02 fluorophore molecules per sugar residue. The linkage is fairly stable in aqueous solution at pH 7.0 (DeBelder and Granath, 1973). This method is convenient and suitable for studies in which the degree of substitution doesn't need to be controlled accurately and detectability of small amounts is the limiting factor, for example, in microcirculation studies, or fluorescence activated cell sorting.

When the method is used to prepare tetramethylrhodamine (TRTC-) derivatives, a minor complication results, apparently due to the hydrophobicity of the fluorophore. TRTC-dextran is slightly soluble in ethanol, so that repeated washing in ethanol sometimes results in progressive loss of labeled material. After two washes, it is recommended that the labeled dextran be dialyzed against distilled water and then desalted on a short column of Sephadex G-25 to remove unreacted dye. If necessary, chromatography on a hydrophobic resin can also be used to remove unreacted dye (for example, see Materials and Methods: Meigs and Wang, 1986).

## 2. AMINOETHYLCARBOXYMETHYL METHOD (Inman, 1975)

This procedure is preferable to DeBelder and Granath for preparing derivatives with a low degree of substitution for studies in which it is necessary to alter the charge and hydrophilicity of the polysaccharide as little as possible. We have used fluorescent Ficolls prepared in this way as inert tracer particles to probe the structure of cytoplasm in living cells by FRAP (Luby-Phelps *et al.*, 1987). Fluorescent dextrans prepared by this method have been used in a recent multiparameter fluorescence study of locomoting Swiss 3T3 cells (DeBiasio *et al.*, 1987).

The procedure involves conversion of hydroxyl groups to O-carboxymethyl ethers and subsequent amidation of these groups by ethylene diamine in the presence of a water soluble carbodiimide (Fig. 1). This aminoethylcarboxymethyl (AECM) derivative can then be reacted directly with amino-selective reagents, such as FITC and TRITC, to obtain fluorescent derivatives.

The number of carboxyls introduced into the polymer at the first step is a function of time and temperature (Fig. 2). The final degree of substitution can thus be controlled by varying these parameters at the first step and using excess ethylene diamine and excess fluorophore at subsequent steps. The procedure detailed in Table II introduces about 11 mol of amino groups/mol of Ficoll 400, or 0.01 mol/mol of sucrose (0.005/sugar residue). Adjustment of the duration of the carboxymethylation reaction may be necessary to obtain the desired degree of substitution when the method is applied to polysaccharides of molecular weights other than 400,000. Carboxymethylation can also be carried out at 40°C in order to introduce large numbers

$$H \overline{\phantom{x}} OH \quad OH \overline{\phantom{x}} H \xrightarrow{(1)} H \overline{\phantom{x}} O\text{-}CH_2 \overset{O}{\overset{\|}{C}}OH \quad OH \overline{\phantom{x}} H \xrightarrow{(2)} H \overline{\phantom{x}} O\text{-}CH_2 \overset{O}{\overset{\|}{C}}NCH_2CH_2NH_2 \quad OH \overline{\phantom{x}} H$$

CM-Derivative          AECM-Derivative

FIG. 1. Preparation of AECM-polysaccharides. Reaction (1): The polysaccharide is combined with chloroacetate in strongly alkaline aqueous solution with the elimination of NaCl and formation of a stable carboxymethyl ether linkage. Reaction (2): A monoamide is formed with ethylenediamine, present in large excess. Carboxyl groups are activated with a water-soluble carbodiimide in aqueous solution, pH-stated at 4.7. [Redrawn with permission from J. K. Inman. Thymus-independent antigens: the preparation of covalent, hapten-Ficoll conjugates. *J. Immunol.*, **114**, 704–709, © by American Association of Immunologists (1975).]

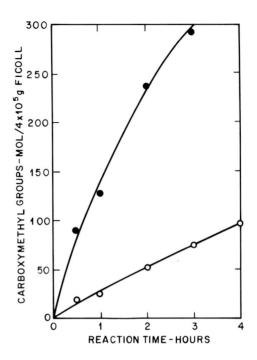

FIG. 2. Time course of the carboxymethylation reaction [Reaction (1), Fig. 1] at 25°C (O) and 40°C (●). Initial concentrations were as follows: Ficoll 53.2 g/liter, sodium chloroacetate 1 $M$, and NaOH 2 $M$; the solvent was water. Reaction was terminated at various times by neutralization of samples with 5 $N$ HCl. Following exhaustive dialysis against water, samples were analyzed for carboxymethyl content by acid–base titration and calculation of hydrogen ion-binding capacity. [With permission from J. K. Inman. Thymus-independent antigens: the preparation of covalent, hapten-Ficoll conjugates. *J. Immunol.*, **114**, 704–709, © by American Association of Immunologists (1975).]

of carboxyl groups while minimizing exposure to harsh conditions. As pointed out by Inman (1975), this may be desirable when the polysaccharide is to be used as a carrier for immunization with a hapten.

The number of free carboxymethyl groups at any step can be determined by titrating a known dry weight of material between pH 2.2 and 7.0, since the carboxymethyl groups have a $pK_a$ of 3.6. The number of free amino groups of either the AECM or the fluorescent derivatives can be determined by titration from pH 5.6 to 11.0, since the amino groups have a $pK_a$ of 9.2 (Inman, 1975). When titrating heavily labeled material, the $pK_a$ of the fluorophore must be taken into account. Free amines can also be determined by the ninhydrin method, outlined in Table IV (see also below).

The labeling procedure detailed in Table III, which uses a concentration of FITC or TRITC in 10-fold molar excess of the number of free amines on

TABLE IV

NINHYDRIN TEST FOR FREE AMINES

---

*Materials*
  SnCl$_2$ · 2H$_2$O (80 mg)
  Ninhydrin (2 g)
  0.2 *M* citrate buffer, pH 5.0
  50% aqueous *n*-propanol (500 ml)
  1 m*M* glycine or glucosamine (10 ml)
  Methyl cellulsolve (ethylene glycol monomethyl ether) (50 ml)
*Procedure*
  1. Ninhydrin solution (100 ml):
     a. Dissolve 80 mg reagent grade SnCl$_2$ · 2H$_2$O in 50 ml of 0.2 *M* citrate buffer (4.3 g citric acid, 8.7 g Na$_3$ citrate · 2H$_2$O in 250 ml distilled water, pH to 5.0).
     b. Add this solution to 50 ml methyl cellusolve containing 2 g of dissolved ninhydrin. Store in cold.
  2. Take 50 $\mu$l aqueous samples of material to be tested.
  3. Add 150 $\mu$l ninhydrin solution with vortexing.
  4. Heat in a boiling water bath for 20 minutes, then cool in a room temperature water bath.
  5. Add 800 $\mu$l of 50% aqueous *n*-propanol while vortexing.
  6. Let color develop for 10 minutes at room temperature. Blue color indicates a positive test.

---

the AECM-polysaccharide, produces a final fluorescent product with a negligible content of free carboxyls or free amines. These derivatives have a very low net surface charge, as shown by flat bed electrophoresis at pH 7.0 in agarose gels. Fluorescein- or rhodamine-labeled polysaccharides with a low degree of substitution show negligible migration in the electric field during a time in which both fluorescein isothiocyanate and fluorescein-labeled BSA migrate several centimeters toward the cathode.

In principle, AECM-Ficoll or -dextran can be labeled with any amino-selective reagent. However, we have found that lissamine rhodamine B sulfonyl chloride (Molecular Probes, Eugene, Oregon), apparently labels sites other than amino groups since we obtained very high substitution ratios before all the amino groups were blocked. This excess dye was not removed by boiling in SDS, suggesting that it was covalently bound to the polysaccharide.

While all the linkages involved in the synthesis are reported to be relatively stable (Inman, 1975), we have found that a low molecular weight, fluorescent contaminant appears when the labeled polysaccharides are stored in aqueous solution at 4°C for extended periods. We have not determined whether this is due to release of dye or to digestion of the polysaccharide by microorganisms. The labeled polysaccharides have proved to be quite stable when stored in aqueous solutions at ≤−20°C or when lyophilized to a powder and stored with desiccant at ≤4°C.

# B.   Characterization of Fluorescent Polysaccharides

### 1.   Degree of Substitution

The degree of substitution can be determined with reasonable accuracy by reading the absorbance of the fluorophore for a known concentration of fluorescent polysaccharide (dry wt/vol). For FTC-derivatives, the absorbance is read at 495 nm in a buffer of pH 8.0, and a molar extinction coefficient ($\varepsilon$) of $7.5 \times 10^4$ $M^{-1}$ cm$^{-1}$ (Haugland, 1985) is used to calculate the concentration of fluorescein. For TRTC-derivatives, absorbance is read at 554 nm at pH 7.2, and an $\epsilon$ of $5.5 \times 10^4$ $M^{-1}$ cm$^{-1}$ (Haugland, 1985) is used. The approximate molar ratio of fluorophore to polysaccharide can be determined by reference to the nominal $\overline{M}_w$ of the polysaccharide. If this is unknown, as, for example, after size-fractionation, or if greater precision is desired, the anthrone reaction (Table V) can be used to determine the glucose or sucrose content of a given dry weight of sample.

### 2.   Determination of Free Amino Groups by Ninhydrin Assay (Spackman et al., 1958)

Details of a method for determining the content of free amino groups on AECM-polysaccharides by the ninhydrin assay are given in Table IV. For

TABLE V

Anthrone Determination of Sugars (Jermyn, 1975)

*Materials*

| Conc. HCl | Anthrone |
|-----------|----------|
| 90% Formic acid | 80% (v/v) Sulfuric acid |
| Glucose or sucrose | |

*Procedure*

1. Place a 1 ml aqueous sample of the material to be assayed in a 25-ml test tube.
2. Add 1 ml of concentrated HCl and 0.1 ml of 90% formic acid.
3. Add slowly to avoid excessive foaming, 8 ml of freshly prepared anthrone reagent (200 $\mu$g/ml anthrone dissolved at room temperature in 80% (v/v) sulfuric acid).
4. Mix thoroughly, heat tubes for 12 minutes in a boiling water bath, and then cool immediately in a cold-water bath.
5. When cool, mix on a Vortex mixer and allow to stand for 5 minutes to allow bubbles to disperse.
6. Read absorbance at 630 nm and compare to a standard curve (glucose for dextran, sucrose for Ficoll) to determine micrograms of monosaccharide. The assay is linear from 20 to 100 mg of total carbohydrate in a 1-ml sample. For assay of smaller amounts, the entire assay may be scaled down by a factor of ten. Under these conditions it is linear from 2 to 10 $\mu$g in 0.1 ml of sample.

quantitative tests, the absorbance of each sample at 570 nm can be read and compared to a standard curve of glycine or glucosamine. When performed as described here, the standard curve is linear over the range of 0–50 nmol of amino groups.

A complication arises when using the ninhydrin test to quantitate free amines on rhodamine derivatives because the absorbance maximum of the fluorophore ($\approx$ 550 nm) is very close to the wavelength at which the assay is read (570 nm). In this case, it is better to quantitate free amines on the AECM-polysaccharide before reacting it with the fluorophore and then subtract the number of fluorophores (determined from the absorbance of a known dry weight of sample, as described above) to obtain the number of unreacted amino groups. Alternatively, the content of free amino groups may be determined by the fluorescamine assay (Udenfriend et al., 1972; Weigele et al., 1972). In cases where quantitation is not important, the ninydrin test can also be used qualitatively as a colorimetric indicator of the presence of unreacted amine groups on fluorescent polysaccharide derivatives. Fluorescein and rhodamine do not appear to interfere with the production of color in the ninhydrin test, or to give false positives. However, other fluorophores should be tested for these possibilities before applying the ninhydrin test in their presence.

## 3.   SIZE-FRACTIONATION

Polysaccharides can be size fractionated by gel permeation chromatography (Granath and Flodin, 1961; Laurent and Granath, 1967). Residual free dye is removed at the same time. In general, the best choice of a resin for size-fractionation can be determined based on the fractionation range of the resin (for polysaccharides) relative to the $\overline{M}_w$ of the unfractionated polysaccharide. This information can usually be obtained from the manufacturer of the chromatography resin. The column can be eluted with any dilute aqueous buffer, provided a small amount of salt ($\geq$ 10 m$M$) is present to minimize interaction of the polysaccharide with the column resin.

To obtain adequate amounts of narrowly fractionated material, it is necessary to load a relatively large volume of fairly concentrated material on a long column. For example, we routinely fractionate FTC-Ficoll 400 by loading 100 mg of material at 25 mg/ml on a 3 $\times$ 150-cm column of Sepharose CL-6B (Pharmacia), and collecting 5-ml fractions (Luby-Phelps et al., 1987). The upper limits of volume and concentration that can be applied to the column to maximize the yield will depend on the dimensions of the column and the incremental viscosity of the polysaccharide. As for gel permeation chromatography in general, a sample volume that is too large relative to the elution volume of the column, or a sample viscosity that is too high relative to the viscosity of the eluant, will result in poor fractionation (Pharmacia, 1981). The relative viscosity for dextran and Ficoll at a given concentration can be estimated by reference to the product literature provided by Pharmacia for these polysaccharides. Selected column fractions should be dialyzed extensively against distilled water, lyophilized to a powder, and stored with desiccant.

### 4.  MEASUREMENT AND MEANING OF HYDRODYNAMIC RADIUS

Size-fractionated polysaccharides are not necessarily monodisperse. Although polydispersity will eventually become negligible with repeated chromatography (Laurent and Granath, 1967; Basedow and Ebert, 1979; Bohrer et al., 1984), this is not recommended for applications where yield is important. Thus, polydispersity may have to be taken into account when using size-fractionated dextran or Ficoll molecules as "molecular rulers" to measure the size of pores or channels in permeability studies (for more extensive discussion, see Luby-Phelps et al., 1988).

The average molecular size of the distribution of particles in each fraction can be described by a mean hydrodynamic radius ($\bar{R}_H$). The $\bar{R}_H$ of the particles contained in each column fraction can be determined from the aqueous diffusion coefficient, $D$, which, for fluorescent derivatives, is measured most easily by FRAP (Axelrod et al., 1976; Yguerabide et al., 1982). The Einstein equation is used to define $\bar{R}_H = kT/6\pi\eta D$, where $k$ is Boltzmann's constant, $T$ is absolute temperature, and $\eta$ is the viscosity of the solvent. When 100 mg of FTC-Ficoll 400 at 25 mg/ml is fractionated on Sepharose CL-6B, as described above, it is found that $\bar{R}_H$ decreases by about 2 Å between successive 5-ml fractions. Thus, from unfractionated Ficoll 400, size-fractions ranging in $\bar{R}_H$ from 30- to 230 Å can be obtained. Size-fractions with a slightly larger $\bar{R}_H$ can be obtained by pooling and concentrating the void volumes from several column runs and chromatographing them together on Sepharose CL-4B.

For permeability studies, it is necessary to know how the hydrodynamic radius ($R_H$) of a monodisperse tracer particle is related to the size parameter that determines exclusion of the particle from a pore or channel. According to Tanford (1961) the radius of gyration, $R_G$, determines the distance of closest approach of a macromolecule to a barrier. Recently, Cassasa (1985) has proposed the mean external diameter ($\bar{X}$) as a more appropriate size parameter, based on thermodynamic considerations. The relationship of $\bar{X}$ to $R_H$ will depend on the conformation and flexibility of the macromolecule in question. For a hard sphere, $\bar{X} \approx 2R_H$. For a random coil, $\bar{X}$ is only slightly larger (13%) than $2R_G$. Thus, for random coils, radius of gyration remains a convenient and reasonably precise descriptor of molecular dimensions. In aqueous solution, dextran molecules larger than 2000 D appear to be almost ideal statistical coils in spite of their branching (Basedow and Ebert, 1979). According to the Kirkwood-Riseman result, for an ideal statistical coil, $R_H = 0.665 R_G$. (Tanford, 1961). By including a correction for the degree of branching, Granath (1958) calculated that $R_H = 0.65 R_G$ for the type of dextran provided by Pharmacia. This value was closely approximated by her experimental data. Thus, $R_G$ probably best describes the dimensions of a dextran molecule, and can be approximated by $R_H/0.65$. In contrast to dextran, Ficoll has a much more compact configuration and a well-defined size that is more accurately described by $R_H$ (Laurent and Granath, 1967; Deen et al., 1981; Bohrer et al., 1984).

A possible drawback to using dextrans in permeability studies is that they are highly flexible molecules. Thus, although their average configuration is described by the radius of gyration, their deformability may enable them to fit through pores or channels of diameter $\leq 2 R_G$ (e.g., Bohrer et al., 1984). In addition, dextrans of $\leq 2000$ D apparently have a rigid, rodlike conformation (Granath, 1958; Basedow and Ebert, 1979) and should not be used in permeability studies unless this is taken into account.

# III. Further Considerations in the Use of Fluorescent Dextrans and Ficolls

In working with polysaccharides, it should be remembered that dextran is susceptible to attack by the dextranases secreted by some microorganisms, and that the glycosidic linkages of polysaccharides are readily hydrolyzed at pH $\leq 3$, especially at high temperature. If antimicrobial agents are employed during the preparation or storage of the fluorescent derivatives, they should be dialyzed out of the preparation before its use with living cells.

The procedures described in this chapter are aimed at obtaining fluorescently labeled polysaccharides with minimal net charge at pH 7.0. For studies where it is of interest to compare the behavior of electroneutral particles with the behavior of charged particles, these procedures can be adapted to leave some of the carboxyl or amino groups unmodified, thus giving a final product with either a net negative or a net positive charge.

The labeling procedures can also be adapted so that each polysaccharide molecule is lableled with two different fluorophores. This would be very useful, for example, for mapping intracellular pH by the ratio of fluorescein to rhodamine fluorescence. This method would have a better signal-to-noise ratio than the current method involving measurement of the ratio of fluorescein fluorescence at two excitation wavelengths, and having the two fluorophores on the same carrier would eliminate the potential problem of differential partitioning of the fluorescein and rhodamine probes within the cell. Similarly, each polysaccharide molecule could be labeled with a fluorophore and a photoactivatable cross-linker in order to create a fluorescent derivative that could be fixed in place within cells.

## ACKNOWLEDGMENTS

The author gratefully acknowledges Dr. Paul McNeil, Dr. Lauren Ernst, Dr. Fred Lanni, Dr. D. Lansing Taylor, and Phil Castle for their advice and assistance in the labeling and characterization of polysaccharides. Thanks also to Dr. Lauren Ernst and Dr. Fred Lanni for critical reading of the manuscript. The author's research is supported by NSF grant DCB86-16089.

## REFERENCES

Arfors, K.-E., and Hint, H. (1971). *Microvasc. Res.* **3**, 440.

Axelrod, D., Koppel, D. E., Schlessinger, J., Elson, E., and Webb., W. W. (1976). *Biophys. J.* **16**, 1005–1069.

Basedow, A. M., and Ebert, K. H. (1979). *J. Polym. Sci. Polym. Symp.* **66**, 101–115.

Berlin, J. M., and Oliver, J. M. (1980). *J. Cell Biol.* **85**, 660–671.

Bohrer, M. P., Patterson, G. D., and Carroll, P. J. (1984). *Macromolecules* **17**, 1170–1173.

Bright, G. R., Fisher, G. W., Rogowska, J., and Taylor, D. L. (1987). *J. Cell Biol.* **104**, 1019–1033.

Casassa, E. F. (1985). *J. Polym. Sci. Polym. Symp.* **72**, 151–160.

DeBelder, A. N., and Granath, K. (1973). *Carbohydr. Res.* **30**, 375–378.

DeBiasio, R., Bright, G. R., Ernst, L. A., Waggoner, A. S., and Taylor, D. L. (1987). *J. Cell Biol.* **105**, 1613–1622.

Deen, W. M., Bohrer, M. P., and Epstein, N. B. (1981). *Am. Inst. Chem. Eng. J.* **27**, 952–959.

Farquhar, M. G. (1982). *In* "Cell Biology of Extracellular Matrix" (E. D. Hay, ed.), pp. 335–378. Plenum, New York.

Fechheimer, M., Denny, C., Murphy, R., and Taylor, D. L. (1986). *J. Cell Biol.* **40**, 242–247.

Gimlich, R. L., and Braun, J. (1985). *Dev. Biol.* **109**, 509–514.

Gingell, D., Todd, I., and Bailey, J. (1985). *J. Cell Biol.* **100**, 1334–1338.

Goren, M. B., Swendsen, C, L., Fiscus, J., and Miranti, C. (1984). *J. Leukocyte Biol.* **36**, 273–292.

Granath, K. A. (1958). *J. Colloid Sci.* **13**, 308–328.

Granath, K. A., and Flodin, P. (1961). *Makromol. Chem.* **48**, 160–171.

Grotte, G. (1956). *Acta Chir. Scand. Suppl.* **11**, 1–84.

Hara, Y., and Nakao, M. (1986). *J. Biol. Chem.* **261**, 12655–12658.

Haugland, R. P. (1985) "Handbook of Fluorescent Probes and Research Chemicals," Molecular Probes, Inc., Eugene, Oregon.

Heiple, J., and Taylor, D. L. (1982). *In* "Intracellular pH: Its Measurements, Regulation and Utilization in Cellular Function," (R. Nuccitelli and D. W. Deamer, eds.), pp. 22–54. Liss, New York.

Holter, H., and Moller, K. M. (1956). *Exp. Cell Res.* **15**, 631–632.

Inman, J. K. (1975). *J. Immunol.* **114**, 704–709.

Jermyn, M. A. (1975). *Anal. Biochem.* **68**, 332–335.
Jiang, L.-W., and Schindler, M. (1986). *J. Cell Biol.* **102**, 853–858.
Lang, I., Scholz, M., and Peters, R. (1986). *J. Cell Biol.* **102**, 1182–1190.
Lanni, F., Waggoner, A. S., and Taylor, D. L. (1985). *J. Cell Biol.* **100**, 1091–1102.
Larm, O., Lindberg, B., and Svensson, S. (1971). *Carbohydr. Res.* **20**, 39–48.
Laurent, T. C., and Granath, K. A. (1967). *Biochem. Biophys. Acta* **136**, 191–198.
Luby-Phelps, K., Lanni, F., and Taylor, D. L. (1985). *J. Cell Biol.* **101**, 1245–1256.
Luby-Phelps, K., Taylor, D. L., and Lanni, F. (1986). *J. Cell Biol.* **102**, 2015–2022.
Luby-Phelps, K., Castle, P. E., Lanni, F., and Taylor, D. L. (1987). *Proc. Natl. Acad. Sci. U.S.A.* **84**, 4910–4913.
Luby-Phelps, K., and Taylor, D. L. (1988). *Cell Motil.* **10**, 28–37.
Luby-Phelps, K., Lanni, F., and Taylor, D. L. (1988). *Annu. Rev. Biophys. Chem.* **17**, 369–396.
McNeil, P. L., Tanasugarn, L., Meigs, J., and Taylor, D. I. (1983). *J. Cell Biol.* **97**, 692–702.
McNeil, P. L., Murphy, R. F., Lanni, F., and Taylor, D. L. (1984). *J. Cell Biol.* **98**, 1556–1564.
Mayhan, W. G., and Heistad, D. D. (1985). *Am. J. Physiol.* **248**, H712–H718.
Meigs, J. B., and Wang, Y-L. (1986). *J. Cell Biol.* **102**, 1430–1438.
Mosier, D. E., Johnson, B. M., Paul, W. E., and McMaster, P. R. B. (1974). *J. Exp. Med.* **139**, 1354–1360.
Murphy, R. F. (1985). *Proc. Natl. Acad. Sci. U.S.A.* **82**, 8523–8526.
Murphy, R. F., Powers, S., and Cantor, C. R. (1984). *J. Cell Biol.* **98**, 1757–1762.
Nugent, C. J., and Jain, R. K. (1984). *Cancer Res.* **44**, 238–24.
Okhuma, S., and Poole, B. (1978). *Proc. Natl. Acad. Sci. U.S.A.* **75**, 3327–3331.
Paradiso, A. M., Tsien, R. Y., and Machen, R. E. (1987). *Nature (London)* **325**, 447–450.
Pharmacia Fine Chemicals (1981). "Gel Filtration: Theory and Practice." Rahms i Lund, Stockholm.
Rothenberg, P., Glaser, L., Schlesinger, P., and Cassel, D. (1983). *J. Biol. Chem.* **258**, 12644–12653.
Simionescu, N., and Palade, G. E. (1971). *J. Cell Biol.* **50**, 616–624.
Sowers, A. E. (1986). *J. Cell Biol.* **102**, 1358–1362.
Spackman, D. H., Stein, W. H., and Moore, S. (1958). *Anal. Chem.* **30**, 1190–1206.
Stutzin, A. (1986). *FEBS Lett.* **197**, 274–280.
Tanasugarn, L., McNeil, P. L., Reynolds, G., and Taylor, D. (1984). *J. Cell Biol.* **98**, 717–724.
Tanford, C. (1961). "Physical Chemistry of Macromolecules." Wiley, New York.
Thorball, N. (1981). *Histochemistry* **71**, 209–233.
Tycko, B., and Maxfield, F. (1982). *Cell* **28**, 643–651.
Udenfriend, S., Stein, S., Böhler, P., Dairman, W., Leimgruber, W., and Weigele, M. (1972). *Science* **178**, 871–872.
Waggoner, A. S. (1986). *In* "Applications of Fluorescence in the Biomedical Sciences" (D. L. Taylor, A. S. Waggoner, R. F. Murphy, F. Lanni, and R. R. Birge, eds.), pp. 3–28. Liss, New York.
Weigele, M., De Bernardo, S. L., Tengi, J. P., and Leimgruber, W. (1972). *J. Am. Chem. Soc.* **94**, 5927–5928.
Yguerabide, J., Schmidt, J. A., and Yguerabide, E. E. (1982). *Biophys. J.* **39**, 69–75.

# Chapter 5

# A Fluorescent Derivative of Ceramide: Physical Properties and Use in Studying the Golgi Apparatus of Animal Cells

RICHARD E. PAGANO

*Department of Embryology*
*Carnegie Institution of Washington*
*Baltimore, Maryland 21210*

## I. Introduction

The synthesis, molecular sorting, and intracellular transport of lipids in animal cells are important processes to the cell biologist interested in membrane assembly and membrane dynamics in animal cells. We have developed a series of fluorescent lipid derivatives which are useful in studying these processes (reviewed in Pagano and Sleight, 1985). Because of their fluorescent properties, we can examine the distribution of these lipids and their metabolites within *living* cells and then correlate that distribution

75

with the metabolism of these compounds as studied by conventional lipid biochemical procedures.

One lipid, a fluorescent analog of ceramide, is particularly interesting because of its ability to prominently "stain" the Golgi apparatus of living cells (Lipsky and Pagano, 1983, 1985a,b; Pagano and Martin, 1988) and to provide a means for studying the traffic of sphingolipids through this organelle (Lipsky and Pagano, 1983, 1985a; van Meer *et al.*, 1987). When cells are treated with $C_6$-NBD-ceramide at low temperature, washed, and warmed to 37°C, the Golgi apparatus and later the plasma membrane become intensely fluorescent (Lipsky and Pagano, 1983, 1985a). During this redistribution of intracellular fluorescence, the $C_6$-NBD-ceramide is metabolized to fluorescent sphingomyelin and glucosylceramide. Consistent with the increasing fluorescence at the plasma membrane over time, increasing amounts of each fluorescent metabolite can be removed from the cell surface by back-exchange to nonfluorescent acceptor liposomes. Furthermore, the ionophore monensin, which is known to inhibit the transport of newly synthesized glycoproteins to the cell surface, also inhibits the delivery of these fluorescent lipid metabolites to the cell surface and causes accumulation of fluorescence in the region of the Golgi apparatus. These results suggest that both the fluorescent sphingomyelin and glucosylceramide analogs are synthesized intracellularly from $C_6$-NBD-ceramide and then translocated through the Golgi apparatus to the plasma membrane.

In this chapter, some biophysical properties of this fluorescent lipid and its use in staining the Golgi apparatus in both living and fixed preparations are presented.


## II.  Synthesis and Purification of $C_6$-NBD-ceramide

The structure of $N$-{6-[(7-nitrobenzo-2-oxa-1,3-diazol-4-yl)amino]caproyl}-sphingosine ($C_6$-NBD-ceramide) is shown in Fig. 1. It is formed by N-acylation of the long chain base, sphingosine, with the short chain fluorescent fatty acid, $N$-[7-(4-nitrobenzo-2-oxa-1,3-diazole)]-6-aminocaproic acid ($C_6$-NBD-FA). Because there are two chiral carbons on the sphingosine base, four possible stereoisomers of sphingosine and the resulting ceramides synthesized from it are possible. Only the D-*erythro* isomer is found in nature.

$C_6$-NBD-ceramide is readily synthesized from sphingosine and $C_6$-NBD-FA by oxidation–reduction condensation with triphenylphosphine and 2,2'-dipyridyldisulfide (Kishimoto, 1975), or by reacting sphingosine with $N$-succinimidyl $C_6$-NBD-FA (Schwarzmann and Sandhoff, 1987). The

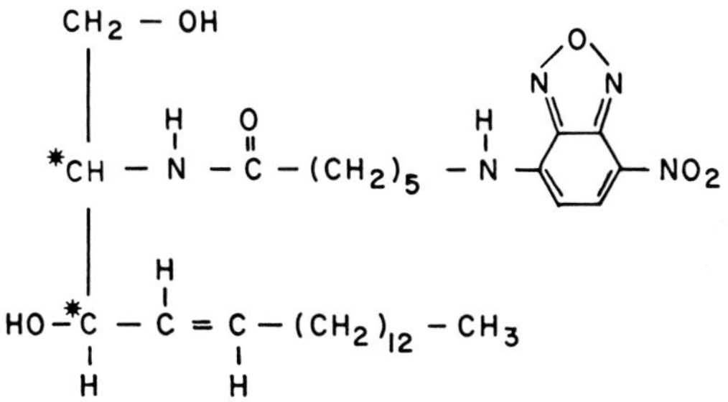

FIG. 1. Structure of $C_6$-NBD-ceramide. Asterisks indicate chiral carbon atoms.

product can be purified by preparative thin-layer chromatography (TLC) using $C:M:H_2O$ 65:25:4 ($R_f = 0.72$) or $C:M:NH_4OH:H_2O$ 160:40:1:3 ($R_f = 0.60$) as the developing solvent. $C_6$-NBD-ceramide is also commercially available from Molecular Probes (Eugene, Oregon).

## HPLC Analysis of $C_6$-NBD-ceramide

Commercial sources of D-*erythro*-sphingosine can be contaminated with significant amounts of L-*threo*-sphingosine, resulting in two isomers of $C_6$-NBD-ceramide being produced during synthesis. The two isomers are not easily resolved by TLC, and the presence of the L-*threo* isomer can complicate metabolic studies. For example, some biosynthetic pathways discriminate between the two isomers while others do not (Pagano and Martin, 1988). Furthermore, the D-*erythro* and L-*threo* isomers of some of the *metabolites* of $C_6$-NBD-ceramide may be difficult to identify by TLC without authentic standards. For these reasons it is recommended that each preparation of $C_6$-NBD-ceramide is subjected to high-performance liquid chromatography (HPLC) analysis, and, if necessary, purified by HPLC before use.

For HPLC analysis and purification of $C_6$-NBD-lipids we use a Waters Model 510 HPLC equipped with a McPherson Model 749 spectrofluorometric detector and a Shimadzu C-R3A plotter–integrator (Martin and Pagano, 1986). Fluorescence is detected using a 24-$\mu$l (illuminated volume) flow cell. Maximum sensitivity is obtained by excitation at the 437-nm mercury line which is on the excitation peak of NBD, and emission is detected using a 515-nm cutoff filter. The $C_6$-NBD-ceramide isomers are separated using an analytical reversed-phase ODS-II column (Regis Chemical Company, Morton Grove, Illinois) and methanol/water/$H_3PO_4$ (850/150/1.5, v/v/v) as the mobile phase. Samples are injected in 20 $\mu$l of

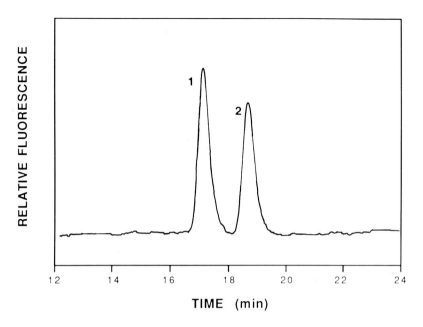

FIG. 2.    HPLC chromatogram of a mixture of the D-*erythro* (peak 1) and L-*threo* (peak 2) isomers of $C_6$-NBD-ceramide.

the HPLC solvent using a Rheodyne 7125 injector and eluted at a flow rate of 1 ml/minute. Figure 2 shows an HPLC chromatogram of a mixture of the D-*erythro* (peak 1) and L-*threo* (peak 2) isomers of $C_6$-NBD-ceramide.

# III.    Use of Resonance Energy Transfer (RET) to Study $C_6$-NBD-ceramide in Liposomes

## A.    Spontaneous Transfer between Liposomes

Resonance energy transfer (RET) between $C_6$-NBD-labeled lipids (the energy donor) and (lissamine) Rhodamine B-labeled phosphatidylethanolamine (*N*-Rh-PE) (energy acceptor) can be used to monitor the rate of transfer of $C_6$-NBD-labeled lipids between two populations of lipid vesicles (Nichols and Pagano, 1982). When both fluorescent lipid analogs are incorporated into the same donor vesicle at a concentration of 1 mol%, NBD fluorescence is quenched by the *N*-Rh-PE due to RET. Addition of nonfluorescent acceptor vesicles to these donor vesicles results in an imme-

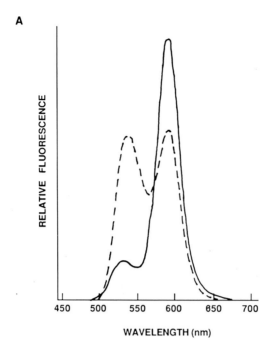

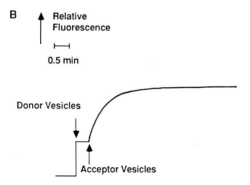

FIG. 3. Use of resonance energy transfer to measure the kinetics of NBD-labeled lipid transfer between vesicles. (A) Emission spectra before and after transfer of $C_6$-NBD-ceramide. Emission spectrum ($\lambda_{ex} = 475$ nm) of donor vesicles (——). Emission spectrum after addition of an approximately equal concentration of DOPC vesicles (---). (B) Kinetic measurement of $C_6$-NBD-ceramide transfer between DOPC vesicles. Donor vesicles were added to HCMF and allowed to equilibrate in the fluorimeter at 25°C. Fluorescence was recorded at 525 nm ($\lambda_{ex} = 475$ nm). At time zero, an equal amount of acceptor vesicles was added.

diate and continuous increase in NBD fluorescence. This is due to the
spontaneous transfer of the NBD lipids (but not the nonexchangeable
N-Rh-PE) to the acceptor vesicles and concomitant relief of RET quenching. Thus, by continuously monitoring NBD fluorescence intensity, both
the rate of transfer and the equilibrium distribution of the NBD-labeled
lipid can be determined.

A typical experiment is shown in Fig. 3 in which donor vesicles
comprised of dioleoylphosphatidylcholine (DOPC)/N-Rh-PE/$C_6$-NBD-
ceramide (98/1/1, mol/mol/mol) were mixed with an equal amount of
nonfluorescent DOPC acceptor vesicles. In Fig. 3A, solid line, the emission
spectrum of the donor vesicles is shown ($\lambda_{ex} = 475$ nm). Here we see
significant quenching of the NBD fluorescence at 525 nm and emission of
rhodamine fluorescence at 590 nm which would not normally be excited at
475 nm. The emission spectrum of the sample after addition of the acceptor
vesicles is shown in Fig. 3A, dashed line. After equilibration of the $C_6$-
NBD-ceramide between donor and acceptor vesicles, the intensity of emitted light at the NBD peak increased and that of the rhodamine peak

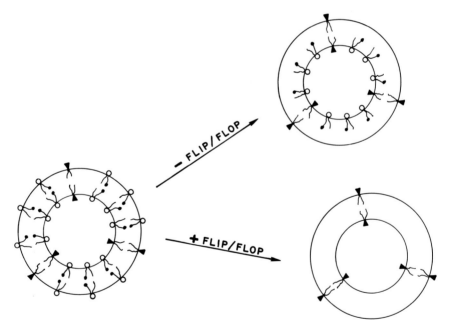

FIG. 4.    Resonance energy transfer test for transbilayer movement (flip/flop). Schematic
illustration shows donor vesicles containing the nonexchangeable lipid, N-Rh-PE (▶⊂), and
an exchangeable $C_6$-NBD-lipid (○⊂ ) *before* (left side) and *after* (right side) incubation with an
*excess* of acceptor vesicles. If no transbilayer movement occurs, only the $C_6$-NBD-lipid present
in the outer leaflet is depleted from the vesicles (upper right), whereas if transbilayer movement
occurs, all of the $C_6$-NBD-lipid is removed from the donor vesicles (lower right).

decreased due to a reduction in RET in the donor vesicles. Figure 3B shows the time course of this process which was monitored by following the "relief" of NBD quenching at 525 nm ($\lambda_{ex}$ = 475 nm). Elsewhere we have shown that the half time to reach equilibrium can be measured directly from the recorded fluorescence trace (Nichols and Pagano, 1982). Using this approach we have measured a half time for spontaneous transfer of $C_6$-NBD-ceramide among DOPC vesicles of 0.42 ± 0.02 minutes at 25°C.

## B. Transbilayer Movement

RET can also be used to test the ability of a $C_6$-NBD-lipid to undergo transbilayer movement in unilamellar vesicles (Pagano and Longmuir, 1985). Donor vesicles containing the NBD-lipid and small amounts of the nonexchangeable lipid $N$-Rh-PE are prepared and the RET between NBD and rhodamine is measured. This measurement is then repeated after adding increasing amounts of nonfluorescent acceptor vesicles (up to a 15- to 20-fold excess). As illustrated in Fig. 4, if no transbilayer movement occurs, addition of an excess of acceptor vesicles will remove all of the NBD-lipid from the *outer* leaflet of the donor vesicles and reduce the RET signal by approximately 50%. However, if transbilayer movement occurs, the NBD-lipid should be removed from *both* leaflets, eliminating the RET signal.

Figure 5 shows such an experiment using donor vesicles containing

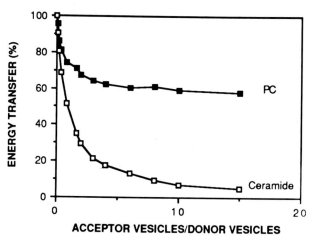

FIG. 5. Transbilayer movement of $C_6$-NBD-ceramide in lipid vesicles. Large unilamellar DOPC/$N$-Rh-PE/$C_6$-NBD-ceramide or -PC (98:1:1) donor vesicles were incubated with various amounts of DOPC acceptor vesicles and the resonance energy transfer (RET) between NBD and rhodamine was measured. The ability of a $C_6$-NBD-lipid to undergo transbilayer movement is inferred from the decrease in RET when an excess of acceptor vesicles is used (see Fig. 4). □, $C_6$-NBD-ceramide; ■, $C_6$-NBD-PC.

DOPC/$N$-Rh-PE/$C_6$-NBD-PC (upper curve) or DOPC/$N$-Rh-PE/$C_6$-NBD-ceramide (lower curve). About 50% of the fluorescent PC is transferred to acceptor vesicles, while essentially all of the $C_6$-NBD-ceramide is transferred. Assuming that the NBD-lipids were approximately equally distributed on the two leaflets of the donor vesicle bilayer, this result indicates that $C_6$-NBD-ceramide, but not $C_6$-NBD-PC, can undergo transbilayer movement.

# IV.   Properties of $C_6$-NBD-ceramide in Cells

## A.   Vital Staining of the Golgi Apparatus (Lipsky and Pagano, 1985b).

To vitally stain the Golgi apparatus of cells, a stock solution of $C_6$-NBD-ceramide complexed with bovine serum albumin (BSA) is first prepared. Cells are then briefly incubated with this solution, washed, and observed in the fluorescence microscope.

### 1.   PREPARATION OF $C_6$-NBD-CERAMIDE/BSA COMPLEX

An aliquot of a chloroform–methanol stock solution of $C_6$-NBD-ceramide containing 50 nmol of the NBD-lipid is dried, first under a stream of nitrogen and then *in vacuo*. The dried lipid is dissolved in 200 $\mu$l ethanol and injected into 10 ml of 10-m$M$ Hepes-buffered minimal essential medium (HMEM) containing 0.34-mg defatted BSA (Sigma Chemical Company, St. Louis, Missouri)/ml while vortexing. This solution is dialyzed overnight at 4°C against 500 ml of HMEM, aliquoted into plastic tubes, and stored frozen at −20°C. The complex therefore contains approximately a 5-$\mu M$ concentration of both the $C_6$-NBD-ceramide and BSA.

### 2.   INCUBATION WITH CELLS

Cells are grown on 25-mm diameter, #1 glass cover slips in 35-mm diameter tissue culture dishes. The cover slips are washed several times with HMEM to remove the culture medium and then incubated for 5 to 10 minutes at 37°C with the $C_6$-NBD-ceramide/BSA complex. The cells are then washed with HMEM, the cover slip is mounted on a depression slide, and the specimen observed in the fluorescence microscope. Prominent staining of the Golgi apparatus similar to that shown in Fig. 6 is generally seen, but it may be necessary to alter incubation conditions with different cell types to optimize the labeling pattern.

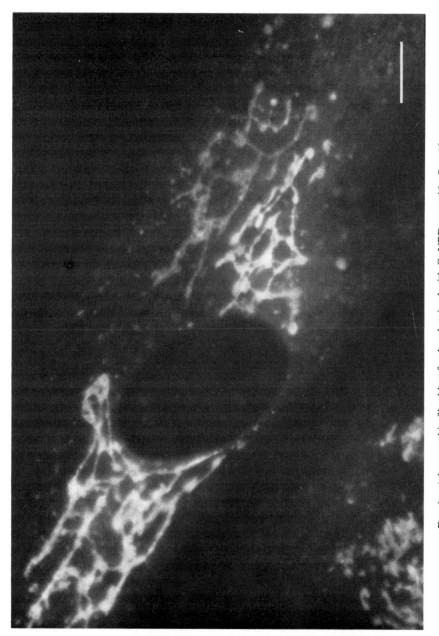

FiG. 6. A human skin fibroblast fixed and stained with C$_6$-NBD-ceramide. Bar, 10 $\mu$m.

## B.   Staining of the Golgi Apparatus in Fixed Cells

Appropriately fixed cells also accumulate $C_6$-NBD-ceramide at the Golgi apparatus. The molecular basis for this labeling is not yet known, but could be due to binding of the fluorescent ceramide to specific proteins or lipids in the Golgi apparatus, or to a physical property of the membrane of this organelle (e.g., the phase state of its lipids).

Cells grown on glass cover slips are rinsed in HMEM and fixed for 5 to 10 minutes at room temperature in glutaraldehyde or paraformaldehyde. We routinely use 0.5% glutaraldehyde/10% sucrose/100 m$M$ PIPES, pH 7.0 as the fixative, but in our experience, fixation time, temperature, and buffer composition are generally not critical. (We did find that brief treatment of fixed cells with detergents or fixation with methanol–acetone at $-20°$C eliminates labeling of the Golgi apparatus.) After fixation, the fixed cells are washed in a simple balanced salt solution (e.g., 10 m$M$ Hepes-buffered calcium- and magnesium-free Puck's saline; HCMF). For glutaraldehyde-fixed cells, the culture dishes containing the glass cover slips are transferred to an ice–water bath and incubated (3 × 5 minutes) with freshly prepared $NaBH_4$ (0.5 mg/ml) in ice-cold HCMF (Weber *et al.,* 1978). The cells are then rinsed several times over a 20- to 30-minute period in cold HCMF, allowed to warm to room temperature, and incubated for 60 minutes with $C_6$-NBD-ceramide/BSA prepared as described above. The fixed cells are then washed in HCMF, and incubated 30–90 minutes at room temperature with 10% fetal calf serum or 2 mg/ml BSA to back-exchange excess $C_6$-NBD-ceramide from the preparation and enhance the staining of the Golgi apparatus. The glass cover slips are then mounted on depression slides and observed under the fluorescence microscope. A typical fluorescence micrograph of a glutaraldehyde-fixed human skin fibroblast, showing prominent labeling of the Golgi apparatus, is shown in Fig. 6.

## V.   Prospectus

The Golgi apparatus is currently the focus of much attention because of its central role in the processing, sorting, and targetting of glycoproteins and glycolipids in eukaryotic cells. The fluorescent ceramide analog reviewed in this chapter provides cell biologists with a tool for readily visualizing this organelle within intact cells, and, in conjunction with new techniques in fluorescence microscopy discussed elsewhere in these volumes, now make it possible to study the dynamics of this organelle and to elucidate its three-dimensional structure. In addition, the physical properties of $C_6$-NBD-ceramide may be useful in designing other ceramide derivatives which can be metabolized and which accumulate in the Golgi apparatus. A ceramide derivative having these properties and bearing an appropriate cross-linking moiety may be particularly useful in selectively labeling proteins within the Golgi apparatus which are involved in sphingolipid metabolism, sorting, and transport.

## ACKNOWLEDGMENTS

The author thanks Mr. J. Maslanski for work on transbilayer movement of $C_6$-NBD-ceramide in the author's laboratory, and Dr. J. Wylie Nichols (Emory University) for measuring the kinetics of monomer transfer of $C_6$-NBD-ceramide between liposomes. Supported by United States Public Health Service grant GM-22942.

## REFERENCES

Kishimoto, Y. (1975). *Chem. Phys. Lipids* **15**, 33–36.
Lipsky, N. G., and Pagano, R. E. (1983). *Proc. Natl. Acad. Sci. U.S.A.* **80**, 2608–2612.
Lipsky, N. G., and Pagano, R. E. (1985a). *J. Cell Biol.* **100**, 27–34.
Lipsky, N. G., and Pagano, R. E. (1985b). *Science* **228**, 745–747.
Martin, O. C., and Pagano, R. E. (1986). *Anal. Biochem.* **159**, 101–108.
Nichols, J. W., and Pagano, R. E. (1982). *Biochemistry* **21**, 1720–1726.
Pagano, R. E., and Longmuir, K. J. (1985). *J. Biol. Chem.* **260**, 1909–1916.
Pagano, R. E., and Martin, O. C. (1988). *Biochemistry* **27**, 4439–4445.
Pagano, R. E., and Sleight, R. G. (1985). *Science* **229**, 1051–1057.
Schwarzmann, G., and Sandhoff, K. (1987). *In "Methods in Enzymology"* (V. Ginsburg, ed.), Vol. 138, pp. 319–341. Academic Press, Orlando, Florida.
van Meer, G., Stelzer, E. H. K., Wijnaendts-van-Resandt, R. W., and Simons, K. (1987). *J. Cell Biol.,* **105**, 1623–1635.
Weber, K., Rathke, P. C., and Osborn, M. (1978). *Proc. Natl. Acad. Sci. U.S.A.* **75**, 1820–1824.

# Chapter 6

## Fluorescent Labeling of Cell Surfaces

### MICHAEL EDIDIN

*Department of Biology*
*The Johns Hopkins University*
*Baltimore, Maryland 21218*

## I. Introduction

Like other powerful tools, fluorescence microscopy has been used and abused in the study of cell surfaces. Surface organization, physical state, dynamics, and function are all accessible through appropriate labels of the surface, but these labels must be applied intelligently if they are to yield valid information. In this chapter, rather than detail techniques of labeling, I want to emphasize the importance of determining the location and specificity of fluorescent surface labels. Whatever the application, a label must localize to the surface and in the surface should bind specifically to the structure to be probed. These ideals are in practice hard to reach, particularly when labeling nucleated cells. Probes readily go astray and some that

87

behave impeccably in synthetic membranes create monster artifacts when inserted into cells. They may label multiple (internal) sites as well as the surface or perturb the sites that are labeled, or react with a different chemistry at the membrane than they display in reaction with pure proteins. Even that most tractable of probes, fluorescent IgG antibodies, may bear sites specific for more than one species of cell surface protein and hence label multiple sites when only one is intended for labeling. This is why I want to approach the problems of cell labeling in a general way, illustrating advantages and disadvantages with examples, instead of laying down hard and fast protocols which are certain to lead to artifacts if thoughtlessly applied.

## II.   Direct Chemical Labels of Cell Surfaces

Many cells are tough enough to withstand direct chemical modification with fluorophores. The fluorophores used either form covalent bonds with membrane proteins or insert into the hydrophobic regions of the bilayer. The former typically are reagents, such as fluorescein isothiocyanate, used for protein modification in solution; and the latter are lipid-soluble molecules, ranging from hydrophobic fluorescent compounds such as pyrene or diphenylhexatriene, to proteins and carbohydrates doubly modified by conjugation with both a fluorophore and one or more hydrophobic tails.

## A.   Covalent Chemical Labels

Most fluorophores reacting with proteins in solution ought to react with cell surfaces. The coupling of commonly used fluorophores to cells is limited then, by the reaction conditions required and by the hydrophobicity of the fluorophore. Clearly, reaction conditions must be compatible with cell survival, requiring aqueous, isotonic medium, and a limited, though surprisingly large, range of pH. Fluorescein isothiocyanate, FITC, and the related (4,6-dichlorotriazinyl) aminofluorescein (DTAF) have been used to label surfaces of cultured fibroblasts and of erythrocytes, with a high degree of localization to the cell surface, while the more hydrophobic rhodamine isothiocyanate, though it may react with surface proteins, can penetrate the membrane and heavily label cell cytoplasm.

Our work with FITC labeling of cells highlights a number of the problems with this approach to cell surface labeling (Edidin et al., 1976). When we attempted to label L cells with FITC at pH 7.5–8.0, the bulk of the fluorescence was found in the cytoplasm. The cells on cover slips did survive labeling at pH 9.5 for 10 to 20 minutes, and cells so labeled appeared intensely ring stained when examined after labeling. Though no systematic analysis was made of labeled proteins, it appears that a significant amount of extracellular materials were conjugated, since the label was immobile, by fluorescence recovery after photobleaching (FRAP) (Jacobson *et al.,* 1983), for several hours after the reaction. Suspension of the cells with trypsin and replating eliminated this immobile component, strongly suggesting that it represented extracellular membrane-adherent molecules. The remaining fluorescence was insoluble in chloroform – methanol, was not removed from cells by trypsin, but was almost entirely removed by papain, indicating that the label remained on the surface on membrane integral proteins.

In the experiments just described, the "surface" label was partitioned between integral and extracellular peripheral proteins, and only a fraction was suitable for photobleaching studies. In contrast, Schlessinger *et al.* (1978) could label L-6 cultured myoblast cells with FITC at pH 7.6, though it appears that some of the population were internally labeled. From 30 to 50% of the label was mobile by FRAP, typical for surface membrane proteins.

Reactive fluorescein derivatives have proven very useful and specific in labeling erythrocyte surfaces. The reaction of FITC, DTAF, or eosinITC is largely with band 3 and glycophorin, and there is little penetration of any of the labels to the cytoplasm. These conclusions are firmly founded on quantitative chemical analysis of ghost and supernatant preparations (Peters *et al.,* 1974; Fowler and Branton, 1977; Golan and Veatch, 1980; Schindler *et al.,* 1980).

Even sperm and eggs have been successfully labeled with FITC. Shapiro and co-workers (Gabel *et al.,* 1979) labeled membranes of sea urchin *(Strongylocentrotus pupuratus)* and golden hamster sperm with 30 $\mu M$ or 13 $\mu M$ FITC at very slightly alkaline pH (7.7). The labeled sperm were motile and could fertilize eggs. By a variety of extraction criteria the label was covalently bound to sperm surface proteins, which persist for days in the developing embryos. In a later paper (Gundersen and Shapiro, 1984), the persistence of fluorescent sperm surface proteins was documented in later pluteus larvae, using intensified video microscopy. This paper also shows the range of sperm membrane proteins, at least 15, labeled by FITC, or a radioiodinated derivative of FITC. Peters and co-workers (1981) were able to label sea urchin *(Paracentrotus lividus)* eggs with FITC at pH 10 to give labeled surfaces, with much of the label on integral membrane proteins, but with a significant fraction on the vitelline membrane. Despite this, lateral diffusion measurements could be made of the mobility of the label within the membrane. Label was biologically relevant — labeled eggs fertilized normally and raised fertilization membranes.

FITC and other isothiocyanates react mainly with amino groups. Proteins of surface membranes and their relatives may also be labeled with -SH-specific fluorophores. Our own experience with such labels, for disk membranes of vertebrate rods, points to further subtleties in covalent labeling. Wey and co-workers (1981) labeled freshly isolated bullfrog *(Rana catesbiana)* rods with iodoacetamido tetramethylrhodamine (ITMR) in isotonic buffer, pH 7.4. A molar ratio of ITMR to rhodopsin of about 10:1 labeled rods on both plasma and disk membranes after 20 minutes incubation. The bulk of the label could be shown by gel electrophoresis to be in rhodopsin. The labeled rhodopsin was mobile in the disk membranes. FRAP measurements gave $D = (3 + 1.2) \times 10^{-9}$ cm$^2$ second$^{-1}$, comparable to $D$ determined for endogenously labeled rhodopsin, $D = (5.3 + 2.4) \times 10^{-9}$ cm$^2$ second$^{-1}$ (Poo and Cone, 1974; Liebman and Entine, 1974). Philips (1985) labeled the same rod preparation with iodoacetamido fluorescein (IF) and measured approximately the same diffusion coefficients as previously reported (5). However, electrophoretic analysis of the labeled proteins clearly showed that the label was not in rhodopsin, but in a membrane protein of slightly lower molecular weight (R.A. Cone, personal communication). Though the reactive group is the same in the two experiments, the fluorophores differ in their water solubility and this difference apparently targets the labels to different membrane proteins.

## B.   Hydrophobic Labels

Hydrophobic fluorescent molecules such as pyrene, diphenylhexatriene, and the carbocyanine dyes readily label lipid bilayers, in liposomes or in cells. There is an enormous literature on the use of these molecules to report on membrane organization (reviewed by Waggoner, 1986). Their attraction for cell biologists lies in the ease of and intensity of labeling and in the enormous knowledge of the probes' behavior in model membranes. Intact cells may be labeled at intensities that allow more sophisticated measurements, for example, determination of fluorescence lifetimes on intact cells, and properties of probes, for example, fluorescence intensity, lifetime, and polarization, as well as lateral diffusion, can be tentatively ascribed to their environments by reference to their behavior in liposomes of known composition and physical properties. In principle then these measurements can report on the presence of gel domains in the cell surface (Wolf *et al.,* 1981; Klausner *et al.,* 1980), on local order of lipids (Shinitzky and Yuli, 1982), and even on transmembrane potential (reviewed by Szollosi *et al.,* 1987). In practice, while they are good reporters of these and other membrane properties in bilayers or in membrane fractions, including plasma membrane fractions derived from cells, they are not easily localized to the cell surface in intact cells. Hydrophobic dyes partition between all accessible membranes and readily label internal as well as cell surface membranes. Some of them, for example, pyrene and diphenylhexatriene, also emit fluorescence at wavelengths overlapped by cell autofluorescence. It is therefore important when using these and other lipophilic labels, even antibodies labeled with lipophilic dyes such as tetramethylrhodamine, to look at the dyes' distribution in cells before making measurements of fluorescence from cell populations. Only if labels are confined to the surface does one have any possibility of interpreting their fluorescence signals.

Hydrophobic dyes also may perturb membranes to the point that cells are irreversibly damaged. Care should be taken to use these lipid labels at the lowest possible concentration. Even then, viewing the labels may cause photodamage to membranes, affecting their function (for an example, see Pagano *et al.,* 1981).

A good example of the problems and successes with lipophilic membrane labels is given in the work of Packard and co-workers (1984). They attempted to label cultured epithelial cells with carbocyanine dyes and found multiple lifetimes of fluorescence for the labeled cells. The label could be seen to be internal as well as on the surface, and it was suggested that each lifetime represented dye in a different membrane. Had the cells not be visualized, the lifetimes could have been interpreted in terms of multiple environments within a single, surface membrane, an incorrect interpretation in this case. Packard then synthesized collarein, a fluorescent phospholipid analog. She measured a single lifetime for collarein in cells and fluorescence microscopy showed only surface labeling (ring staining) of the cells.

Our own work with the indocarbocyanin dyes, the so-called diIs (Wolf *et al.,* 1981), measured their lateral diffusion and compared this to lateral diffusion of the dyes in liposomes containing known gel and fluid phases. We could see internal fluorescence in labeled cells (sea urchin eggs) whose intensity varied with the particular diI used and with time after labeling, but the contribution of this internal fluorescence to our measurements could be reduced by optical sectioning of the image using high numerical aperature objectives. The use of confocal laser microscopy (Brackenhoff *et al.,* Volume 30, this series; White *et al.,* 1987), or through-focus, deconvolution methods of three-dimensional analysis (Agard and Sedat, Volume 30, this series), ought to further reduce the contribution by internal fluorescence to signals from the cell surface. The localization of lipid labels to inner or outer leaflets of the cell surface may also be determined using impermeant quenchers of fluorescence (Wolf, 1985) or by using unlabeled liposomes to extract label from the cell surface (Sleight and Pagano, 1984).

Amphipilic molecules also make good surface labels. Another chapter in this volume discusses fluorescent phospholipid analogues (Pagano,

Chapter 5). Here we mention use of fatty acyl tails to integrate fluorescent dextrans (Wolf *et al.,* 1980) or proteins (Peacock *et al.,* 1986) into membranes. This area has been little explored, but it is clear that the implanted macromolecules can be useful probes of the cell surface, and that they can even function as specific antigen receptors. Implanted proteins may also be useful as controls for proximity of membrane molecules to one another. In our laboratory we measure the proximity between viral antigens on the cell surface and the class I major histocompatibility complex (MHC) class I antigens. Resonance energy transfer experiments indicated proximity of the two sorts of labeled molecules, but left open the question of its specificity. Rather than label another membrane protein and examine its proximity to the viral antigens, we prepared ribonuclease, tailed with an average of 1 palmitic acid per ribonuclease, and fluoresceinated. This amphiphile-labeled cell membrane well appeared, from later diffusion measurements, to be integrated into the bilayer, rather than stuck to its surface. (Lateral mobility of the label, with diffusion coefficients $\geq 10e - 10$ is a good indication of its integration into, rather than absorption on, the cell surface.) However, no energy transfer was detected between it and the viral antigens, reinforcing the idea that observed energy transfer between the class I antigens and viral antigens was specific and biologically relevant (Hochman and Edidin, 1987).

There is one report that a physiological, amphiphile, rhodamine-conjugated, $\beta$-adrenergic receptor has been successfully integrated into unlabeled cell surfaces (Cherksey *et al.,* 1985). Rotational mobility and functional activity of the receptor could both be measured in this preparation.

One other caution must be stated before leaving the subject of hydrophobic and amphiphilic labels. All these are delivered to cells in traces of detergent or organic solvent. These are membrane-active molecules and may be present at deceptively high molar concentrations in apparently dilute solutions. Thus, 1% v/v ethanol solution is approximately 0.2 $M$ ethanol, a substantial concentration. The onus is on the investigator to show, by varying labeling methods and conditions, that the solvent present does not significantly perturb the cell surface.

## III.   Fluorescent Ligands as Labels of Cell Surfaces

The diverse proteins of cell surfaces bind or are bound by ligands ranging from peptide hormones, toxins, through antibodies and their fragments and lectins, to large particles – low density lipoprotein particles, and even viruses (see Maxfield, Chapter 2 and Angelides, Chapter 3, this volume). In principle, all of these ligands bind reversibly to specific sites, and, given time, the population of ligand molecules should reach an equilibrium between bound and free forms. We are concerned here with fluorescent labels, but the quantitative study of ligand associations at cell surfaces has been most thoroughly done using radiolabels. The principles and results of such studies are well discussed and reviewed in a recent monograph (Limbird, 1986).

## A.  Binding of Labeled Ligand

Any labeled ligand must be specific for its receptor, and must be applied in such a manner that its binding can approach equilibrium. This is readily measured by two experiments, a time course at constant ligand concentration and a titration of labeled ligand at constant time. The rigor of these experiments depends upon the application. Brief periods of binding may be sufficient if cell surfaces exhibit a large signal when only a fraction of receptors are occupied. For example, we find that for qualitative work only 15 minutes incubation with antibodies of moderately high affinity ($K_a \sim 10^8$ $M^{-1}$) is sufficient time for labeling. Such short times also reduce the possibilities for internalization or degradation of label. These unwanted events are also reduced by labeling, when possible, at low temperature (0–4°C).

Cells labeled for microscopy must be washed after labeling, otherwise the high concentration on fluorescent ligand in the medium will swamp signals from the surface. This washing also removes some nonspecifically bound ligand and, by disrupting the equilibrium between free and bound ligand, some specifically bound ligand as well. Most cell-bound ligands, whether antibodies and their fragments or peptide hormones, are surprisingly resistant to dissociation by dilution. Washing may be avoided and fluorescence quantitated if labeled cells are examined by flow cytometry instead of microscopy. The geometry of the flow cytometer ensures that most fluorescence is detected from the cell surface, rather than from the small sheath of fluid surrounding it. We find increases of only 20–25% of the peak channel for washed cells if instead the sample is applied in a solution that is $\sim 10^{-6}$ $M$ in fluorescein. Flow cytometry of ligands in equilibrium with cells has been discussed extensively by Bohn (1980).

## B.  Specificity of Labeling

Specificity of ligand labeling is classically determined by competing the labeled ligand with an excess of unlabeled ligand which should largely or entirely block labeling. The effectiveness of blocking can often be estimated by eye in the fluorescence microscope. However, quantitative microscopy or flow cytometry is more informative, since it allows estimates of the fraction of specific binding and may also detect unexpected effects of the unlabeled antibody. For example, in unpublished work we found that the monoclonal antibody to class I MHC antigens, 28.14.8 (Shiroishi *et al.*, 1985) creates more sites for its own binding when it binds to the cell surface. This was detected in terms of the shift to higher channels of a flow cytogram while attempting a blocking control for specificity of fluorescein-labeled 28.14.8. In the event, the labeled antibody did prove to be specific, but much higher than expected levels of unlabeled antibody were required to block all binding sites for the labeled antibody.

When labeling cells it may be possible to perform an additional specificity control by measuring the labeling of cells negative for the receptor of interest. Thus, in work on expression of mutant genes in transfected cells, the nontransfected cells are an excellent control. This control needs be used with care. In one instance we prepared a rabbit antiserum to the light chain of class I MHC antigens, $\beta_2$-microglobulin. By immunofluorescence, the antiserum labeled hamster and human MHC class I antigen-positive cells as expected, but it also reacted well with human Daudi cells, which lack a functional $\beta_2$-microglobulin gene and fail to express any surface class I antigens. Immunoprecipitation and blocking experiments showed that the reactivity was with the Fc receptor of these cells; the rabbit antibody had in fact bound specifically, but irrelevantly for our purposes (S. Barbour and M. Edidin, unpublished).

## C.   Sensitivity of Labeling

Detection of labeled molecules depends upon both the intensity of fluorescence from the cell surface and upon the ability to resolve specific fluorescence from background, that is, upon the contrast between label and background. Intensity of fluorescence depends upon the surface concentration of molecules to be labeled and upon the number of fluorophores associated with each. Resolution of a small number of molecules will depend upon the extent to which cell background from autofluorescence, from scattered light, or from nonspecifically bound label, can be subtracted from the signal. This subtraction may be quantitative, for example, when comparing peak positions in cell populations analyzed by flow cytometry, or "before and after" digital video images of cells, or may be qualitative, when comparing positive and negative controls by eye in the fluorescence microscope. We find that by flow cytometry we can detect a few thousand molecules of fluorescein above a background signal equivalent to about 10,000 molecules, but that this requires titration of the fluorescein–ligand

($\beta_2$-microglobulin) and is not estimated from a single value for peak or mean fluorescence channel. The resolution of a 20% change in fluorescence intensity is much better than that achieved by looking at positive and negative samples. In that case, 2-fold or greater differences are those reliably detected by the observer.

All of the variables affecting fluorescence intensity can be controlled to some extent. Even the surface concentration of the molecule of interest can be increased by choice of cell or by appropriate treatment of cells of interest. For example, epidermal growth factor receptors are far more abundant on A431 epidermoid carcinoma cells than on normal fibroblasts (Fabricant *et al.*, 1977), making the former more useful for studies of fluorescent ligand binding than the latter. Class I MHC antigens are expressed at moderately high levels on most cells, but their expression can be enhanced by treatment of cells with interferon (reviewed by Friedman and Vogel, 1983).

Unfortunately, it seems to be generally true that such exceptional systems are limited in their usefulness and that sooner or later investigators must turn to cells bearing lower numbers of the receptor or antigen of interest. This leads to attempts to raise fluorescence intensity by raising the number of fluorophores on each labeled ligand. Some of these attempts end in failure. Increasing the number of conjugated fluorophores per ligand molecule leads both to reduction of fluorescence due to excited state dimer formation, and to loss of biological activity, or at least to increasingly nonspecific binding of the labeled material. Both of these phenomena are more common with more hydrophobic labels, for example, tetramethylrhodamine, or sulforhodamine (Texas Red), than with fluorescein.

The number of fluorophores per ligand may also be increased by building layers of label. Thus, binding of the Fab fragment of a monoclonal antibody against anti-MHC class I antigen could not be detected when the Fab fragment was conjugated directly to fluorescein, but could be detected using a conjugated antiglobulin. This increases sensitivity both because more fluorophores can be conjugated per molecule of IgG without denaturing it than can be conjugated to Fab, and because there are multiple binding sites for the antiglobulin on each Fab. Such multiplications might be continued, using third fluorescent layers to further increase fluorescence, but a price in nonspecific binding must be paid for each layer. Also, the washing required between each label becomes extensive, and may even elute previously bound ligand, vitiating the attempt.

Multistage or even single-stage multivalent labels exact a further price; if they cross-link their binding sites this may often lead to rapid internalization of label and loss from the surface.

An effective way to raise fluorescence intensity is to use phycobiliprotein labels (Oi *et al.*, 1982). Each phycobiliprotein bears multiple fluorophores.

The extinction coefficients of the molecules are between 1 and $2 \times 10^6$ compared to about 80,000 for fluorescein. The proteins may be directly conjugated to antibodies, or to biotin or avidin for labeling of avidin- or biotin-conjugated antibodies. Their intense fluorescence is also their chief disadvantage. Labeled cells must be thoroughly washed to remove all traces of unbound label. Otherwise, backgrounds are very high indeed. Again, this is less of a problem for flow cytometry than for microscopy. For a given solution of phycoerythrin label we found that three washes were sufficient for flow cytometry while nine or more washes were required before background was reduced sufficiently for fluorescence microscopy.

Even the most favorable labels may not be detected if cells themselves fluoresce. Cell autofluorescence is common in cultured cells and appears to be a function of both cell type and culture conditions. Lymphoid lines, freshly explanted lymphocytes, and transformed fibroblasts commonly have low levels of autofluorescence, but even L cells, transformed fibroblasts, may contain red autofluorescent regions if cultured at high density in some media. Other cell types, for example, hepatocytes and 3T3 adipocytes, in our hands, are brightly fluorescent. Though this is cytoplasmic fluorescence, it swamps dim surface stain.

Cell autofluorescence is not readily dealt with. It appears to arise mainly from intracellular flavins and flavoproteins and the emissions cover all the useful visible range, from 500 to 600 nm (Benson *et al.*, 1979; Aubin, 1979). Some advantage is gained by using fluorophores that excite and emit at longer, rather than shorter wavelengths, for example, Texas Red or tetramethylrhodamine instead of fluorescein (see also Waggoner *et al.*, Volume 30, this series). Exciting light should be carefully and thoroughly filtered to block all wavelengths shorter and longer than those required to excite the fluorophore chosen. These strategies used with a flow cytometer, rather than a microscope, may allow detection of specific surface label despite persistent autofluorescence. Indeed, recently, Alberti *et al.* (1987) have described methods for subtracting autofluorescence from total fluorescence on a cell-by-cell basis. This method could be applied to digital fluorescence microscopy as well.

## D. Localization of Bound Ligand

Having achieved detectable and specific labeling, we want to confirm the surface location of our label. Impermeant quenchers may be used to confirm surface localization, much as they are used to quench hydrophobic or amphiphilic labels (see Section II, B). However, it is often more convenient to strip label from the surface using either buffers of low pH or proteases. The difference in cell fluorescence between fully labeled and stripped cells (corrected for autofluorescence of the cell blank or specificity control) over the (corrected) total cell fluorescence gives the fraction of label inaccessible to stripping and therefore assumed to be internalized. These measurements have been shown to be somewhat sensitive to the agent used to remove the label. The amount of internalized insulin estimated from tryptic digestion of intact cells is higher than that estimated from treatment of the cells with excess unlabeled insulin (Kahn and Baird, 1978). Presumably, the trypsin molecules cannot reach insulin receptors in deeply invaginated coated pits, while insulin molecules can.

Trypsin is most useful for stripping labeled antibodies and lectins from cells, while acidic buffers, pH 3.5–4.5 (depending upon what a particular cell type can tolerate) are useful for stripping smaller molecules, peptide hormones, and $\beta_2$-microglobulin. Work on dissociation of ligands from isolated, cell-free receptors may be a useful guide to other dissociating agents. For example, we find that fluorescent $\beta_2$-microglobulin is effectively dissociated from cell surfaces in 50 m$M$ CuCl$_2$ solution. This treatment has been shown to dissociate $\beta_2$-microglobulin from isolated soluble MHC class I antigens (Revilla et al., 1986).

## E. Useful Fluorescent Ligands

An old dictionary defines both "cow" and "dog" as "an animal well known." The same definition may be used for the most useful and versatile cell surface labels, lectins and antibodies. Antibodies are particularly versatiles fluorescent labels. Their high specificity, large size, and multivalence make them attractive labels. Except for the very important factor of specificity, the same may be said of lectins. Antibodies have the further attraction that they are easily cleaved to form monovalent fragments, useful labels for kinetic studies of surface molecules. Though some lectins may also be cleaved or dissociated to reduce their valence, the chemistry is not as straightforward as that for antibodies. The following section deals primarily with the applications of antibodies to labeling cell surfaces.

### 1. ANTIBODIES

The classic fractionations and cleavages of immunoglobulins were worked out for rabbit immunoglobulins. These, isolated as polyclonal mixtures, are still useful reagents for surface labeling, not least because they are robust and are not readily degraded or damaged by fractionation and labeling. With the advent of monoclonal antibodies produced in rat or mouse, it has become necessary to take more care in antibody handling and labeling. Also, though it ought not be the case, reactions that are designed for milligrams of polyclonal immunoglobulins need to be scaled down for the micrograms of monoclonal Ig that are often all that are available.

The fluorescent conjugation reactions are fairly standardized and extensively described elsewhere (see Haugland, Volume 30, this series). They generally depend upon reaction with protein of an isothiocyanate, sulfonyl chloride, or dichlortriazine derivative of the fluorophore. The specificity of reaction varies with the reactive group and the pH of labeling. Classically, at alkaline pH (>9) isothiocyanates react with epsilon amino groups of lysines. The resulting fluorescent derivative is very stable and is readily fractionated by charge, since each substitution removes one positive charge of the protein. Such fractionation is useful for producing well-defined labels whose fluorescence can often be unambiguously characterized. More importantly, choice of a suitable charge fraction reduces nonspecific labeling and optimizes fluorescence by eliminating highly coupled molecules whose fluorescence is self-quenched.

The reaction specificity of both sulfonyl chlorides and dichlorotriazines is broader than that of isothiocyanates. Even hydroxyl groups may be efficiently substituted by these derivatives. In our experience, sulfonyl chlorides form less stable bonds than do isothiocyanates and shedding of dye is a serious problem when using conjugates made with sulfonyl chloride derivatives of rhodamine, notably, Texas Red (Titus et al., 1982). The most commonly used dichlorotriazine, DTAF, will conjugate most proteins at pH 7.5 (Blakeslee and Baines, 1976). This spares them the harsh effects of the alkaline buffer used for other conjugations.

For most purposes it is necessary to separate unreacted and hydrolyzed dye from conjugated and native protein (see Haugland, Volume 30, this series; Wang, Chapter 1, this volume). A G-25 or P-2 column of sufficient length usually does the job, though it sometimes fails to resolve protein from free rhodamine. In any case, rhodamine conjugates of all types should be further purified by absorption on SM-2 biobeads (Spack et al., 1986). Free dye and overcoupled protein bind to these hydrophobic beads, while moderately coupled and native molecules are not bound. This step eliminates much of the nonspecific staining of rhodamine conjugates. Since some conjugates are not stable, we routinely treat rhodamine-labeled antibodies with SM-2 before use and routinely find free dye bound to the beads.

Conjugates are ideally stored unfrozen at 0°C. We use this approach, storage in an ice–water mixture, if a sample is to be used over the course of a week or so. It becomes inconvenient for longer storage. In our experience, rhodamine conjugates are more stable in frozen storage than are fluorescein conjugates and the latter are best stored at 4°C after sterile filtration.

## 2. ANTIBODY FRAGMENTS

Intact Ig is not suitable for some applications, either because the cells of interest bear Fc receptors which bind the Ig, or because the bulk and/or valence of the Ig affects the membrane property of interest, for example, stimulating or blocking internalization. Rabbit antibodies are classically cleaved to monovalent Fab by overnight incubation with about 1 part/100 w/w papain (Porter, 1959). This tactic is guaranteed to cleave mouse Ig to peptides, if not to amino acids. Fab can be made from mouse IgG by papain cleavage for a short period, 10–20 minutes, at optimum pH (5.5) (see Edidin and Zuniga, 1984) or for longer periods (a few hours) by reaction at suboptimal pH, 7.5. Alternatively, divalent $(Fab')_2$ fragments can be prepared from the Ig and then reduced to give monovalent $Fab'$ (Mason and Williams, 1980). Both preparations should be freed of intact Ig, either by sizing or on protein A columns and the product should be checked for purity by gel electrophoresis. The results of papain treatment of IgG vary significantly from antibody to antibody of the same isotype. Hence the procedures given can only be taken as a rough guide to preparation of fragments.

It is sometimes possible to produce Fab fragments of IgM antibodies by cleavage with trypsin (A. Schreiber and M. Edidin, unpublished, following the work of Putnam *et al.,* 1973). These may be of such low affinity that they fail to bind to cell surfaces. However, in some instances useful monovalent labels have been produced from IgM.

## 3. LABELS FOR SPECIFIC RECEPTORS

This topic is largely covered in other chapters in this volume. Here we mention several labels of biological interest as surface labels.

Not all specific receptors bind small ligands. Low-density lipoprotein (LDL) receptors have been successfully labeled with fluorescent LDL (Barak and Webb, 1982). This is an intense label, whose fluorophore is a lipophilic dye absorbed to the lipid core of the LDL. Fluorescence intensity is so high that movements of single LDL molecules can be followed at the cell surface.

Labeling receptors for peptide hormones with fluorescent hormones present numerous and formidable difficulties (see Maxfield, Chapter 2, this volume). The hormones themselves are easily damaged by labeling, and the addition of a fluorophore, even when it does not denature the hormone, may hinder its binding to the receptor. There are few surface receptors for most peptide hormones, and so even if a specific analog is achieved its signal may be hard to detect. Epidermal growth factor is most tractable to conjugation and highly specific rhodamine EGF conjugates have been described and employed (Schechter *et al.*, 1978; Schlessinger *et al.*, 1978). Several cultured cell lines, notably A431, bear large numbers (millions) of receptors, so detection of bound ligand is easy.

In contrast, insulin is hard to derivatize and hard to visualize. Both directly labeled and lactalbumin–rhodamine-labeled insulins bind nonspecifically to cells. Only 1–10% of these derivatives could be displaced by unlabeled insulin (Schechter *et al.*, 1978; Murphy *et al.*, 1982). The directly labeled insulin could not be visualized on cells unless the cells were warmed and the label internalized. Recently, Due and co-workers (1985) prepared fluorescein insulin and purified the derivative by HPLC. Forty percent of the best preparation was not displaceable by native insulin. This was sufficient to label cells for flow cytometry, but one suspects that it would not do for fluorescence microscopy of insulin receptors.

## 4. A SMALL "UNIVERSAL" SURFACE LABEL

My laboratory works on a number of aspects of the class I major histocompatibility complex antigens, class I MHC antigens. The MHC antigens consist of a polymorphic heavy chain, a transmembrane protein of ~45 kDa, noncovalently associated with a highly conserved light chain, $\beta_2$-microglobulin. MHC class I antigens molecules are displayed on the surfaces of most cells of the body, including liver, kidney epithelium, skin, connective tissue, lymphoid tissue, and adipocytes, and their expression is maintained in cultured derivatives of these tissues. They are found in all mammals examined, in birds, and in fish. Thus, if reagents are available, they represent a near universal site for surface labeling, expressed at levels of 1000s to 100,000s per cell (reviewed by Kimball and Coligan, 1983; Klein, 1986). Class I MHC antigens are readily labeled by alloantibodies, made within a species, and there are some cross-species reagents as well, for example, rabbit anti-human MHC. However, there are no reagents commonly available for labeling cells of many species.

We have overcome this problem by preparing fluorescent derivatives of the $\beta_2$-microglobulin light chain of the molecule (Hochman and Edidin, 1988). Careful choice of stoichiometries yields a population of fluorescein-conjugated $\beta_2$-microglobulin molecules with an average of 1 fluorescein/protein. The conjugate fractionates on hydroxylapatite to yield derivatives that exchange specifically with endogenous. We find that these conjugates label the surfaces of dog, chinese hamster, human, and rodent cells. The label can be stripped almost completely (85–90%) by $CuCl_2$ solutions or by acid buffers. The fluorescein–$\beta_2$-microglobulin may be used to label for simple visualization of the surface distribution of MHC antigens or for studies in membrane dynamics of the antigens. Based on results with radiolabeled $\beta_2$-microglobulin we expect the fluorescein derivative to label cells of any vertebrate species.

This success story also has its cautionary side. Overcoupled and unfractionated $\beta_2$-microglobulin does not bind specifically to class I MHC antigen heavy chains. Also, we find that several different rhodamine derivatives of $\beta_2$-microglobulin bind specifically, to the extent that they do not bind to MHC class I antigen-negative cells, but are rapidly internalized. We do not understand the basis of this difference from fluorescein–$\beta_2$-microglobulin but we know that it is not due to aggregation induced by the coupling.

## REFERENCES

Alberti, S., Parks, D. R., and Herzenberg, L. A. (1987). *Cytometry,* in press.

Aubin, J. E. (1979). *J. Histochem. Cytochem.* **27**, 36–43.

Barak, L. S., and Webb, W. W. (1982). *J. Cell Biol.* **95**, 846–852.

Benson, R. C., Meyer, R. A., Zaruba, M. E., and McKhann, G. M. (1979). *J. Histochem. Cytochem.* **27**, 44–48.

Blakeslee, D., and Baines, M. G. (1976). *J. Immunol. Methods* **13**, 305.

Bohn, B. (1980). *Mol. Cell. Endocrinol.* **20**, 1–15.

Cherksey, B. D., Mendelsohn, S. A., Zadunaisky, J. A., and Altszuler, N. (1985). *J. Membr. Biol.* **84**, 105–116.

Due, C., Linnet, K., Langeland-Johansen, N., and Olsson, L. (1985). *Diabetologia* **28**, 749–755.

Edidin, M., and Zuniga, M. (1984). *J. Cell Biol.* **99**, 2333–2335.

Edidin, M., Zagyansky, Y., and Lardner, T. (1976). *Science* **191**, 466–468.

Fabricant, R. N., Delarco, J. E., and Todaro, G. J. (1977). *Proc. Natl. Acad. Sci. U.S.A.* **74**, 565–569.

Fowler, V., and Branton, D. (1977). *Nature (London)* **268**, 23–26.

Friedman, R. M., and Vogel, S. N. (1983). *Adv. Immunol.* **34**, 97–140.

Gabel, C. A., Eddy, E. M., and Shapiro, B. M. (1979). *J. Cell Biol.* **82**, 742–754.

Golan, D. E., and Veatch (1980). *Proc. Natl. Acad. Sci. U.S.A.* **77**, 2537–2541.

Gundersen, G. G., and Shapiro, B. M. (1984). *J. Cell Biol.* **99**, 1343–1353.

Hochman, J. H., and Edidin, M. (1987). *In* "Major Histocompatibility Genes and Their Role in Immune Function" (C. S. David, ed.). Plenum, New York, in press.

Hochman, J., and Edidin, M. (1988). *J. Immunol.* **140**, 2322–2329.

Jacobson, K., Elson, E., Koppel, D., and Webb, W. W. (1983). *Fed. Proc., Fed. Am. Soc. Exp. Biol.* **42**, 72–79.

Kahn, C. R., and Baird, K. (1978). *J. Biol. Chem.* **253**, 4900–4906.

Kimball, E. S., and Coligan, J. (1983). *Contemp. Top. Mol. Immunol.* **9**, 1–64.

Klausner, R. D., Kleinfeld, A. M., Hoover, R. L., and Karnovsky, M. J. (1980). *J. Biol. Chem.* **255**, 1286–1295.

Klein, J. (1986). "Natural History of the Major Histocompatibility Complex." Wiley, New York.

Kleinfeld, A. M., Klausner, R. D., Hoover, R. L., and Karnovsky, M. J. (1980). *J. Biol. Chem.* **255**, 1286–1295.

Liebman, P., and Entine, G. (1974). *Science* **185**, 457.

Limbird, L. (1986). "Cell Surface Receptors, A Short Course on Theory and Methods." Kluwer, Boston.

Mason, D. W., and Williams, A. F. (1980). *Biochem. J.* **187**, 1–20.

Murphy, R. F., Powers, S., Verderame, M., Cantor, C. R., and Pollack, R. (1982). *Cytometry* **6**, 402–406.

Oi, V. T., Glazer, A. N., and Stryer, L. (1982). *J. Cell Biol.* **93**, 98–100.

Packard, B. S., Karukstis, K. K., and Klein, M. P. (1984). *Biochim. Biophys. Acta* **769**, 201–208.

Pagano, R. E., Longmuir, K. J., Martin, O. C., and Struck, D. K. (1981). *J. Cell Biol.* **91**, 872–877.

Peacock, J. S., Londo, T. R., Roess, D. A., and Barisas, B. G. (1986). *J. Immunol.* **137**, 1916–1923.

Peters, R., Peters, J., Tews, K. H., and Bahr, W. (1974). *Biochim. Biophys. Acta* **367**, 282–294.

Peters, R., Brunger, A., and Schulten, K. (1981). *Proc. Natl. Acad. Sci. U.S.A.* **78**, 962–966.

Phillips, E. S. (1985). Ph.D. thesis, The Johns Hopkins University, Baltimore.

Poo, M.-M., and Cone, R. A. (1974). *Nature (London) New Biol.* **247**, 438–441.

Porter, R. R. (1959). *Biochem. J.* **73**, 119–126.

Putnam, F. W., Florent, G., Paul, C., Shindoa, T., and Shimuzu, A. (1973). *Science* **182**, 287.

Revilla, Y., Ferreira, A., Villar, M. L., Bootello, A., and Gonzalez-Porque, P. (1986). *J. Biol. Chem.* **261**, 6486–6491.

Schecter, Y., Schlessinger, J., Jacobs, S., Chang, K.-S., and Cuatrecasas, P. (1978). *Proc. Natl. Acad. Sci. U.S.A.* **75**, 2135–2139.

Schindler, M., Koppel, D. E., and Sheetz, M. P. (1980). *Proc. Natl. Acad. Sci. U.S.A.* **77**, 1457–1461.

Schlessinger, J., Axelrod, D., Koppel, D. E., Webb, W. W., and Elson, E. L. (1977). *Science* **195**, 307–309.

Schlessinger, J., Schechter, Y., Willingham, M. C., and Pastan, I. (1978). *Proc. Natl. Acad. Sci. U.S.A.* **75**, 2659–2663.

Shinitzky, M., and Yuli, I. (1982). *Chem. Phys. Lipids* **30**, 261.

Shiroishi, T., Evans, G. A., Appella, E., and Ozato, K. (1985). *J. Immunol.* **134**, 623.

Sleight, R. G., and Pagano, R. E. (1984). *J. Cell Biol.* **99**, 742–751.

Spack, E. G., Jr., Packard, B., Wier, M. L., and Edidin, M. (1986). *Anal. Biochem.* **158**, 233–237.

Szollosi, J., Damjanovich, S., Mulhern, S. A., and Tron, L. (1987). *Prog. Biophys. Mol. Biol.* **49**, 65–87.

Titus, J. A., Haugland, R., Sharrow, S. O., and Segal, D. M. (1982). *J. Immunol. Methods* **50**, 193–204.

Waggoner, A. (1986). *In* "Applications of Fluorescence in the Biomedical Sciences" (D. L. Taylor, A. S. Wagonner, R. S. Murphy, F. Lanni, and R. R. Birge, eds.), pp. 3–28. Liss, New York.

Wey, C.-L., Cone, R. A., and Edidin, M. (1981). *Biophys. J.* **33**, 225–232.

White, J. G., Amos, W. B., and Fordham, M. (1987). *J. Cell Biol.* **105**, 41–48.

Wolf, D. E. (1985). *Biochemistry* **24**, 582–586.

Wolf, D. E., Henkart, P., and Webb, W. W. (1980). *Biochemistry* **19**, 3893–3904.

Wolf, D. E., Kinsey, W., Lennarz, W., and Edidin, M. (1981). *Dev. Biol.* **81**, 133–138.

# Chapter 7

# Fluorescent Labeling of Mitochondria

## LAN BO CHEN

*Dana-Farber Cancer Institute*
*Harvard Medical School*
*Boston, Massachusetts 02115*

## I. Introduction

Success in mitochondrial research in the last 50 years represents a major triumph in biochemistry and cell biology. Few compartments in the cell are better understood than mitochondria. However, there remain unsolved problems, for example, the role of mitochondria in cellular differentiation, the molecular basis of mitochondrial motility, the role of mitochondria in diseases including cancer, obesity, and atherosclerosis, the signal transduction from plasma membrane to mitochondria, the interaction of mitochondria with endoplasmic reticulum, lysosomes and other organelles, the coordination of gene expressions in mitochondria and in nuclei, the assembly

METHODS IN CELL BIOLOGY, VOL. 29

and maintenance of mitochondria, and the role of mitochondria in calcium homeostasis.

Some of these unsolved problems are difficult to investigate by the conventional approach of using isolated mitochondria *in vitro*. Genetic approaches and recombinant DNA technology are essential and have already altered the prospect of mitochondrial research. Applying these approaches to yeast, mitochondriologists are rapidly solving the problems of the biogenesis of mitochondria and the nuclear–mitochondrion communication. However, yeast cells are too small for the analysis of mitochondrial motility and the interactions of mitochondria with other organelles. In addition, the presence of a cell wall makes yeast different from mammalian cells in signal transduction from the plasma membrane to mitochondria. The role of mitochondria in differentiation and human disease are also very difficult to study in yeast.

Thus, it is imperative to develop complementary approaches to study mitochondria, in particular, in its native context: the living cells. For example, we need a simple, unambiguous technique to detect mitochondria in living cells; a convenient, reliable technique to monitor respiration, electrochemical gradient, oxidative phosphorylation in a single living cell; to measure mitochondrial pH in living cells; to visualize and follow mitochondrial DNA; to micromanipulate mitochondria; and to monitor calcium flux across mitochondria.

Toward these goals, several laboratories, most prominently Britten Chance's lab, have been developing new approaches. One of Chance's early successes was to probe mitochondrial bioenergetics in living cells by monitoring NADH fluorescence (Chance, 1970a,b). He has shown that pyridine nucleotide is an excellent indicator for the oxygen requirements of energy-linked functions of mitochondria (Chance, 1976). He has applied this technique to monitor bioenergetics in tumor, liver, heart, brain, skeletal muscle, even in sperm that contain few mitochondria. Chance and others have also recognized early the need of applying exogenous fluorescent dyes to probe mitochondrial membrane bioenergetics. In one of Chance's reviews (Chance, 1976), he predicted, "Ultimately, a fluorescent probe that binds the mitochondrial membrane exclusively may be developed." Pursuing such a goal, he and others have used fluorescent dye, such as ANS and oxonol V, to reflect the energized state of isolated mitochondria *in vitro*. Fortuitously, these dyes happened to not be positively charged. Otherwise, Chance's group would have discovered "a fluorescent dye that binds the mitochondrial membrane exclusively" in living cells as early as 1955. Interestingly, later Chance's group did collaborate with Waggoner who has pioneered the use of lipophilic, cationic, fluorescent dyes for monitoring membrane potential (Waggoner, 1976, 1979a,b). Had epifluorescence microscopy been available at the time, fluorescent dyes that "bind the mitochondrial membrane exclusively" in living cells could have been discovered in that collaboration.

## II.   Unique Property of Mitochondria: High Membrane Potential

Mitochondria are known to have a proton electrochemical gradient generated by proton pumps which in turn are driven by respiratory electron transport chains utilizing NADH and succinate. According to Mitchell (1966), energy stored in this gradient is primarily responsible for the conversion of $ADP + P_i$ to ATP by ATP synthetase (widely called $F_0F_1ATPase$). This gradient has two components: the electric component (the membrane potential) and the chemical component (the pH gradient). The rate of proton pumping is mostly dictated by the rate of respiration but the expression of proton gradient as membrane potential or pH gradient might be linked to cellular programs other than respiration.

Since the mitochondrion is the only organelle known to have a significant membrane potential with negative charge inside, lipophilic compounds with delocalized positive charge may be taken up by mitochondria to a much greater extent than that by other organelles. Although plasma membrane potential would also allow the equilibration of such compound into cytoplasm, the concentration as predicted by the Nernst equation would be much lower than that in mitochondria. If cells have a plasma membrane potential of 60 mV and a mitochondrial membrane potential of 180 mV, 1× concentration of a lipophilic cation in culture medium may lead to 10× in cytoplasm and 10,000× in mitochondria at equilibrium as calculated by the Nernst equation under ideal conditions (Weiss and Chen, 1984; Davis et al., 1985). Although cells are not in ideal conditions to allow 10,000× accumulation in mitochondria, 1000× are readily reached experimentally (recalculated from data presented in Davis et al., 1985; and Emaus et al., 1986). Moreover, before reaching equilibrium, as a result of the much higher membrane potential across mitochondria than that of the plasma membrane, most of the lipophilic cation would be electrophoresed first into the more negative sink of mitochondria, not the cytoplasm or other organelles. If such a compound happened to be a fluorescent dye, under epifluorescence microscopy, mitochondria in living cells should be brightly stained.

## III.   Fluorescent Dyes with Delocalized Positive Charges

Indeed, mitochondria in living cells are intensely stained against a dark cytoplasmic background by fluorescent lipophilic cations such as rhodamines 123, 3B, and 6G (Figs. 1 and 2; Johnson *et al.,* 1980) and cyanines developed by Waggoner and others (Waggoner, 1976, 1979a,b; Cohen *et al.,* 1981; Johnson *et al.,* 1981). The selective staining is not dependent on the specific structural feature other than lipophilicity and delocalized, positive charge. After testing 3000 lipophililc cations including rhodamines, cyanines, acridines, pyryliums, and quinoliniums, it now seems safe to infer that all lipophilic compounds with delocalized positive charge ($pK_a > 8.2$) are accumulated by mitochondria in accordance with the Nernst potential.

Among all lipophilic, cationic, fluorescent dyes tested thus far, the least toxic has been rhodamine 123. It has a relatively stable yellowish–green fluorescence under blue excitation. Where phase-contrast and Normaski optics fail to reveal mitochondria, rhodamine 123 staining in conjunction with epifluorescence excels. Rhodamine 123 is effective in revealing mitochondria cluster, granular mitochondria difficult to be distinguished from lysosomes and other vesicular structure; fused mitochondria in three-dimensional networks; and mitochondria in perinuclear region and other thickened areas of cells (Johnson *et al.,* 1980, 1981, 1982; Chen *et al.,* 1982, 1984; Summerhayes *et al.,* 1982; Hedberg and Chen, 1986).

Although rhodamine 123 staining is developed for mitochondria in living cells, mitochondria in glutaraldehyde-fixed cells may also be stained (Terasaki *et al.,* 1984; Terasaki, Chapter 8, this volume). Whether glutaraldehyde-fixed mitochondria maintain a membrane potential, which is not unlikely in view of the early report that such mitochondria can respire

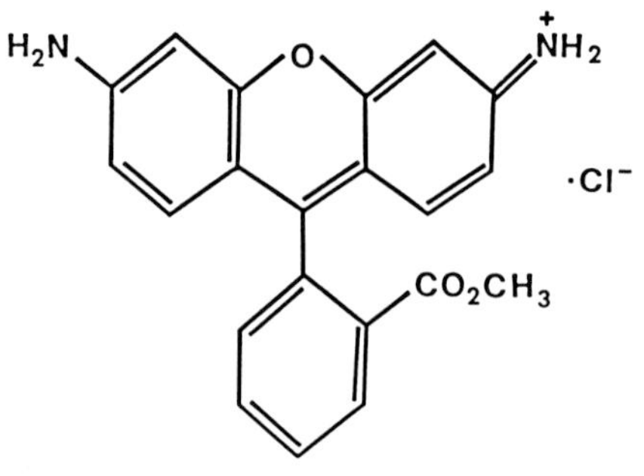

FIG. 1.   Structure of rhodamine 123.

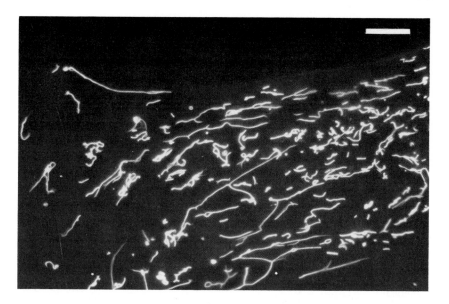

FIG. 2. Mitochondria in a portion of a large living cell stained with rhodamine 123 (10 μg/ml for 10 minutes). Bar, 10 μm.

(Deamer *et al.,* 1967; Seibert *et al.,* 1971), or, after fixation an entirely different mechanism is invoked, is still unknown. As discussed by Terasaki, the partitioning of lipophilic cations into lipid bilayer after glutaraldehyde fixation should be considered. Moreover, since rhodamine 123 inhibits $F_0F_1$ATPase (discussed below), the possibility of direct binding to this enzyme after glutaraldehyde treatment should also be considered.

## IV. Rhodamine 123 as a Conditional Supravital Dye

When cells are exposed to rhodamine 123 at 10 μg/ml for 10 minutes at 37°C, no toxic effect on cells has been detected thus far. This conclusion is based on nearly 10 years experience with this dye involving numerous experiments in this lab, and those reported by others (Darzynkiewicz *et al.,* 1981, 1982; James and Bohman, 1981; Albertini, 1984; Gundersen *et al.,* 1982; Arslan *et al.,* 1984; Collins and Foster, 1983; DeMartinis *et al.,* 1987). In this lab, we find that the standard staining procedure does not affect cell growth, DNA synthesis, RNA synthesis, protein synthesis, glucose transport, organization of endoplasmic reticulum and Golgi apparatus, fibronectin synthesis and expression on the cell surface, organizations of microfilaments, microtubules and intermediate filaments, cell structures revealed by both transmission and scanning electron microscopy, or cell locomotion monitored by time-lapse videomicroscopy. And, most important of all, at 10 $\mu$g/ml for 10 minutes, rhodamine 123 does not affect mitochondrial respiration, membrane potential, pH gradient, ATP synthesis, or cristae examined by thin-section electron microscopy. Thus, it seems reasonable to consider that at 10 $\mu$g/ml for 10 minutes, rhodamine 123 is a supravital dye for all cells.

However, under two conditions, rhodamine 123 cannot be classified as a supravital dye. One is the exposure of stained cells to intense blue or green light. Like most fluorescent dyes, rhodamine 123 is susceptible to photobleaching which leads to the formation of free radicals, singlet oxygen, or heat. In the environment of mitochondria where oxygen is abundant, the electron transport chain is imbedded in lipid bilayers, and the electrochemical gradient can be dissipated as heat, light sensitive lipophilic cations have an excellent opportunity to exert photoinduced cytotoxicity. The conditions normally used in epifluorescence are not photoactive toward rhodamine 123 except for the combination of a 100-W mercury light source and 63X objective lens with numerical aperture >1.2. Phototoxicity may be dramatically reduced by inserting a neutral density filter, or a polarizer in the incident light path to reduce the total dose of light. While phototoxicity should be avoided for mitochondrial localization, a combination of high doses of light and lipophilic cations including rhodamine 123 and cyanines has been exploited for the phototherapy of cancer (Oseroff et al., 1986; Powers et al., 1986; Beckman et al., 1987).

The other condition that can lead to cytotoxicity without light is the prolonged exposure (>16 hours) to rhodamine 123 at >5 $\mu$g/ml. Such a condition is particularly detrimental to certain carcinoma cells (Lampidis et al., 1983), and myocardiac cells (Lampidis et al., 1984) that take up 5- to 20-fold more dye than do normal epithelial cells or lymphocytes. Like phototherapists, chemotherapists are also exploiting this effect for the purpose of killing carcinoma cells without light (Bernal et al., 1982b, 1983). The basis for cytotoxicity of prolonged exposure of cells to rhodamine 123 has not been fully established but the following inhibitory activities of rhodamine 123 may all contribute: mitochondrial $F_0F_1$ATPase (Mai and Allison, 1983; Modica-Napolitano et al., 1984; Lampidis et al., 1984; Emaus et al., 1986; Modica-Napolitano and Aprille, 1987), mitochondrial protein synthesis (Abou-Khalil et al., 1986), transport of carbamoyl-phosphate synthetase I and ornithine transcarbamylase (Morita et al., 1982), and transhydrogenase (Lubin et al., 1987) into mitochondria.

## V.  Methods of Labeling and Quantitation

To stain mitochondria in living cells with rhodamine 123 cells are grown on glass cover slips (12-mm round or square, from Bradford Scientific, Epping, New Hampshire) and two drops of rhodamine 123 (Eastman Kodak, 10 $\mu$g/ml) in culture medium added to the cells. After 10 minutes at 37°C, cells are washed in dye-free medium for 1 to 2 minutes at room temperature. Because rhodamine 123 staining is dependent on mitochondrial membrane potential which in turn is critically dependent on oxygen, a living cell chamber was developed. This chamber is made of a piece of silicon rubber (North American Reiss Co., Belle Mead, New Jersey) with three holes (10 mm in diameter) made by a rubber stopper hole puncher. The rubber is pressed onto a microscope slide by hand to create a seal. Culture medium is then added into the hole and cover slips mounted. The excess medium is removed with layers of Kimwipe gently placed on top of cover slips and tapped with fingers. Before microscopy, the surface of cover slips may be further cleaned with alcohol. This living cell chamber allows a prolonged observation since medium can readily be changed (by lifting cover slips with a tweezer) and oxygen and carbon dioxide can diffuse through silicon rubber. Moreover, since the thickness of the rubber is <0.7 mm, it allows the use of a high power objective lens including 100×. Without grease, agar, nail polish, or clumsy mechanical accessories, this silicon rubber chamber for living cells has served us well for studying living cells in the last 10 years.

The excitation spectrum of rhodamine 123 has a maximum at 500 nm and the molar extinction coefficient is $7.5 \times 10^4$ cm$^{-1}$ mol$^{-1}$ in water (Darzynkiewicz *et al.*, 1981). The maximum of the emission spectrum varies, depending on concentrations. At low concentrations ($10^{-8}$ $M$), the maximum is at 525 nm; at a higher concentration ($10^{-3}$ $M$), 545 nm. However, the absorption spectrum does not change with dye concentration. Using computer simulation of spectral changes, Darzynkiewicz *et al.* (1982) suggested that concentration-dependent red shift of the emission is the result of an inner filter effect rather than excimer formation. The emission spectrum of the rhodamine 123 localized in mitochondria has a red shift of 12 nm compared with dye in water (Darzynkiewicz *et al.*, 1982; Emaus *et al.*, 1986). This shift is more likely a result of complex formation, instead of inner filter effect or excimer formation.

For visualizing rhodamine 123 by fluorescence microscopy, the filter system should be that for fluorescein (blue excitation), not rhodamine (green excitation). The most suitable objective lens for detecting mitochondria in the absence of correction for out-of-focus light is the Zeiss Planapo 40× or its equivalent since it has a smaller depth of field. Because of plasticity of living mitochondria, it is essential to use the shortest possible exposure time for photography. We use Kodak Tri-X or preferably, Kodak T-Max 400 at Exposure Index 1600 or 3200 and developed in Kodak HC-110 dilution B for 12 to 15 minutes at 25°C. The printing is simply made on Kodak Polycontrast resin-coated (RC) paper with #3 filter and processed by Kodak Ektamatic Processor designed for Ektamatic paper. The prints are fixed in Kodak Rapid Fixer for 2 minutes, rinsed in running water, and air-dried in a warm room. For videomicroscopy, in order to prolong observation time, mitochondria stained with rhodamine 123 are best visualized under weak blue light from a tungsten or halogen lamp (with proper neutral density filters). In some cell types, continuous monitoring of rhodamine 123-stained mitochondria by videomicroscopy is possible for 2 to 10 hours.

Fluorescence intensity of rhodamine 123 in mitochondria of living cells may be quantitated by various techniques described in other chapters in this volume (i.e., Wampler and Kutz, Chapter 14; Spring and Lowy, Chapter 15; and Aikens et al., Chapter 16). Rhodamine 123 uptake at a single cell level may be determined (Wiseman et al., 1985). In principle, fluorescence of a single mitochondrion may also be determined by various techniques described in Volume 30, in particular, those involving image processors. However, due to dye quenching (Emaus et al., 1986), it is essential to quantitate the rhodamine 123 concentration by extracting it with organic solvents and measuring it outside the cells in the monomeric state by fluorescence spectrophotometry.

In a typical 6 hour-uptake experiment, forty 60-mm cell culture dishes of cells are evenly seeded and grown to subconfluence. Prior to the experiment, 18 of the dishes are aspirated free of medium, wrapped in plastic, and placed at −20°C for 12 to 16 hours. On the day of the experiment, these control dishes are allowed to thaw at room temperature for 10 minutes, then each received 3 ml of freshly prepared rhodamine 123 (1 $\mu$g/ml) in culture medium supplemented with 5% calf serum. At the same time, 18 dishes of live cells are aspirated free of medium and also received dye-containing medium. All dishes are incubated at 37°C. At hourly intervals over a 6 hour-period, three live-cell and three frozen-cell dishes are removed from the incubator, quickly washed three times with 3 ml of phosphate-buffered saline (PBS), and extracted with 2.5 ml butanol. After 20 minutes at room temperature, the butanol extracts are transferred to disposable cuvettes (Evergreen Scientific), sealed with Parafilm, and stored in the dark at 4°C. Samples are measured within 48 hours against freshly prepared standards of 0 to 100 ng/ml rhodamine 123 in butanol. Control experiments showed less than 1% variation during this storage interval. Measurements are made using a Hitachi Perkin-Elmer model MPF-2A or MPF-4 fluorescence spectrophotometer with bandwidths set at 10 nm, an excitation wavelength of 485 nm, and an emission wavelength of 532 nm. The total uptake of rhodamine 123 per dish of live cells per time point is calculated from the butanol measurements as ng rhodamine 123/dish, and the values from the three live-cell dishes are averaged. Nonspecific uptake is calculated for each time point in the same manner using cells frozen before labeling. Specific rhodamine 123 uptake on a "per cell" basis is calculated by subtracting the nonspecific uptake from the total uptake and dividing the result by the average number of cells per dish. Uptake is graphed as a function of time in 1 $\mu$g/ml of rhodamine 123.

# VI. Mitochondrial Localization

Mitochondria in the following cell types have been very difficult to detect in living cells, but stand out unambiguously after staining with rhodamine 123: sperm (Johnson *et al.*, 1980), Sertoli cells (Tanphaichtr *et al.*, 1984), oocytes (Albertini, 1984), fertilized egg (Gundersen *et al.*, 1982), lymphocytes (Arslan *et al.*, 1984), granulocytes (Collins and Foster, 1983), macrophages (Johnson *et al.*, 1980), adipocytes (DeMartinis *et al.*, 1987), cardiac muscle cells (Johnson *et al.*, 1980; Siemens *et al.*, 1982; Lampidis *et al.*, 1984), myoblasts (Summerhayes *et al.*, 1982; Chen *et al.*, 1982), chondrocytes (Benel *et al.*, 1986), glomerular cells (Oberley *et al.*, 1982), presynaptic nerve terminal (Yoshikami and Okun, 1984), *Plasmodium yoelii*-infected mouse erythrocytes (Tanabe, 1983), *Toxoplasma gondii*-infected mouse fibroblasts (Tanabe, 1985), reovirus-infected cells (Sharpe *et al.*, 1982), interferon-treated cells (Brouty-Boye *et al.*, 1981), senescence cells (Martinez *et al.*, 1986), *Trypanosoma cruzi* (Wolfson *et al.*, 1987), plant cells (Wu, 1987), and yeast (L. B. Chen, unpublished results).

# VII. Mitochondrial Morphology

Since the seminal observations by Lewis and Lewis (1915), mitochondria are known to have a capacity to change morphology rapidly, as much as 15–20 different shapes in 10 minutes. The origin and consequence of the morphological transformation of mitochondria remain totally unknown. It seems unlikely that these events take place without biochemical meanings. However, mitochondrial morphology is not totally autonomous and appears to be sensitive to the surrounding environment as well. The long,

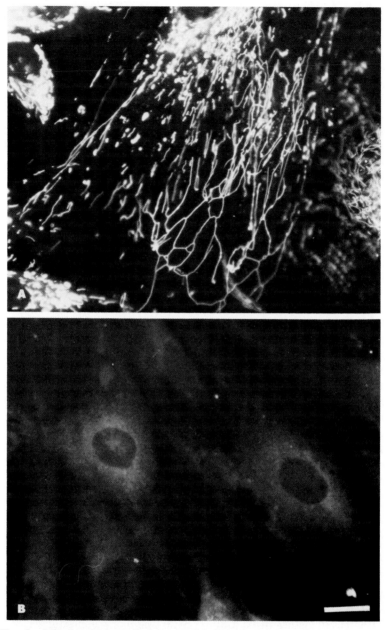

FIG. 3. Mitochondrial membrane potential dependent uptake of rhodamine 123. (A) Human fibroblasts FS-2 pretreated with 137 m$M$ K$^+$ for 30 minutes to abolish plasma membrane potential (Davis *et al.*, 1985), followed by the incubation of rhodamine 123 at 0.1 $\mu$g/ml in 137 m$M$ K$^+$ medium for 30 minutes. (B) Same as above except oligomycin (5 $\mu$g/ml) and sodium azide (1 m$M$) were included in 137 m$M$ K$^+$ medium. Bar, 15 $\mu$m.

straight mitochondria clearly depend on the presence of intact microtubules. Once microtubules are disrupted by colchicine, the straight filamentous mitochondria become bent and twisted (Johnson et al., 1980; Summerhayes et al., 1983).

Filamentous versus granular forms are interchangeable and can be externally controlled. For example, phorbol ester, TPA/PMA, converts all filamentous mitochondria into granular form between 2 to 5 hours (Chen et al., 1984). Once TPA is removed, the granular form returns filamentous. If indeed protein kinase C is the predominant target for TPA/PMA, certain substrates for this enzyme could be responsible for the filamentous to granular conversion. Interestingly, Weinstein's group has described mitochondrial proteins (69K, 37K, and 17K) whose phosphorylation is calcium dependent and markedly stimulated by the presence of protein kinase C and TPA/PMA (Backer et al., 1986). These phosphoproteins may provide the first clue to the molecular basis of mitochondrial morphology. A similar but irreversible change in mitochondrial morphology occurs during myogenesis. When myoblasts fuse into myotubes, the filamentous mitochondria in the myoblasts gradually change into the granular form and increase in size over time (Chen et al., 1982). In 3T3 cells, platelet derived growth factor (PDGF) tends to promote granular form of mitochondria instead of filamentous form (Chen et al., 1984).

Another interesting enigma observed is the fused mitochondrial network (Johnson et al., 1980). It has been detected in yeast, fibroblasts, certain kideny epithelial cells, glial cells, and neuronal cells, but rarely in other epithelial cells, myotubes, cardiac muscle cells, lymphocytes, macrophage, or endothelial cells. Whether mitochondria network has significant biochemical origin or consequence is unknown. Although some extremes of mitochondrial morphology are observed in normal muscle and fat cells, the most dramatic cases are seen in human tumor cells (Summerhayes et al., 1982; Chen et al., 1984; DeMartinis et al., 1987).

## VIII. Mitochondrial Motility and Distribution

Saltatory motion of mitochondria has been well documented for many years (Tzagoloff, 1982). Rhodamine 123 stained mitochondria have been used for further understanding of mitochondrial motility (Salmeen et al., 1985; Herman and Albertini, 1984). Whether it is an active motion propelled by its own motor, a passive movement regulated by forces generated by cytoskeleton or Brownian random walk, is still unresolved. Using rhodamine 123 and a silicon intensified target (SIT) video camera, Salmeen et *al.* (1985) reported that the predominant translational motion of the mito-chondria satisfied the formal conditions of a Brownian motion for a free particle, although sometimes there was a slow drift superimposed on the random motion. They determined the apparent diffusion coefficients were about $5 \times 10^{-12}$ cm$^2$/second, and the drift speeds about $2 \times 10^{-3}$ $\mu$m/sec-ond. Although this study infers that the mitochondrion itself might not have a motor, it does not preclude the involvement of other forces in mitochondrial motility, in particular where they should inhabit.

That microtubule is intimately associated with mitochondrial motility–distribution has been repeatedly suggested (Raine *et al.,* 1971; Allen, 1975; Smith *et al.,* 1975, 1977; Wang and Goldman, 1978). Colchicine is a potent inhibitor of mitochondrial movement and the distribution of microtubules correlates well with that of mitochondria (Heggeness *et al.,* 1978; Ball and Singer, 1982; Summerhayes *et al.,* 1983). Whether certain mitochondrial outer membrane proteins might interact with microtubules or, whether certain proteins might bind both microtubules and mitochondria (such as kinesin), remains to be determined. Recent success in maintaining move-ment of organelles along microtubules *in vitro* (Vale *et al.,* 1985a–c; Allen *et al.,* 1985) should open a new avenue to study mitochondrial motility at a molecular level.

However, the microtubule may not be the only structure involved in mitochondrial motility–distribution. Although microfilaments may not play a major role in mitochondrial motility–distribution, it is difficult to rule out a role of intermediate filaments. With regard to distributions, microtubules correlate with mitochondria, and intermediate filaments cor-relate with microtubules (Geiger and Singer, 1980; Ball and Singer, 1981), therefore, mitochondria should correlate with intermediate filaments, at least by the first approximation. Indeed, there were earlier reports consist-ent with this contention (Starger and Goldman, 1977; Wang and Goldman, 1978; Lee *et al.,* 1979; Toh *et al.,* 1980). However, Ball and Singer (1982) showed that in fibroblasts transformed with Rous sarcoma virus or treated with cycloheximide, the mitochondria are only associated with microtu-bules but not intermediate filaments. Lin and Feramisco (1981) reported that when intermediate filament distribution was disrupted by microinjec-tion of a monoclonal antibody recognizing a 95K protein, neither motility nor distribution of mitochondria was affected, which was subsequently confirmed (Chen *et al.,* 1982; Summerhayes *et al.,* 1983). However, the notion that intermediate filament is not involved in mitochondrial distri-bution has become less unambiguous with time. By use of rhodamine 123, a renewed effort was made to resolve the issue of intermediate filaments in mitochondrial motility–distribution.

First, in several human carcinoma cell lines, mitochondria revealed by rhodamine 123 are clustered near the nucleus, although these cells are well spread and the microtubule distribution appears normal (Chen *et al.,* 1984; Hedberg and Chen, 1986). Despite abundant cytoplasmic space available for migration, mitochondria are trapped in the perinuclear region. Why does the correlation between mitochondria and microtubules break down in these human carcinoma cells? Unexpectedly, when these carcinoma cells were stained for intermediate filaments, both keratin and vimentin distributions correlate closely with mitochondrial distribution (Chen *et al.,* 1984).

Second, in many cell types (but not all, exception shown in Mose-Larsen *et al.,* 1982), when microtubules are disrupted by colchicine or vinblastine, the distribution of coalesced intermediate filaments correlates with that of mitochondria (Starger and Goldman, 1977; Summerhayes *et al.,* 1983).

Third, in CV-1 cells (African green monkey kidney), cycloheximide disrupts the organization of intermediate filament as well as the distribution of mitochondria but leaves microtubules totally intact (Summerhayes *et al.,* 1983; Sharpe *et al.,* 1980).

Fourth, in reovirus-infected cells where the distribution of microtubules is similar to that of uninfected cells, a concommitant disruption in the distribution of mitochondria and intermediate filament is observed (Sharpe *et al.,* 1982).

Fifth, most significantly, Celis and colleagues reported that when cultured fibroblasts were extracted with Triton, the distribution of one of the proteins (called IEF 24, MW 56,000), tenaciously associated with a subset of intermediate filaments, correlates closely with mitochondrial distribution (Mose-Larsen *et al.,* 1982). IEF 24 was suggested as a linking protein between mitochondria and intermediate filaments.

The relationships between cytoskeleton and mitochondria are probably far more complex than initially realized. It appears that both microtubules and intermediate filaments might influence mitochondrial distribution, and although neither appears to play a dominant role when both are present, in the absence of one, the other seems to compensate (Summerhayes *et al.,* 1983). The interactions between mitochondria and cytoskeleton may even vary from one cell type to the other (after all, keratin is not identical to vimentin). And, as suggested by Celis and colleagues, different associations may be revealed depending on the cell type and/or the experimental conditions used (Mose-Larsen *et al.,* 1982).

## IX.  Monitoring Mitochondrial Membrane Potential

A variety of mitochondrial activity is influenced by membrane potential: ATP synthesis, calcium uptake–release, transport of precursor enzymes into mitochondria, and, possibly mitochondrial protein synthesis (Tzagoloff, 1982). It is of interest to study whether the magnitude of mitochondrial membrane potential can be determined by the amount of dye uptake. At equilibrium, the Nernst equation predicts a positive correlation between membrane potential and dye uptake. Data collected thus far strongly suggest that the uptake of rhodamine 123 is indeed membrane potential dependent (Fig. 3; Johnson et al., 1981; Emaus et al., 1986; Modica-Napolitano and Aprille, 1987). Since the dye must first pass through plasma membrane, the dye concentration in the mitochondria is a function of both the plasma membrane and mitochondrial membrane potentials. In order to monitor only the mitochondrial membrane potential, it is necessary to abolish the plasma membrane potential. This can be achieved, without deleterious effect on the cells, in a medium in which the $K^+$ concentration is adjusted to that of the cytoplasm, about 137 m$M$ (Davis et al., 1985).

By the use of rhodamine 123, uptake at a single cell level may be monitored. Cells grown on cover slips are mounted in the silicon-rubber living cell chamber with a modified culture medium containing rhodamine 123 (50 ng/ml) and 137 m$M$ $K^+$. At equilibrium, the fluorescence in a single cell may be measured by a fluorescence spectrophotometer attached on microscope or a low light level camera in conjunction with an image processor (see Volume 30 and Wampler and Kutz, Chapter 14, this volume). Concentration of the dye in mitochondria may be calculated based on the volume of mitochondria measured by image analysis, by electron microscopy and/or by the content of mitochondria-specific enzymes (Weibel, 1969; Loud, 1968). Based on the Nernst equation, mitochondrial membrane potential may then, in principle, be expressed in millivolts for a single cell or a single mitochondrion.

The effect of differentiation on mitochondrial membrane potential is most pronounced in myogenesis. Prior to fusion, myoblasts have a low membrane potential. As soon as two myoblasts fuse, the membrane potential increases dramatically (Chen et al., 1982). Bladder epithelial cells active in migration have a higher mitochondrial membrane potential than those in a confluent, resting state (Johnson et al., 1981).

A significant difference in mitochondrial membrane potential is found between normal epithelial cells and carcinogen-transformed epithelial cells or certain carcinoma-derived cells (Summerhayes et al., 1982). Whereas most normal epithelial cells have a low mitochondrial membrane potential, differentiated adenocarcinoma, squamous cell carcinoma, transitional cell carcinoma, and melanoma-derived cells express a high potential. This phenotype has also been detected in primary explants of surgical specimens of human colon adenocarcinoma, bladder transitional cell carcinoma and

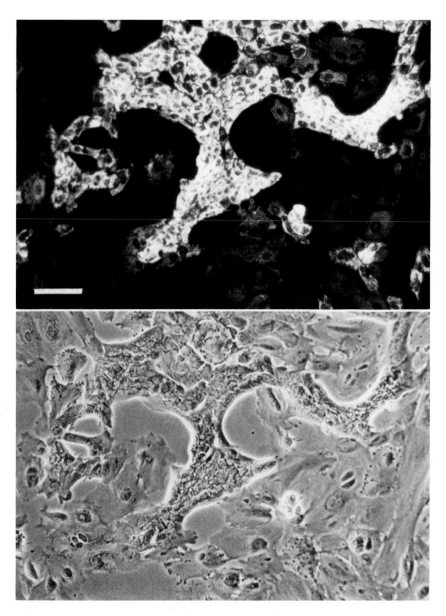

FIG. 4. Primary explant culture of human breast carcinoma incubated with rhodamine 123 (10 μg/ml for 10 minutes) and left in dye-free medium for 24 hours. Normal cells present in tumor retain less dye relative to carcinoma cells. Bar, 60 μm.

melanoma, as well as DMBA-induced rat mammary carcinoma, DMH-in-
duced rat colon carcinoma, and BBN-induced mouse bladder transitional
cell carcinoma, demonstrating that it is not an artifact of immortalization.
An example of differential uptake–retention of rhodamine 123 between
human breast carcinoma cells and human normal breast epithelial cells is
demonstrated in a primary mix culture derived from a surgical specimen
shown in Fig. 4. However, the mitochondrial phenotype of higher membrane potential is not detected in poorly differentiated carcinoma, large cell
carcinoma, oat cell carcinoma, leukemia, lymphoma, neuroblastoma, sarcoma, or osteosarcoma. The molecular basis for these observations are still
unknown but it is of interest to note that fos oncogene has been implicated
in increased rhodamine 123 uptake and retention (Zarbl et al., 1987).

# X.  Monitoring Mitochondrial Total Electrochemical Gradient

It is essential to explore whether the total electrochemical gradient (pH
gradient and membrane potential) across mitochondria can be determined
in living cells since pH gradient alone can also drive ATP synthesis.
Nigericin, a $K^+/H^+$ exchange ionophore which dissipates pH gradient but
leads to compensatory increase in membrane potential with continuous
respiration, satisfies such a need (Johnson et al., 1981; Davis et al., 1985).
Since nigericin also induces hyperpolarization of the plasma membrane, it
is necessary to include ouabain in the medium. The differential in dye
uptake before and after nigericin–ouabain is indicative of the magnitude of
pH gradient. Thus, the total electrochemical gradient across mitochondria
in living cells may be monitored.
    Some cell types–lines such as CV-1 (monkey kidney) have lower mito-
chondrial membrane potential, but nigericin treatment leads to increased
rhodamine 123 uptake to levels near those for cells expressing high mitochondrial membrane potential, a finding indicative of a high pH gradient
but a normal respiratory rate and electrochemical gradient in these cells
(Davis et al., 1985). Since it is known that the uptake of substrates such as
glutamate and pyruvate by mitochondria is proportional to the magnitude
of pH gradient, these cells may have an unusual rate of glutamate and
pyruvate accumulation despite normal electrochemical gradient. Interestingly, none of the cells with a high mitochondrial membrane potential
(cardiac muscle and carcinoma cells), show an increase in dye uptake after
applying nigericin, suggesting that these cells express the entire proton
gradient as membrane potential.

A striking example of the effect of oncogenic transformation on mitochondrial electrochemical gradient is found in feline sarcoma virus-transformed mink fibroblasts. Whereas this fes oncogene-induced transformation does not affect the rate of respiration or the rate of proton pumping, it shifts the state of high membrane potential and low pH gradient to a state of low membrane potential and high pH gradient. In addition, we have identified poorly differentiated human colon carcinomas (such as FET and CCL 237) that have a low mitochondrial membrane potential but fail to show an increase in rhodamine 123 uptake upon nigericin treatment as in fes-transformed mink cells. These cells might have an unusually low rate of proton pumping and/or respiration, or uncoupled mitochondria. They may rely on glycolysis as the source of energy, and 2-deoxyglucose, alone normally nontoxic to other cells, exerts potent toxicity (Modica-Napolitano *et al.*, 1987).

## XI.   Flow Cytometry, Cell Sorting, and Other Applications

In addition to fluorescent microscopy, rhodamine 123 stained cells are ideal for analysis by flow cytometry (Shapiro, 1985; Ronot *et al.*, 1986). Cells with varying mitochondrial number or mitochondrial membrane potential may be separated by a fluorescence-activated cell sorter. These methods have been used to study lymphocyte activation and cell cycle (Darzynkiewicz *et al.*, 1981, 1982; James and Bohman, 1981; Evenson *et al.*, 1985); to purify hybrids from fusion of rhodamine 123 stained cytoplasts and unstained cells or karyoplasts (Walker and Shay, 1981, 1983; Clark and Shay, 1982a,b; Hightower *et al.*, 1981; Kliot-Fields *et al.*, 1983); to study the effect of anticancer drugs (Bernal *et al.*, 1982b); to study the effect of oncogenic transformation (Johnson *et al.*, 1982; Zarbl *et al.*, 1987), the differentiation of promyelocytic cells (Collins and Foster, 1983) and erythroleukemia cells (Tsiftsoglou *et al.*, 1983); to study the effect of aging on fibroblasts (Goldstein and Korczack, 1981) and monocytes (Staiano-Coico *et al.*, 1982), the motility of sperm (Evenson *et al.*, 1982), heterogeneity of tumors (Sonka *et al.*, 1983); and to measure mitochondrial content in tumor cells simultaneously with the cell volume, surface area, and DNA content (Steinkamp and Hiebert, 1982).

Rhodamine 123 staining has also been used to mark mitochondria when mitochondria were moved from one cell type into others (Clark and Shay, 1982a,b); to follow mitochondrial viability during subcellular fractionation (Casey and Anderson, 1982); to monitor recovery from uncoupler FCCP treatment (Maro *et al.*, 1982); and to identify mitochondrial proteins in two-dimensional gels (Anderson, 1981). A related analog, rhodamine 6G, has been used to study submicrometer regions of a single mitochondrion (Siemens *et al.*, 1982), to eliminate mitochondria (Ziegler and Davidson, 1981) or to kill cells based on differential rhodamine accumulation (Johnson *et al.*, 1982; Lampidis *et al.*, 1984).

## REFERENCES

Abou-Khalil, W. H., Arimura, G. K., Yunis, A. A., and Abou-Khalil, S. (1986). *Biochem. Biophy. Res. Commun.* **137,** 759–765.
Albertini, D. F. (1984). *Biol. Reprod.* **30,** 13–28.
Allen, R. D. (1975). *J. Cell Biol.* **64,** 497–503.
Allen, R. D., Weiss, D. G., Hayden, J. M., Brown, D. T., Fujiwake, H., and Simpson, M. (1985). *J. Cell Biol.* **100,** 1736–1752.
Anderson, L. (1981). *Proc. Natl. Acad. Sci. U.S.A.* **78,** 2407–2411.
Arslan, P., Corps, A. N., Hesketh, T. R., Metcalfe, J. C., and Pozzan, T. (1984). *Biochem. J.* **217,** 419–425.
Backer, J. M., Arcoleo, J. P., and Weinstein, I. B. (1986). *FEBS Lett.* **200,** 161–164.
Ball, E. H., and Singer, J. S. (1981). *Proc. Natl. Acad. Sci. U.S.A.* **78,** 6986–6990.
Ball, E. H., and Singer, J. S. (1982). *Proc. Natl. Acad. Sci. U.S.A.* **79,** 123–126.
Beckman, W. C., Powers, S. K., Brown, J. T., Gillespie, G. Y., Bigner, D. D., and Camps, J. L. (1987). *Cancer* **59,** 266–270.
Benel, L., Ronot, X., Kornprobst, M., Adolphe, M., and Mounolou, J. C. (1986). *Cytometry* **7,** 281–285.
Bernal, S. D., Lampidis, T. J., Summerhayes, I. C., and Chen, L. B. (1982a). *Science* **218,** 1117–1119.
Bernal, S. D., Shapiro, H. M., and Chen, L. B. (1982b). *Int. J. Cancer* **30,** 219–224.
Bernal, S. D., Lampidis, T. J., McIsaac, R. M., and Chen, L. B. (1983). *Science* **222,** 169–172.
Brouty-Boye, D., Cheng, Y. S. E., and Chen, L. B. (1981). *Cancer Res.* **41,** 4174–4184.
Casey, C. A., and Anderson, P. M. (1982). *J. Biol. Chem.* **257,** 8449–8461.
Chance, B. (1970a). *Proc. Natl. Acad. Sci. U.S.A.* **67,** 560–564.
Chance, B. (1970b). *Behav. Sci.* **15,** 1–22.
Chance, B. (1976). *Circ. Res.* **38** (Suppl. 1), 31–38.
Chen, L. B., Summerhayes, I. C., Johnson, L. V., Walsh, M. L., Bernal, S. D., and Lampidis, T. L. (1982). *Cold Spring Harbor Symp. Quant. Biol.* **LXVI,** 141–155.
Chen, L. B., Lampidis, T. J., Bernal, S. D., Nadakavukaren, K. K., and Summerhayes, I. C. (1983). *In* "Genes and Proteins in Oncogenesis" (I. B. Weinstein and H. J. Vogel, eds.), pp. 369–387. Academic Press, New York.
Chen, L. B., Summerhayes, I. C., Nadakavukaren, K. K., Lampidis, T. J., Bernal, S. D., and Shepherd, E. L. (1984). *Cancer Cells* **1,** 75–86.
Chen, L. B., Weiss, M. J., Davis, S., Bleday, R. S., Wong, J. R., Song, J., Terasaki, M., Shepherd, E. L., Walker, E. S., and Steele, G. D. (1985). *Cancer Cells* **3,** 433–443.
Chen, L. B., Bleday, R., Song, J., Weiss, M. J., and Steele, G. D., Jr. (1986). *Surg. Forum* **36,** 423–425.
Chung, D., Wong, J. R., and Chen, L. B. (1988). Submitted.
Clark, M. A., and Shay, J. W. (1982a). *Nature (London)* **295,** 605–607.
Clark, M. A., and Shay, J. W. (1982b). *Proc. Natl. Acad. Sci. U.S.A.* **79,** 1144–1148.
Cohen, L. B., and Salzberg, B. M. (1978). *Rev. Physiol. Biochem. Pharmacol.* **83,** 35–88.

Cohen, R. L., Muirhead, K. A., Gill, J. E., Waggoner, A. S., and Horan, P. K. (1981). *Nature (London)* **290**, 593–595.

Collins, J. M., and Foster, K. A. (1983). *J. Cell Biol.* **96**, 94–99.

Darzynkiewicz, Z., Traganos, F., and Melamed, M. R. (1980). *Cytometry* **1**, 98–108.

Darzynkiewicz, Z., Staiano-Coico, L., and Melamed, M. R. (1981). *Proc. Natl. Acad. Sci. U.S.A.* **78**, 2383–2387.

Darzynkiewicz, Z., Traganos, F., Staiano-Coico, L., Kapuscinski, J., and Melamed, M. R. (1982). *Cancer Res.* **42**, 799–806.

Davis, S., and Chen, L. B. (1988). Submitted.

Davis, S., Weiss, M. J., Wong, J. R., Lampidis, T. J., and Chen, L. B. (1985). *J. Biol. Chem.* **260**, 13844–13850.

Deamer, D. W., Utsumi, K., and Parker, L. (1967). *Arch. Biochem. Biophys.* **121**, 641–652.

DeMartinis, F. D., Ashkin, K. T., and Lampe, K. T. (1988). *J. Physiol. (London),* **253**, C783–C791.

Emaus, R. K., Grunwald, R., and Lemasters, J. J. (1986). *Biochim. Biophys. Acta* **850**, 436–448.

Evenson, D. P., Darzynkiewicz, Z., and Melamed, M. R. (1982). *J. Histochem. Cytochem.* **30**, 279–280.

Evenson, D. P., Lee, J., Darzynkiewicz, Z., and Melamed, M. R. (1985). *J. Histochem. Cytochem.* **33**, 353–359.

Fields, T. K., Finney, D. A., and Wiseman, A. (1983). *Somatic Cell Genet.* **3**, 375–389.

Geiger, B., and Singer, S. J. (1980). *Proc. Natl. Acad. Sci. U.S.A.* **77**, 4769–4773.

Goldstein, S., and Korczack, L. B. (1981). *J. Cell Biol.* **91**, 392–398.

Grinius, L. L., Jasaitis, A. A., Kadziuskas, Y. P., Liberman, E. A., Skulachey, V. P., Topoi, V. P., Tsofine, L. M., and Vladimirova, M. A. (1970). *Biochim. Biophys. Acta* **216**, 1–12.

Gundersen, G. G., Gabel, C. A., and Sharpiro, B. M. (1982). *Dev. Biol.* **93**, 59–72.

Hedberg, K. K., and Chen, L. B. (1986). *Exp. Cell Res.* **163**, 509–517.

Heggeness, M. H., Simon M., and Singer, S. J. (1978). *Proc. Natl. Acad. Sci. U.S.A.* **75**, 3863–3866.

Herman, B., and Albertini, D. F. (1984). *J. Cell Biol.* **98**, 565–576.

Hightower, M. J., Fairfield, F. R., and Lucas, J. J. (1981). *Somatic Cell Genet.* **7**, 321–329.

James, T., and Bohman, R. (1981). *J. Cell Biol.* **89**, 256–260.

Johnson, L. V., Walsh, M. L., and Chen, L. B. (1980). *Proc. Natl. Acad. Sci. U.S.A.* **77**, 990–994.

Johnson, L. V., Walsh, M. L., Bockus, B. J., and Chen, L. B. (1981). *J. Cell Biol.* **88**, 526–535.

Johnson, L. V., Summerhayes, I. C., and Chen, L. B. (1982). *Cell* **28**, 7–14.

Kliot-Fields, T., Finney, D. A., and Wiseman, A. (1983). *Somatic Cell Genet.* **3**, 375–389.

Kramer, D. L., Zychilinski, L., Wiseman, A., and Porter, C. W. (1983). *Cancer Res.* **43**, 5943–5950.

Kuzela, S., Joste, V., and Nelson, B. D. (1986). *Eur. J. Biochem.* **154**, 553–557.

Lampidis, T. J., Bernal, S. D., Summerhayes, I. C., and Chen, L. B. (1982). *Ann. N.Y. Acad. Sci.* **397**, 299–302.

Lampidis, T. J., Bernal, S. D., Summerhayes, I. C., and Chen, L. B. (1983). *Cancer Res.* **43**, 716–720.

Lampidis, T. J., Salet, C., Moreno, G., and Chen, L. B., (1984). *Agents Act.* **14**, 751–757.

Lee, C. S., Morgan, G., and Wooding, F. B. P. (1979). *J. Cell Sci.* **38**, 125–135.

Levenson, R., Macara, I. G., Smith, R. L., Cantley, L., and Housman, D. (1982). *Cell* **28**, 855–863.

Lewis, M. R., and Lewis, W. H. (1915). *Am. J. Anat.* **17**, 339–401.

Lin, J. J. C. (1981). *Proc. Natl. Acad. Sci. U.S.A.* **7**, 2335–2339.

Lin, J. J. C., and Feramisco, J. R. (1981). *Cell* **24**, 185–193.

Loud, A. V. (1968). *J. Cell Biol.* **37**, 27–46.

Lubin, I. M., Wu, L. N. Y., Wuthier, R. E., and Fisher, R. R. (1987). *Biochem. Biophys, Res. Commun.* **144**, 477–483.

Mai, M., and Allison, W. S. (1983). *Arch. Biochem. Biophys.* **221**, 467–476.

Maro, B., Marty, M. C., and Bornens, M. (1982). *EMBO J.* **1**, 1347–1352.

Martinez, A. O., Vigil, A., and Vila, J. C. (1986). *Exp. Cell Res.* **164**, 551–555.

Mitchell, P. (1966). *Biol. Rev.* **41**, 455–502.

Mitchell, P. (1979). *Science* **206**, 1148–1151.

Modica-Napolitano, J. S., and Aprille, J. R. (1987). *Cancer Res.* **47**, 4361–4365.

Modica-Napolitano, J. S., Weiss, M. J., Chen, L. B., and Aprille, J. R. (1984). *Biochem. Biophy. Res. Commun.* **118**, 717–723.

Modica-Napolitano, J. S. *et al.* (1988). In preparation.

Morita, T., Mori, M., Ikeda, F., and Tatibana, M. (1982). *J. Biol. Chem.* **257**, 10547–10550.

Mose-Larsen, P., Bravo, R., Fey, S. J., Small, J. V., and Celis, J. E. (1982). *Cell* **31**, 681–692.

Nadakavukaren, K. K., Nadakavukaren, J. J., and Chen, L. B. (1985). *Cancer Res.* **45**, 6093–6099.

Nass, M. (1984). *Cancer Res.* **44**, 2677–2688.

Oberley, T. D., Murphy, P. J., Steinert, B. W., and Albrecht, R. M. (1982). *Virchows Arch.* **41**, 145–170.

Oseroff, A. R., Ohuoha, D., Ara, G., McAulieffe, D., Foley, J., and Cincotta, L. (1986). *Proc. Natl. Acad. Sci. U.S.A.* **83**, 9729–9733.

Pedersen, P. L. (1978). *Prog. Exp. Tumor Res.* **22**, 190–274.

Powers, S. K., Pribil, S., Gillespie, G. Y., and Watkins, P. J. (1986). *J. Neurosurg.* **64**, 918–923.

Raine, C. S., Ghetti, B., and Shelanski, M. L. (1971). *Brain Res.* **84**, 386–393.

Ronot, X., Benel, L., Adolphe, M., and Mounolou, J. (1986). *Biol. Cell* **57**, 1–8.

Rosenbaum, R. M., Wittner, M., and Lenger, M. (1969). *Lab Invest.* **26**, 516–528.

Rouiller, C. (1960). *Int. Rev. Cytol.* **9**, 227–292.

Salmeen, I., Zacmanidis, P., Jesion, G., and Feldkamp, L. A. (1985). *Biophys. J.* **48**, 681–686.

Seibert, M., Chance, B., and DeVault, D. (1971). *Arch. Biochem. Biophys.* **146**, 611–617.

Shapiro, H. M. (1985). "Practical Flow Cytometry." Liss, New York.

Sharpe, A. H., Chen, L. B., Murphy, J. R., and Field, R. N. (1980). *Proc. Natl. Acad. Sci. U.S.A.* **77**, 7267–7271.

Sharpe, A. H., Chen, L. B., and Fields, B. N. (1982). *Virology* **120**, 399–411.

Siemens, A., Walter, R., Liaw, L. H., and Berns, M. W. (1982). *Proc. Natl. Acad. Sci. U.S.A.* **79**, 466–470.

Sims, P. L., Waggoner, A. S., Wang, C. H., and Hoffman, J. F. (1974). *Biochemistry* **13**, 3315–3329.

Smith, D. S., Jartfors, U., and Cameron, B. F. (1975). *Ann. N.Y. Acad. Sci.* **253**, 472–502.

Smith, D. S., Jartfors, U., and Cayer, M. L. (1977). *J. Cell Sci.* **27**, 255–272.

Staiano-Coico, L., Darzynkiewicz, Z., Melamed, M. R., and Weksler, M. (1982). *Cytometry* **3**, 79–83.

Starger, J. M., and Goldman, R. D. (1977). *Proc. Natl. Acad. Sci. U.S.A.* **74**, 2422–2426.

Steinkamp, J. A., and Hiebert, R. D. (1982). *Cytometry* **2**, 232–237.

Summerhayes, I. C., Lampidis, T. J., Bernal, S. D., Nadakavukaren, J. J., Nadakavukaren, K. K., Shepherd, E. L., and Chen, L. B. (1982). *Proc. Natl. Acad. Sci. U.S.A.* **79**, 5292–5296.

Summerhayes, I. C., Wong, D., and Chen, L. B. (1983). *J. Cell Sci.* **61**, 87–105.

Tanabe, K. (1983). *J. Protozool.* **30**, 707–710.

Tanabe, K. (1985). *Experientia* **41**, 101–103.

Tanphaichtr, N., Chen, L. B., and Bellve, A. (1984). *Biol. Reprod.* **31,** 1049–1060.

Terasaki, M., Song, J., Wong, J. R., Weiss, M. J., and Chen, L. B. (1984). *Cell* **38,** 101–108.

Toh, B. H., Lolait, S. J., Mathy, J. P., and Baum, R. (1980). *Cell Tissue Res.* **211,** 163–169.

Tsiftsoglou, A. D., Nunez, M. T., Wong, W., and Robinson, S. H. (1983). *Proc. Natl. Acad. Sci. U.S.A.* **80,** 7528–7532.

Tzagoloff, A. (1982). "Mitochondria," pp. 1–342. Plenum, New York.

Vale, R. D., Schnapp, B. J., Reese, T. S., and Sheetz, M. P. (1985a). *Cell* **40,** 449–454.

Vale, R. D., Schnapp, B. J., Reese, T. S., and Sheetz, M. P. (1985b). *Cell* **40,** 559–569.

Vale, R. D., Reese, T. S., and Sheetz, M. P. (1985c). *Cell* **42,** 39–51.

Waggoner, A. S. (1976). *J. Membr. Biol.* **27,** 317–334.

Waggoner, A. S. (1979a). *Annu. Rev. Biophys. Bioeng.* **8,** 47–68.

Waggoner, A. S. (1979b). "Methods in Enzymology" (S. Fleischer and L. Packer, eds.), Vol. 55, pp. 689–695. Academic Press, New York.

Walker, C., and Shay, J. W. (1981). *J. Cell Biol.* **91,** 379a.

Walker, C., and Shay, J. W. (1983). *Somat. Cell Genet.* **9,** 469–476.

Walsh, M. L., Jen, J., and Chen, L. B. (1979). *Cold Spring Harbor Conf. Cell Prolif.* **6,** 513–520.

Wang, E., and Goldman, R. D. (1978). *J. Cell Biol.* **79,** 708–726.

Weakley, B. S. (1976). *Cell Tissue Res.* **169,** 531–550.

Weibel, E. R. (1969). *Int. Rev. Cytol.* **26,** 235–302.

Weiss, M. J., and Chen, L. B. (1984). *Kodak Lab. Chem. Bull.* **55,** 1–4.

Weiss, M. J., Wong, J. R., Ha, C. S., Bleday, R., Salem, R. R., Steele, G. D., Jr., and Chen, L. B. (1987). *Proc. Natl. Acad. Sci. U.S.A.* **84,** 5444–5448.

Wiseman, A., Fields, T. K., and Chen, L. B. (1985). *Somat. Cell Mol. Genet.* **11,** 541–556.

Wolfson, J. S., McHugh, G. L., Swartz, M. N., Ng, E. Y. W., and Hooper, D. C. (1987). *J. Parasitol.,* in press.

Wu, F. S. (1987). *Planta* **171,** 346–357.

Yoshikami, D., and Okun, L. M. (1984). *Nature (London)* **310,** 53–56.

Zarbl, H., Latreille, J., and Jolicoeur, P. (1987). *Cell,* **51,** 357–369.

Ziegler, M. L., and Davidson, R. L. (1981). *Somat. Cell Genet.* **7,** 73–88.

# Chapter 8

# *Fluorescent Labeling of Endoplasmic Reticulum*

### MARK TERASAKI

*Laboratory of Neurobiology, IRP, N.I.N.C.D.S.*
*National Institutes of Health at the Marine Biological Laboratory*
*Woods Hole, Massachusetts 02543*

## I.  Introduction

A large fraction of the intracellular membranes is endoplasmic reticulum (ER). The ER is an extensive network of tubular and cisternal membranes which is known to have several different functions, principally protein synthesis, lipid synthesis, and calcium sequestration.

In the thin, spread periphery of many types of cultured cells, the ER has essentially a two-dimensional distribution, parallel to the substrate. Due to

125

this arrangement, the pattern of the ER can be carefully inspected in living cultured cells, as well as in unsectioned, fixed cells. In fact, the first clear observations of the ER were made by "whole-mount" electron microscopy of osmium-fixed fibroblasts grown on formvar-coated EM grids (Porter *et al.*, 1945; Porter, 1953). In these cells, a "lace-like reticulum" was noted, with a distribution primarily in the endoplasm (or nonmoving regions) of the cells; this distribution led to its naming as the "endoplasmic reticulum" (Porter and Kallman, 1952). This same reticular network has been observed by other whole mount electron microscopy techniques (Buckley and Porter, 1975; Song *et al.*, 1985), by phase contrast microscopy (Buckley, 1964), by labeling with fluorescent lipid analogs (Pagano *et al.*, 1981), and by immunofluorescence (Munro and Pelham, 1987; Saga *et al.*, 1987). Recently, we have found that immunofluorescence staining using an antibody to BiP (Bole *et al.*, 1986), a protein thought to be in the ER, produces an identical pattern as staining by $DiOC_6(3)$ in the same cell (Terasaki and Reese, 1988).

The subject of this chapter is the use of the fluorescent dye $DiOC_6(3)$ to observe the ER In living cells (Terasaki *et al.*, 1984). At certain concentrations, this dye stains all, or almost all, intracellular membranes, most probably by a simple partitioning process. The dye therefore does not specifically label the ER, but the ER is easily recognized by its distinctive morphology in the stained cells. The method is relatively simple and gives a bright clear image in optimum conditions. The major drawback of the use of $DiOC_6(3)$ is that it has some toxic effects on live cells. For certain applications, it may be preferable to use $DiOC_6(3)$ to stain glutaraldehyde-fixed cells, where the staining is somewhat clearer and there is no problem with toxic effects (Terasaki *et al.*, 1986).

$DiOC_6(3)$ is a dicarbocyanine dye. The conventions for naming the dicarbocyanine dyes are as follows: "Di" refers to the two identical halves of the molecule; O refers to the two oxygen atoms in the rings; $C_6$ refers to the pair of 6 carbon chains; and 3 refers to the 3 carbon chain in the middle connecting the two halves of the molecule.

The fluorescence is produced by the conjugated double bonds in the middle of the molecule. The middle portion and the carbon ring structures form a planar structure, so that the molecule is essentially flat. The molecule has mostly hydrophobic characteristics but it also has a positive charge, giving it hydrophilic characteristics as well.

In the 1970s, the dicarbocyanine dyes (Sim *et al.*, 1974) were introduced to biological studies as membrane potential sensitive dyes. They have been used to make many measurements of membrane potential in blood cells and nerve cells (see Haugland, 1984, for references). At a later time, when they were used on thinly spread cultured cells, many of these molecules were found to stain mitochondria (Johnson *et al.,* 1981). They were used, along with rhodamine 123 and other dyes, to monitor relative changes in mitochondrial membrane potentials.

It was subsequently discovered, by Micheal Weiss in Dr. Lan Bo Chen's laboratory, that higher doses of dicarbocyanine dyes stained a reticular structure. Evidence was gathered that this structure was the endoplasmic reticulum and various dyes were screened to find one most useful for experimentation (Terasaki *et al.,* 1984). Of the dyes tested, $DiOC_6(3)$ was found to give the brightest image with the slowest bleaching rate.

# II. Materials

## A. Cells

A significant practical limitation of the technique lies not in the labeling itself, but rather in the fact that ER is densely distributed in the cell. In regions where the ER has much three-dimensional distribution, the overlapping of the ER tubules in the microscope field obscures all details of organization. Thus, the best material for this technique is cells which have regions with two-dimensional distribution of the ER. The peripheral regions of many well-spread cells in culture have such ER, and the dye labels the ER in these regions very well. Such thin, spread cells include CV-1 (monkey kidney epithelial cell line), PtK-2 (kangaroo rat kidney epithelial cell line), 3T3 (mouse fibroblast cell line), chick embryo fibroblasts, and bovine aortic endothelial cells, as well as numerous other cell types, including nerve cells with growth cones (Dailey and Bridgman, 1987). On the other hand, many cell types in culture are not well spread on the substrate and have little or no thin regions; it is usually difficult to make observations in these cells without confocal microscopy or three-dimensional microscopy (see Brackenhoff *et al.* and Agard and Sedat, Volume 30, this series).

Thick regions of cytoplasm are generally present in the central regions of all cells. These regions contain the Golgi apparatus and microtubule organizing center; details here are obscured by the overlapping membranes stained by $DiOC_6(3)$. Perhaps the processes occurring in this region are so complex that they require a three-dimensional organization, while the processes in the periphery are of a type that can occur with an essentially two-dimensional organization.

The ER also tends to become more dense and three dimensional as cells become crowded, so it is usually easier to observe the ER in sparse cell cultures.

With epifluorescence microscopy, it is possible to observe cells at surfaces of tissues (though only when these cells are thin) and we have attempted to do this with some preparations. A common problem is a higher fluorescence background in such preparations. Cells on the inside of the goldfish scale, which have been used for immunofluorescence (Byers *et al.*, 1980; Byers and Fujiwara, 1982) were adequately observable, but we had difficulty in getting a good image from endothelial cells of the bovine artery. Recently, some workers have used the dye to obtain images in onion bulb epidermis cells *in situ* (Quader *et al.*, 1987).

It should be pointed out that confocal scanning microscopy and three-dimensional microscopy with a cooled CCD camera (Brakenhoff, Volume 30, this series; White *et al.*, 1987; and Agard and Sedat, Volume 30, this series) may allow better observations of thick regions of cells in culture or surfaces of tissues.

## B.  Dye

$DiOC_6(3)$ is listed as dihexaoxacarbocyanine iodide in the Kodak Laboratory and Research Products catalog (Rochester, New York). It is also obtainable from Molecular Probes (Eugene, Oregon) and Polysciences (Warrington, Pennsylvania). It is relatively inexpensive, in that 1 g will last a very long time in normal usage. A convenient volume and concentration for use as a stock solution is 10 ml of 0.5 mg $DiOC_6(3)$/ml 100% ethanol. This stock should be kept protected from light at room temperature, and it appears to be stable for at least a year.

$DiOC_6(3)$ is a "lipophilic" substance. To demonstrate this, make a dilution of the stock solution to 2.5 $\mu$g/ml $DiOC_6(3)$ in water and layer it on top of a nonpolar solvent. The dye will transfer into the nonpolar solvent either by contact with the aqueous solution or after shaking the mixture.

## C.  Mounting Cells for Microscopy

There are many ways to mount cells in chambers for observation by fluorescence microscopy, and they should all be compatible with this technique. A silicon rubber chamber is described in Chapter 7 by Chen, this volume.

## D.  Fluorescence Characteristics

$DiOC_6(3)$ is observable with optics for fluorescein and has no spillover into rhodamine type filters. When cells are stained well, the fluorescence is bright and distinct. Compared to the fluorophores commonly used for immunofluorescence, the rate of bleaching is noticeably slower than fluorescein and is similar to that of rhodamine. When using high intensity illumination, the rate of bleaching also appears to be slower with illumination from xenon arc lamps than from mercury arc lamps.

## E. Optics

The ER is dense, and in order to see individual tubules, it is best to use high magnifications. Of the lenses from Zeiss which we have experience with, the 40X objective lens [numerical aperture (NA) 1.0], combined with the usual 10–12X eyepiece, is barely adequate, while the 63X (NA 1.4) and 100X (NA 1.3) objectives lenses are best. The 100X lens may be preferable for direct viewing, because more details are apparent while viewing even though the 63X lens has a higher NA.

Photography has been quite easy using this dye. The 100X lens is preferred by this author for photography also. The staining is also well suited for video recordings with a silicon intensified target (SIT) camera and image processing (Lee *et al.,* 1987; Sanger *et al.,* 1987; Dailey and Bridgman, 1987). Low light level imaging microscopy techniques can be very useful for reducing exposure of the cells to light (see Spring and Lowy, Chapter 15 and Aikens *et al.,* Chapter 16, this volume). This minimizes the amount of photodynamic damage, which is a serious problem when using $DiOC_6(3)$ to label the ER in living cells.

## III. Labeling Procedures

The labeling procedures using $DiOC_6(3)$ are simple, so that it should be relatively easy to begin working with the technique. Since the procedures are straightforward, one can also experiment with labeling periods, materials, and equipment in order to identify the optimal conditions for labeling.

Two ways to label the cells with $DiOC_6(3)$ will be described here. In the first method, cells are incubated continuously in the presence of the dye, and in the second, cells are labeled with a higher concentration of $DiOC_6(3)$ for a given length of time, then washed and mounted in dye-free media. Both methods have relative advantages and disadvantages.

For the first method, make a stock solution of 0.5 mg $DiOC_6(3)$/ml ethanol. Then, make a staining solution using culture media (preferably with low serum, such as 1%) with 0.5 $\mu$g $DiOC_6(3)$/ml. Mount the cells directly in this media and observe them with fluorescein optics at the highest magnification available.

The population of cells on a cover slip usually has a variation in levels of staining. However, if the dye concentration is too low, most or all cells will only be stained for mitochondria. In this case, the mitochondria will maintain their normal worm-like appearance, and the outline of the cell will be invisible. If the concentration is too high, the dye will be toxic to most of the cells and they will start to round up and detach from the dish. In either case, one can mount another cover slip using a staining solution with lower or higher concentration than 0.5 $\mu$g/ml.

In well-stained cells, the outline of the cell will be clearly visible. Mitochondria will be swollen and throughout the cell will be the fine reticular network of the ER. Vesicles of varying sizes will also be stained, including lysosomes and in some cases, small rapidly transported vesicles.

For the second method, using the stock solution prepared previously, make a staining solution with 2.5 $\mu$g DiOC$_6$(3)/ml in media or phosphate-buffered saline (PBS). Stain cells for 1 minute in this solution. Wash the cover slip and mount in dye-free media.

There is usually less general fluorescence background from the media by this method compared to the first method. If the cells are not well stained, one can either vary the concentration of DiOC$_6$(3) or duration of staining. A disadvantage of this method of staining is that DiOC$_6$(3) slowly leaks out of cells, so that after 10 or 20 minutes, a well-stained cover slip may revert to a population of cells mostly stained only for mitochondria. On the other hand, some cell types retain the dye better than others, so this should be tested empirically.

As for the general behavior of the ER, in most cells the reticular network of the labeled ER will appear to be completely stationary. In a few cells, there will be small regions of the cell where there is rapid movements of ER tubules of several different types. Extension and retraction of tubules, along with "contraction" of polygons of ER tubules, are commonly seen in these regions. Often, the ER tubules in these regions appear to be undergoing Brownian motion. Occasionally, one can also see cisternae which have very active edges. Recently, detailed observations using low light levels and video recording have been made (Lee and Chen, 1988).

It may be interesting to observe the toxic effects induced by interactions of light and dye (photodynamic damage). To observe the photodynamic damage, expose a stained cell to full excitation irradiation for approximately 20 seconds. Without moving the microscopy stage, block the excitation light to the sample and wait for 10 minutes. Then, open the shutter for excitation, and one should see a rounded, retracted, dying or dead cell. Then move the microscopy stage to other regions and one should see that the other cells are seemingly fine. Photodynamic damage may be minimized by controlling the dose of excitation by decreasing the light level, and the time of illumination by using the methods developed for low light level imaging (see Spring and Lowry, Chapter 15 and Aikens *et al.*, Chapter 16, this volume).

Membrane fractions in various kinds of cell homogenates or cell-free preparations can be stained "live" by $DiOC_6(3)$ (Terasaki *et al.*, 1987; Dabora and Sheetz, 1987). There is less variability in staining periods and a shorter exposure to the dye is sufficient (2.5 $\mu$g/ml for 10 seconds followed by two brief washes and mounting for observation).

$DiOC_6(3)$ interacts with some materials found outside the cells. It binds to components in serum, probably albumin, so that a low concentration of serum is sometimes desirable to reduce the background fluorescence from the media. The dye stains nonpolar substances, such as droplets of microscope lens immersion oil, or material which is present on uncleaned cover slips or other glass surfaces. $DiOC_6(3)$ also stains taxol crystals, which form in taxol solutions greater than 20 $\mu M$. These crystals have a very superficial resemblance to centrosomes. Since the dye was being used to stain a cell-free extract, for a short time, this author thought that the extract was forming centrosomes with attached ER tubules!

# IV.   Analysis of the Staining

## A.   Interactions of $DiOC_6(3)$ and Living Cells

Interactions of $DiOC_6(3)$ with living cells are not completely understood at the molecular level. However, considerations are somewhat simplified by the lack of evidence for enzymatic modification of dicarbocyanine dyes in cells. Given this, one can concentrate on the behavior of the single molecule with respect to the cell.

$DiOC_6(3)$ is positively charged and at low concentrations, its distribution in cells seems to be governed by a combination of this charge and its lipophicity. $DiOC_6(3)$ accumulates in mitochondria, due to the negative mitochondrial membrane potential which is in turn generated by the metabolic activities of the mitochondria (Johnson *et al.*, 1981). At higher concentrations of $DiOC_6(3)$, the mitochondria lose their normal morphology and the $DiOC_6(3)$ stains other intracellular membranes. A possible explanation for this is that excess $DiOC_6(3)$ interferes with mitochondrial metabolism, causing a loss of the membrane potential. This would then release excess $DiOC_6(3)$, allowing it to partition into the other membranes by virtue of its lipophilicity.

## B.  Variability

Within a given cell, the degree of labeling of intracellular membranes is constant throughout the cell. However, individual cells within a population show variations in the degree of labeling. There is also a day to day variation in the $DiOC_6(3)$ concentration required to stain a population of cells to the same degree. These variations are due to intrinsic properties of the cells rather than properties of the $DiOC_6(3)$ molecule, and are most probably due to variations in physiological state among cells.

It may be interesting to note that this variability in living cells is not present in glutaraldehyde fixed cells. In fixed cells, $DiOC_6(3)$ stains intracellular membranes (e.g., mitochondria and ER) equally well even at low concentrations. $DiOC_6(3)$ also stains all cells in a population equally and shows much less day to day variation. In addition, $DiOC_6(3)$ seems to stain equally well various membrane fractions from homogenized cells. These differences in interaction of $DiOC_6(3)$ with living versus nonliving cells are another example of complex properties of individual, whole living cells which are not evident when one studies fixed cells, fractions of disrupted cells, or even averaged measurements of a population of living cells.

## C.  Toxic Effects of $DiOC_6(3)$

The variability discussed above and time dependence of staining are, in practice, moderately troublesome problems of the technique. These problems have to do with how well the cells are stained. However, the most difficult problems of the technique are those of toxicity, which have to do with the normality of the behavior and distribution of the ER as it is observed by the technique. For some applications, these problems may even be insoluble. We can distinguish three kinds of toxicity, the first of which is avoidable, while the other two are more formidable.

$DiOC_6(3)$ is clearly toxic to cells if used in excess concentration. Cells will round up, leaving retraction fibers, and eventually detach from the dish. However, this can easily be avoided because moderate concentrations of $DiOC_6(3)$ leaves cells apparently unaffected, i.e., with normal morphology by phase contrast microscopy.

Of the two more problematic toxic effects of $DiOC_6(3)$, the first is that mitochondria become swollen at the concentrations required to stain the ER. This effect is not dependent on light. As discussed above, $DiOC_6(3)$ is accumulated, via membrane potential, in mitochondria. When the accumulated $DiOC_6(3)$ is too high, the mitochondria may become "poisoned," losing their membrane potential, thereby releasing $DiOC_6(3)$ which then could partition according to its lipophilicity. If this is true, then it is not possible to observe the ER in cells with normal functioning mitochondria using this technique. It should be pointed out, however, that vesicle movement can be observed in some labeled cells (indicating that ATP levels are not depleted), and cultured cells can survive and appear healthy for at least several days in the presence of 2.5 $\mu$g/ml DiOC$_6$(3), a concentration which causes mitochondria to become swollen (unpublished observations). Perhaps the affected mitochondrial functions are not required for many cell processes, or the cellular metabolism is able to adapt and compensate for the impaired functions.

The second type of toxic effect of the dye is photodynamic damage. When the intense light sometimes used for fluorescence interacts with fluorescent molecules, free radical formation causes cell damage and eventually cell death. With DiOC$_6$(3), this type of damage is relatively rapid (see demonstration in Section III).

If one observes labeled cells with continuous illumination, there are no obvious short-term changes in the distribution or behavior of the stained ER. However, toxic processes must be occurring in the cells because prolonged exposure leads eventually to gross morphological changes and cell death. Photodynamic damage is rather nonspecific in nature and its effects on the many interacting processes and structures in living cells cannot be determined at the present time. There are therefore many ways in which photodynamic damage during the early stages of exposure could affect the distribution and behavior of the ER. It could disrupt interactions of the ER with microtubules (Terasaki et al., 1986), or it could disrupt physiological processes in the ER which are involved in ER distribution and behavior. For instance, local phospholipid synthesis could cause expansion of the ER in those regions, or ribosome binding could cause changes in the form of the membranes.

With the existence of many possible interactions, and with the certainty that damaging light-dependent processes are occurring, it is difficult to know if the behavior of the ER will be "normal" in a particular given situation.

One approach to this problem is to minimize the photodynamic damage by reducing the intensity of the light levels and/or reducing the amount of time that the cell is exposed to illumination. Video microscopy techniques, such as use of a SIT camera, image processing, and time-lapse recording, could be of great use in this.

Another approach is to test if there is an adverse effect on the cellular process that is being studied. For instance, if the distribution of the ER during cell movement is being investigated, one can see if the cell does indeed move in a normal fashion during the observations. If the process being studied does occur normally, then there should be much greater confidence that the observed distribution and behavior of the ER is also normal.

# V.  Summary

The fluorescent molecule, $DiOC_6(3)$, can be used to label the ER in living cultured cells. The labeling procedures are simple and rapid, and in optimum conditions, the staining is bright and clear and bleaches slowly. The main disadvantage of the technique is toxicity. Photodynamic damage is probably the most serious of the toxic effects, because the damage can be relatively rapid and because the extent and nature of the damage during exposure to the light cannot be determined. To lessen the damage, the exposure of cells to light should be minimized. For many applications, it would be best to verify that cell function is normal during the experimental observations.

## REFERENCES

Bole, D. G., Hendershot, L. M., and Kearney, J. F. (1986). *J. Cell Biol.* **102,** 1558–1566.
Buckley, I. K. (1964). *Protoplasma* **59,** 569–588.
Buckley, I. K., and Porter, K. R. (1975). *J. Microsc.* **104,** 107–120.
Byers, H. R., and Fujiwara, K. (1982). *J. Cell Biol.* **93,** 804–811.
Byers, H. R., Fujiwara, K., and Porter, K. R. (1980). *Proc. Natl. Acad. Sci. U.S.A.* **77,** 6657–6661.
Dabora, S. L., and Sheetz, M. P. (1987). *J. Cell Biol.* **105,** 89a.
Dailey, M. E., and Bridgman, P. C. (1987). *Neurosci. Meet. Abstr.* **13,** 1477.
Haugland, R. (1984). Molecular Probes catalogue.
Johnson, L. V., Walsh, M. L., and Chen, L. B. (1980). *Proc. Natl. Acad. Sci. U.S.A.* **77,** 990–994.
Johnson, L. V., Walsh, M. L., Bockus, B. J., and Chen, L. B. (1981). *J. Cell Biol.* **88,** 526–535.
Lee, C., and Chen, L. B. (1988). In press.
Lee, C., Greisman, H., Shepherd, E. L., Terasaki, M., and Chen, L. B. (1987). *J. Cell Biol.* **105,** 262a.
Munro, S., and Pelham, H. R. B. (1987). *Cell* **48,** 899–907.
Pagano, R. E., Longmuir, K. J., Martin, O. C., and Struck, D. K. (1981). *J. Cell Biol.* **91,** 872–877.
Porter, K. R. (1953). *J. Exp. Med.* **97,** 727–750.
Porter, K. R., and Kallman, F. L. (1952). *Ann. N.Y. Acad. Sci.* **54,** 882–891.
Porter, K. R., Claude, A., and Fullman, E. (1945). *J. Exp. Med.* **81,** 233–241.
Quader, H., Hoffman, A., and Schnepf, E. (1987). *Eur. J. Cell Biol.* **44,** 17–26.
Saga, S., Nagata, K., Chen, W. T., and Yamada, K. M. (1987). *J. Cell Biol.* **105,** 517–528.
Sanger, J. M., Dome, J. S., Somlyo, A. V., and Sanger, J. W. (1987). *J. Cell Biol.* **105,** 87a.
Sims, P. J., Waggoner, A. S., Wang, C. H., and Hoffman, J. F. (1974). *Biochemistry* **13,** 3315–3330.

Song, J., Terasaki, M., and Chen, L. B. (1985). *J. Cell Biol.* **101,** 57a (Abstr.).
Terasaki, M., and Reese, T. S. (1988). In preparation.
Terasaki, M., Song, J. D., Wong, J. R., Weiss, M. J., and Chen, L. B. (1984). *Cell* **38,** 101–108.
Terasaki, M., Chen, L. B., and Fujiwara, K. (1986). *J. Cell Biol.* **103,** 1557–1568.
Terasaki, M. Gallant, P., and Reese, T. S. (1987). *J. Cell Biol.* **105,** 128a.
White, J. G., Amos, W. B., and Fordham, M. (1987). *J. Cell Biol.* **105,** 41–48.

# Chapter 9

## Fluorescent Labeling of Endocytic Compartments

JOEL SWANSON

*Department of Anatomy and Cellular Biology*
*Harvard Medical School*
*Boston, Massachusetts 02115*

---

## I. Introduction

Endocytosis is that cellular process in which plasma membrane invaginates to form a closed vesicle within the cytoplasm. Extracellular solutes and surface-bound molecules enclosed by endocytic vesicles may encounter a variety of fates, depending upon both the engulfed molecule and the engulfing cell. Some endocytic vesicles simply return to the cell surface, recycling their contents in the process. Others fuse with organelles, which may include other endocytic vesicles, Golgi cisternae, lysosomes, or nonlysosomal acidic compartments called endosomes or receptosomes (Silverstein *et al.*, 1977; Pastan and Willingham, 1985). Many of these organelles can also recycle contents via vesicular delivery to the cell surface (Steinman *et al.*, 1983). Here I define the endocytic compartment broadly as any

METHODS IN CELL BIOLOGY, VOL. 29

vesicular organelle which receives solutes from extracellular medium via endocytosis.

The usual target of endocytosed solutes is the lysosome, an acidic compartment of degradative hydrolysis, and the repository for many nondigestible and membrane-impermeant molecules. Early methods for labeling this compartment for fluorescence microscopy used acridine orange, which readily diffuses across cellular membranes at neutral pH, but becomes membrane impermeant at low pH, and therefore concentrates in lysosomes (Robbins and Marcus, 1963; Hart and Young, 1975). At high concentrations the acridine orange fluorescence emission spectrum exhibits a red shift, consequently making acidic organelles appear red. However, not all acidic organelles which might be labeled by this method necessarily take part in endocytosis.

A more direct method for labeling the endocytic compartment for fluorescence microscopy is by endocytosis. Cells are incubated for 60 minutes or more in medium containing a fluorescent, nontoxic, membrane-impermeant solute, then washed in buffer or unlabeled medium and observed in a fluorescence microscope. Fluorescent vesicles inside the cells are part of the endocytic compartment.

Although such a procedure may be delightful, it is scarcely informative. Fluorescent vesicles could be any of the variety of organelles that receive endocytosed solutes, directly or indirectly. They may have been labeled at the beginning of the incubation, or just minutes before the cells were washed. Most importantly, the fluorescent probe molecule may have altered the natural process, either by directing flow into vesicles otherwise not involved in endocytosis, or by inducing vesicle formation.

In this chapter I describe some of the fluorescent probes which have been used in living cells to delineate movements through the endocytic compartments. I also outline methods for determining where a particular probe has traveled and what it may have done to the cell in getting there. Techniques which employ fixed cells, such as electron microscopic cytochemistry and immunofluorescence microscopy, can be very useful for identifying components of the endocytic pathways. Nonetheless, endocytosis involves the intracellular movement of sometimes delicate vesicular structures, and this makes observation of the process in living cells not only instructive but often necessary.

## II. Fluorescent Probes

There are several ways an extracellular solute might arrive in an endocytic vesicle. If it does not adsorb to the cell, and cannot simply diffuse across the plasma membrane, it may be endocytosed as part of the fluid contents of a pinosome. If the molecule does adsorb to the cell surface, either by nonspecific binding or by interaction with a membrane receptor for that molecule, its concentration at the cell surface increases and its rate of endocytosis may consequently increase above the rate of bulk solute uptake. A fluorescent ligand which enters cells by receptor-mediated endocytosis can also enter by fluid-phase pinocytosis, especially when added to cells at concentrations which exceed receptor capacity. On the other hand, at low concentrations, many fluorescent indicators of fluid phase pinocytosis exhibit a detectable adsorptive uptake.

One should know as much as possible about a fluorescent probe and its effects on cells before using it to explain processes occurring in unperturbed cells. A fluorescent probe which is not visibly toxic to cells may nonetheless alter endocytosis. Normal endocytic flow is most often accompanied by degradation of ingested macromolecules in lysosomes (Ehrenreich and Cohn, 1967). Therefore, it is possible that nondegradable probes used for measuring pinocytosis distort constitutive flow through the endocytic compartment by their extensive accumulation in lysosomes. Conversely, fluoresceinated proteins can be hydrolyzed in lysosomes to produce fluorescent degradation products which may either accumulate in lysosomes or diffuse into cytoplasm. These products could have adverse effects on cell health. The fluorescence properties of the fluorophore may also be altered by ionic conditions in the endocytic compartment (Okuma and Poole, 1978). Finally, a probe that properly reports normal endocytic movements in one type of cell may alter pinocytosis in another cell type. The enzyme, horseradish peroxidase, reports fluid phase pinocytosis well for Chinese hamster ovary cells (Adams et al., 1982), but stimulates pinocytosis in thioglycollate-elicited mouse peritoneal macrophages (Swanson et al., 1985). Carboxyfluorescein and Lucifer Yellow are small fluorescent molecules which have been successfully used to measure fluid phase pinocytosis in some cell types (Goldmacher et al., 1986; Swanson et al., 1985). In other cells, however, these molecules may not be restricted to membrane-bound compartments.

## A.   Probes of Fluid Phase Pinocytosis

For reasons that are not yet understood, constitutive pinocytosis is more active in some cells than in others. Fluorescent probes which label endocytic compartments by fluid phase pinocytosis are not ideal for labeling cells with low rates of constitutive pinocytosis. However, many free-living amoebae (Aley et al., 1984; Thilo and Vogel, 1980), as well as macrophages (Swanson et al., 1985) and fibroblasts (van Deurs et al., 1984), readily accumulate fluorescent solutes such as fluorescein isothiocyanate-labeled dextran (FITC-dextran) or Lucifer Yellow by fluid phase pinocytosis. An advantage of labeling endocytic compartments with such probes is that under appropriate conditions they do not disturb the constitutive flow through the cells, and it is possible to determine independently that flow is minimally disturbed. Given enough time, such probes usually label the entire endocytic compartment, which can be especially useful when combined with other fluorescent probes that report the more restricted movements of receptors or ligands.

To qualify as a fluorescent probe for fluid phase pinocytosis, a molecule must satisfy several criteria. First, it should be water soluble and should not diffuse across cellular membranes. Second, it should not be toxic to cells at concentrations above those added to culture medium, since nondegradable molecules often accumulate in lysosomes and attain high concentrations (Steinman et al., 1976). Third, if the molecule is degradable by lysosomal hydrolases, it should not in its degraded form be such that the fluorophore can cross lysosomal membrane into cytoplasm. Fourth, it must not adsorb to cell membranes or otherwise stimulate endocytosis. That is, probe accumulation should be directly proportional to the concentration of probe in the medium, and should not be inhibited by the presence in the medium of an excess of nonfluorescent but otherwise similar molecules. Finally, as most endocytic processes are energy dependent (Silverstein et al., 1977), the accumulation of probe molecules should be inhibitable by low temperature (4°C) or by metabolic poisons. Lucifer Yellow, FITC-dextran, and carboxyfluorescein have satisfied many of these criteria in certain cell types.

## 1. Lucifer Yellow Carbohydrazide

Lucifer Yellow CH (LY) is a low molecular weight, water-soluble, and membrane-impermeant marker for fluid phase pinocytosis. Its fluorescence is constant from pH 2.0 to 12.0 (Stewart, 1978). Its excitation and emission maxima, 430 and 540 nm, respectively, are similar enough to those of fluorescein that it can be visualized with broad bandpass fluorescein filter sets. However, more appropriate filter combinations are available and of course preferable.

Although LY has proven a valid indicator of fluid phase pinocytosis in macrophages, and has been used to label endocytic compartments in other cells as well (Miller et al., 1983; Riezman, 1985), it may not be universally applicable. Before interpreting microscopic studies with a new type of cell one should characterize pinocytosis of LY in cell populations, satisfying as much as possible the criteria outlined above.

*Microscopic Study of Lucifer Yellow Pinocytosis.* A flat cell makes the best optical specimen. Macrophages plated onto 12-mm circular cover slips, No. 1 thickness, spread on the glass to varying degrees. Increased spreading can be induced by treatment of mouse peritoneal macrophages with 10 ng/ml phorbol myristate acetate, a potent tumor promoter (Phaire-Washington et al., 1980), or by opsonizing the cover slip with immunoglobulins to induce a phagocytic response against the surface (Michl et al., 1979; Takemura et al., 1986). Other tissue culture cells, such as fibroblasts or endothelial cells, also spread well when cultured on cover slips.

Cover slip cultures of macrophages are maintained in 24-well tissue culture plasticware, with one cover slip per well and 1 ml of Dulbecco's modified essential medium with 10% heat-inactivated fetal calf serum (DM10F). Lucifer Yellow CH, potassium salt (Molecular Probes, Inc., Eugene, Oregon) is dissolved at 1–5 mg/ml in DM10F, then added to tissue culture wells and incubated for various time periods at 37°C in a 5% $CO_2$ atmosphere (Swanson et al., 1985). Since fluid phase pinocytosis is an inefficient process, even in macrophages, a small volume (0.25 ml) of LY solution can be added to each well without a significant proportion of the dye being removed from the medium by the cells. Macrophages can internalize a detectable amount of LY in a few minutes of incubation, but may be left in LY-containing medium for as long as 3 days. After incubation, cover slips are removed from the wells using jeweler's forceps, then rinsed in several changes of phosphate-buffered saline (PBS) before microscopic observation. For pulse-chase experiments, rinsed cover slips are returned to new wells containing medium without LY, incubated for varying periods, then washed again and observed.

Preparation of cover slips for microscopic observation can be simple or complex, depending on how long one wishes to study the specimen. For a quick survey, I blot most of the moisture from the rinsed cover slip, then dry the back with a paper towel. This is then laid cell side down onto a glass microscope slide, blotted again to remove excess moisture from around the rim of the cover slip, then mounted in the fluorescence microscope (Zeiss Photomicroscope III). Such a preparation will dry out on the microscope stage, however, especially if it is maintained at 37°C. For longer study, an observation chamber is prepared. A similarly blotted cover slip, with its back dried, is inverted onto small fragments of cover slips, which act as struts to support the cell preparation one cover slip height above the slide. Phosphate-buffered saline is added to wet the whole undersurface of the cover slip, then this is sealed with warm "valap." Valap is a mixture of equal portions of vaseline, anhydrous lanolin (VWR, San Francisco, California), and paraffin (52.5°C mp, tissue mat; Fisher Scientific, Springfield, New Jersey). It is applied by first heating it in a Pasteur pipette, then pouring it between the edge of the cover slip and the slide. With practice one can prepare perfusion chambers which allow one to change periodically the solution under the cover slip (Fig. 1). A portion of the LY pinocytosed by

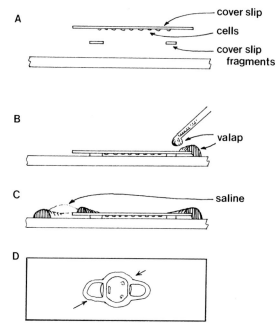

Fig. 1.    A preparation for the study of living cells in the fluorescence microscope. (A) A 12-mm diameter, No. 1 thickness, circular cover slip with attached cells is placed cell side down onto the microscope slide. Fragments of broken cover slips are used to support the cells above the slide. Saline or medium is then added until the entire underside of the cover slip is wet. (B) This chamber is sealed with warm valap applied from a heated Pasteur pipet. (C and D) With careful application of the valap one can assemble a perfusion chamber. D shows the top view of such a chamber, with a circular cover slip resting on three cover slip fragments, and sealed with valap. The arrows in D indicate the plane of section shown in C (side view). The two wells of valap provide ports for the addition and removal of solutions.

macrophages returns to the medium (Swanson *et al.*, 1985). Occasional perfusion with warm PBS can prevent accumulation of exocytosed LY in the microscope preparation.

Many organelles of the endocytic compartments are extremely small. The tubular lysosomes in macrophages are sometimes only 75-nm wide (Swanson *et al.*, 1987). When rendered fluorescent by pinocytosis of LY, they can be detected using a $63\times$ Planapo lens, numerical aperture (NA) 1.4, or a Plan-neofluar $63\times$ lens, NA 1.25. The essential feature of the objective lens for the resolution of fine structure is its high NA.

Intense excitation light will cause some LY-labeled lysosomes to explode, releasing the probe into the cytoplasmic space. This is particularly evident in macrophages loaded for a long time by pinocytosis of LY. It is most likely due to heating of the organelle by excess light, and can be prevented by reducing the excitation light, by insertion of an infrared heat-filter into the excitation light path, and by viewing the cell with the aid of an image intensifying video camera. Such procedures are recommended even if the lysosomes are not explosive, as intense illumination is deleterious to cells in other respects as well.

Does the pattern of fluorescence seen in living macrophages labeled with LY reveal the normal endocytic pathway, or have some of these organelles been formed due to presence of the probe? The answer to this requires independent approaches. Pinocytosis of another fluorescent probe for fluid-phase pinocytosis, such as FITC-dextran, should display a pattern of fluorescence similar to that displayed by LY. Second, the immunofluorescence distribution of antigens specific to the endocytic compartments should appear the same in cells which have pinocytosed LY and in cells which have never been exposed to LY (Swanson *et al.*, 1987).

## 2. FLUORESCENT DEXTRANS

Berlin and Oliver (1980) characterized FITC-dextran as an indicator of fluid phase pinocytosis and used it to show that fluid phase pinocytosis is greatly diminished in J774 macrophages during mitosis. It has since been used as a probe for fluid phase pinocytosis in other cell types and observed not to cross cell membranes or to adsorb appreciably to cell surfaces (vanDeurs *et al.*, 1984). Dextran is a branched polysaccharide which can be prepared in a broad range of molecular weight sizes (see Luby-Phelps, Chapter 4, this volume). FITC-dextran can be tolerated by most cells at concentrations of 10 mg/ml or more. However, commercially available FITC-dextrans sometimes contain free fluorescein or other molecules that adversely affect cell health. It is therefore advisable to pass solutions of FITC-dextran over a gel permeation column (Sephadex G-25, Pharmacia, Inc., Piscataway, New Jersey) before use.

The fluorescence excitation spectrum of fluorescein varies as a function of pH, and this property has been exploited by many investigators to study the pH of endocytic compartments (Okuma and Poole, 1978; Geisow *et al.*, 1981; Heiple and Taylor, 1982; Tycko and Maxfield, 1982). These methods are described in detail elsewhere (see Maxfield, Chapter 2, this volume). For strictly morphological studies, however, this effect of pH on fluorescein fluorescence can be misleading, in that as FITC-dextran enters acidic intracellular compartments its fluorescence decreases. Since much of the endocytic compartment is acidic, this property obviates somewhat the use of FITC-dextran as a morphological marker for this compartment. In addition, fluorescein is easily bleached by intense excitation light, necessitating the use of image intensification for studying FITC-dextran in living cells over extended periods of time.

Dextrans labeled with other fluorophores, such as rhodamine, Texas Red, and Lucifer Yellow, are also available commercially (Molecular Probes, Inc., Eugene, Oregon). These too are useful markers of the endocytic compartments in that bleaching and pH effects are not such significant problems as with FITC-dextran. The different fluorophores confer different adsorptive properties upon the dextrans, however, and one should not assume that they are all legitimate indicators of fluid phase pinocytosis. Likewise, dextrans with conjugated amino groups, called "lysine-fixable" dextrans, are useful in that they can be fixed with aldehydes for immunocytochemical methods, but they may adsorb to cell membranes more than the conventional charge-neutral dextrans.

### 3. CARBOXYFLUORESCEIN

Goldmacher *et al.* (1986) used 5(6)-carboxyfluorescein (CF) as a probe for fluid-phase pinocytosis in lymphocytes and showed that pinocytosis in these cells has a low capacity. They determined that their CF preparation would not diffuse across cell membranes by showing that it did not diffuse out of liposomes. Such precautions are recommended for any application of CF as a probe for fluid-phase pinocytosis. Commercial preparations of CF diffuse across the macrophage plasma membrane into cytoplasm (J. Swanson, unpublished observations). Without further purification of CF this artifact renders it useless for studying fluid-phase pinocytosis. CF may also enter the endocytic pathway by a nonendocytic sequestration from the cytoplasm (see Section III).

## B.   Probes of Adsorptive Pinocytosis

Fluorescent probes which adsorb to cell membranes label endocytic compartments more efficiently. The increased probe concentration at the cell surface consequently increases its concentration in pinosomes. Macromolecules labeled with rhodamine often adsorb to cell surfaces. We have found that ovalbumin labeled with Texas Red (Molecular Probes, Inc.) is endocytosed very efficiently by macrophages and concentrations in the medium as low as 50 $\mu$g/ml quickly render the endocytic compartments brightly fluorescent (T. Steinberg and J. Swanson, unpublished observations). Texas Red can be visualized using conventional rhodamine filter sets. The protein is hydrolyzed within lysosomes, so intracellular fluorescence probably reports the distribution of both Texas Red – ovalbumin and Texas Red conjugated to smaller peptides.

## C.   Probes for Receptor-Mediated Endocytosis

The efficient internalization of many physiologically important macromolecules is mediated by receptors in the plasma membrane. Specific interactions between receptors and their macromolecular ligands increase the concentration of those ligands at the cell surface, and consequently increase their rate of entry into the endocytic compartment. Receptors may be endocytosed and recycled to the cell surface constitutively, transporting a ligand if it happens to bind (Goldstein *et al.*, 1985). Alternatively, some receptors are not internalized unless they are ligated (Carpenter and Cohen, 1976). The subsequent fate of the receptor and ligand, once internalized by the cell, varies with the receptor – ligand combination: transferrin and its receptor both reappear at the cell surface (Ciechanover *et al.*, 1983); the low density lipoprotein receptor returns empty (Brown *et al.*, 1983); and the ligated epidermal growth factor receptor is degraded with its ligand in lysosomes (Carpenter and Cohen, 1976).

A number of fluorescent ligands for receptor-mediated endocytosis have been described. These have been used both to map the routes of these macromolecules into the endocytic compartment (Maxfield *et al.*, 1978; see Maxfield, Chapter 2, this volume) and to determine the pH of the various compartments en route to lysosomes (Tycko and Maxfield, 1982; Yamashiro *et al.*, 1984).

The presence on a cell's surface of specific receptors does not ensure that a fluorescent conjugate of the ligand added to cells will enter by receptor-mediated endocytosis. It could enter nonspecifically if the fluorophore simply causes the ligand to adsorb to the cell surface. Moreover, the concentrations of fluorescent ligand necessary to visualize endocytic compartments in the microscope may saturate or exceed the receptor-mediated process, especially if the receptor – ligand interaction is of low affinity, or if the number of surface receptors per cell is low. As the concentration of fluorescent ligand added to cells increases, so too does the contribution of fluid phase pinocytosis to probe accumulation.

Such concerns are usually addressed experimentally in two ways. First, ligand fluorescence in the microscope is amplified via a low light level camera (Willingham and Pastan, 1978; Spring and Lowry, Chapter 15; and Aikens *et al.*, Chapter 16, this volume). This allows use of lower concentrations of fluorescent ligand. The second approach is a control experiment, which is to incubate cells in the selected concentration of fluorescent ligand plus an excess of unlabeled ligand (Maxfield *et al.*, 1978). Fluorescent ligand entering cells by receptor-mediated endocytosis should be competable, so saturation of the receptors with unlabeled ligand should inhibit cellular accumulation of the fluorescence. Any residual fluorescence accumulation by cells is due to either adsorptive or fluid-phase endocytosis.

Fluorescent proteins which reach lysosomes will very likely be hydrolyzed. Long incubations of cells with fluorescent ligands will, as with Texas Red-labeled ovalbumin, display a mixture of fluorescent ligands and their fluorescent breakdown products.

## III.  Identifying Compartments of the Endocytic Pathway

Many of the various organelles labeled by endocytosis look alike in the fluorescence microscope. Although their separate identities can sometimes be distinguished in fixed cells by immunological methods, the identification of compartments in living cells is more difficult. The pH-dependent fluorescence of fluorescein, or the pH-dependent accumulation of acridine orange, can be used to identify acidic compartments. But since lysosomes, endosomes, and possibly also the trans-Golgi network are all acidic (Anderson and Pathak, 1985), methods which identify low pH compartments may not provide satisfactory distinctions.

Pulse-chase experiments elucidate the pathway of endocytosed molecules from one compartment to another and allow one to devise methods which selectively label "early" compartments (pinosomes and endosomes) and "late" compartments (lysosomes). In macrophages, for example, a 5-minute incubation in 5 mg/ml LY in DM10F labels a cortical population of small vesicles. These are pinosomes (Fig. 2A and B). If after a brief pulse of LY the macrophages are incubated another 30 minutes in unlabeled medium, the cells become much less fluorescent, due to efflux of the LY, and the fluorescence which remains appears in larger and more centrally located organelles, which are lysosomes (Fig. 2C and D; Swanson *et al.*, 1985, 1987). This pulse-chase experiment indicates that some pinocytosed solute enters first into the small cortical population of vesicles, then is transferred to the larger, lysosomal compartment.

The best method at present for labeling lysosomes is to incubate cells for a long period (hours to days) in culture medium containing a nondigestible fluorescent probe, followed by a 1- to 2-hour chase in unlabeled medium. The long chase should deplete probe from compartments such as pinosomes and endosomes, leaving only lysosomes fluorescent. Microscope preparations of macrophage lysosomes labeled by pinocytosis of LY and

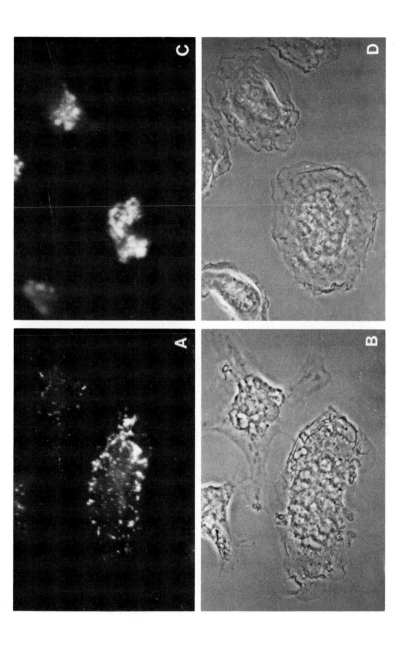

FIG. 2. Thioglycollate-elicited mouse peritoneal macrophages pulse labeled with LY. (A) Cells labeled by a 5-minute incubation in 5 mg/ml LY in DM10F, washed, and mounted as in Fig. 1, then observed and photographed immediately (Kodak PlusX film). Macrophages exhibit a punctate fluorescence at their periphery and margins. (B) Phase micrograph of cell shown in A. (C) Cells labeled as in A, then washed and reincubated in DM10F at 37°C for 30 minutes, washed again, mounted, and photographed. LY has redistributed from the peripheral pinosomes to the larger and more central lysosomes. (D) Phase micrograph of cell shown in C.

maintained at 37°C display dramatic saltatory movements of endocytic compartments. Tubular lysosomes move about within cytoplasm with a worm-like motion, probably along microtubule tracks.

The direct transfer of LY from one organelle to another has not, to my knowledge, been observed in the microscope. Wang and Goren (1987) report that the intercompartmental transfer, from lysosomes to phagosomes, of small sulfonated dyes such as LY or sulforhodamine occurs more rapidly than the transfer of larger fluorescent dextrans. This indicates that the transfer of solutes between compartments may not occur by complete fusion of the intact organelles, but rather by a more limited interconnection.

Lysosomes labeled by pinocytosis can be identified in fixed cells by immunohistochemical localization of lysosomal antigens such as lgp120 or cathepsin L with a primary antibody (Swanson et al., 1987). Pinocytosed LY can be fixed in place, and a rhodamine-labeled secondary antibody will label the lysosomal antigens with a distinguishable fluorophore. Fluorogenic substrates, which become fluorescent when hydrolyzed in lysosomes, should prove useful for identifying lysosomes in living cells (Dolbeare, 1981).

Endosomes can be identified immunologically in few cell types (Wilson et al., 1987). In living cells, it is often difficult to differentiate endosomes from pinosomes. That they are defined differently, pinosomes as the plasma membrane-derived vesicles with enclosed extracellular solutes, and endosomes as the acidic, nonlysosomal compartment in which receptors and ligands part ways, does not mean that they are different organelles, or that one does not gradually become the other. Until such distinctions are possible, one must consider those nonlysosomal compartments labeled by brief incubation in fluorescent probes to represent both endosomes and pinosomes.

There is accumulating evidence that endocytosed macromolecules enter the Golgi apparatus (Snider and Rogers, 1985; Woods et al., 1986; Youle and Colombatti, 1987). Immunofluorescence localization of Golgi markers, in parallel with LY labeling of endocytic compartments, could determine if pinocytosed LY reaches the Golgi.

Presence in a vacuolar compartment of fluorescent molecules such as LY or CF is not by itself evidence that the probes entered the cell by endocytosis. When LY is introduced into macrophage cytoplasm, either by scrapeloading (McNeil et al., 1984) or by ATP-induced permeabilization of the plasma membrane (Steinberg et al., 1987a), it is cleared rapidly into a nonlysosomal vacuolar compartment. Steinberg et al. (1987c) observed that although much of the LY loaded into cytoplasm exits directly from the cell, probably across the plasma membrane, the rest is sequestered into vacuoles, then later delivered into lysosomes (Fig. 3). Other fluorescent organic anions, such as Fura-2 or CF, also exhibit this redistribution after introduction into cytoplasm (DiVirgilio *et al.*, 1988; J. Swanson, unpublished observations). Both clearance and sequestration can be inhibited by addition of the organic anion transport inhibitor probenecid (DiVirgilio *et al.*, 1988; Steinberg *et al.*, 1987b, c). Larger molecules, such as FITC-dextran, are not sequestered from cytoplasm, at least not as quickly. Since this same phenomenon has been observed in other tissue culture cells as well, one should make certain that cells incubated in medium containing LY or CF are labeling endocytic compartments by the conventional route, that is, as solutes enclosed within pinosomes, and not instead by passive diffusion of probe across cell membrane followed by sequestration into vacuoles. This may be tested by examining cells after incubation in medium containing the fluorescent probe plus 5 m$M$ probenecid. If probe enters endocytic compartments by passive diffusion across plasma membrane followed by sequestration, probenecid should block the sequestration and therefore cause the probes to label the cells with a diffuse cytoplasmic fluorescence. Such precautions are unnecessary with larger fluorescent probes such as FITC-dextran, as these are unlikely to diffuse across or be transported across membranes.

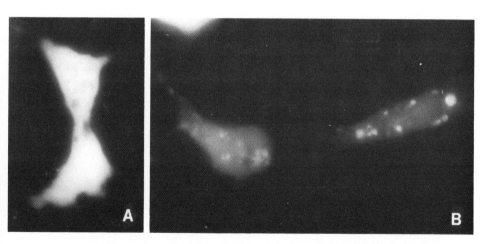

Fig. 3.   Sequestration of LY from the cytoplasm of thioglycollate-elicited mouse peritoneal macrophages. (A) Cells labeled by a 5-minute incubation in 0.5 mg/ml LY, 5 m$M$ ATP$^{4-}$, in DM10F, washed and mounted as in Fig. 1. (B) Cells labeled as in A, then washed and reincubated in DM10F for 15 minutes. Note redistribution of LY from diffuse cytoplasmic labeling into cytoplasmic vesicles. See Steinberg *et al.* (1987c) for a more complete description of this phenomenon.

## References

Adams, C. J., Maurey, K. M., and Storie, B. (1982). *J. Cell Biol.* **93**, 632–637.

Aley, S. B., Cohn, Z. A., and Scott, W. A. (1984). *J. Exp. Med.* **160**, 724–737.

Anderson R. G. W., and Pathak, R. K. (1985). *Cell* **40**, 635–643.

Berlin, R. D., and Oliver, J. M. (1980). *J. Cell Biol.* **85**, 660–671.

Brown, M. S., Anderson, R. G. W., Basu, S. K., and Goldstein, J. L. (1983). *Cold Spring Harbor Symp. Quant. Biol.* **46**, 713–724.

Carpenter, G., and Cohen, S. (1976). *J. Cell Biol.* **71**, 159–171.

Ciechanover, A., Schwartz, A. L., Dautry-Varsat, A., and Lodish, H. F. (1983). *J. Biol. Chem.* **258**, 9681–9689.

DiVirgilio, F., Steinberg, T. H., Swanson, J. A., and Silverstein, S. C. (1988). *J. Immunol.* **140**, 915–920.

Dolbeare, F. (1981). *In* "Modern Fluorescence Spectroscopy" (E. L. Wehry, ed.), Vol. 3, p. 251. Plenum, New York.

Ehrenreich, B. A., and Cohn, Z. A. (1967). *J. Exp. Med.* **126**, 941–958.

Geisow, M. J., Hart, P. D., and Young, M. R. (1981). *J. Cell Biol.* **89**, 645–652.

Goldmacher, V. S., Tinnel, N. L., and Nelson, B. C. (1986). *J. Cell Biol.* **102**, 1312–1319.

Goldstein, J. L., Brown, M. S., Anderson, R. G. W., Russell, D. W., and Schneider, W. J. (1985). *Annu. Rev. Cell Biol.* **1**, 1–39.

Hart, P. D., and Young, M. R. (1975). *Nature (London)* **256**, 47–49.

Heiple, J. M., and Taylor, D. L. (1982). *J. Cell Biol.* **94**, 143–149.

McNeil, P. L., Murphy, R. F., Lanni, F., and Taylor, D. L. (1984). *J. Cell Biol.* **98**, 1556–1564.

Maxfield, F. R., Schlessinger, J., Schechter, Y., Pastan, I., and Willingham, M. C. (1978). *Cell* **14**, 805–810.

Michl, J., Pieczonka, M., Unkeless, J. C., and Silverstein, S. C. (1979). *J. Exp. Med.* **150**, 607–621.

Miller, D. K., Griffiths, E., Lenard, J., and Firestone, R. A. (1983). *J. Cell Biol.* **97**, 1841–1851.

Okuma, S., and Poole, B. (1978). *Proc. Natl. Acad. Sci. U.S.A.* **75**, 3327–3331.

Pastan, I., and Willingham, M. C. (1985). "Endocytosis." Academic Press, New York.

Phaire-Washington, L., Wang, E., and Silverstein, S. C. (1980). *J. Cell Biol.* **86**, 641–655.

Riezman, H. (1985). *Cell* **40**, 1001–1009.

Robbins, E., and Marcus, P. I. (1963). *J. Cell Biol.* **18**, 237–250.

Silverstein, S. C., Steinman, R. M., and Cohn, Z. A. (1977). *Annu. Rev. Biochem.* **46**, 669–722.

Snider, M. D., and Rogers, O. C. (1985). *J. Cell Biol.* **100**, 826–834.

Steinberg, T. S., Newman, A. S., Swanson, J. A., and Silverstein, S. C. (1987a). *J. Biol. Chem.* **262**, 8884–8888.

Steinberg, T. S., Newman, A. S., Swanson, J. A., and Silverstein, S. C. (1987b). *Clin. Res.* **35**, 492A.

Steinberg, T. S., Newman, A. S., Swanson, J. A., and Silverstein, S. C. (1987c). *J. Cell Biol.* **105**, 2695–2702.

Steinman, R. M., Brodie, S. E., and Cohn, Z. A. (1976). *J. Cell Biol.* **68**, 665–687.

Steinman, R. M., Mellman, I. S., Muller, W. A., and Cohn, Z. A. (1983). *J. Cell Biol.* **96**, 1–27.

Stewart, W. (1978). *Cell* **14**, 741–759.

Swanson, J. A., Yirinec, B. D., and Silverstein, S. C. (1985). *J. Cell Biol.* **100**, 851–859.

Swanson, J. A., Bushnell, A., and Silverstein, S. C. (1987). *Proc. Natl. Acad. Sci. U.S.A.* **84**, 1921–1922.

Takemura, R., Stenberg, P. E., Bainton, D. F., and Werb, Z. (1986). *J. Cell Biol.* **102**, 55–69.

Thilo, L., and Vogel, G. (1980). *Proc. Natl. Acad. Sci. U.S.A.* **77,** 1015–1019.

Tycko, B., and Maxfield, F. R. (1982). *Cell* **28,** 643–651.

van Deurs, B., Ropke, C., and Thorball, N. (1984). *Eur. J. Cell Biol.* **34,** 96–102.

Wang, Y.-L., and Goren, M. B. (1987). *J. Cell Biol.* **104,** 1749–1754.

Willingham, M. C., and Pastan, I. (1978). *Cell* **13,** 501–507.

Wilson, J. M., Whitney, J. A., and Neutra, M. R. (1987). *J. Cell Biol.* **105,** 691–703.

Woods, J. W., Doriaux, M., and Farquhar, M. G. (1986). *J. Cell Biol.* **103,** 277–286.

Yamashiro, D. J., Tycko, B., Fluss, S. R., and Maxfield, F. R. (1984). *Cell* **37,** 789–800.

Youle, R. J., and Colombatti, M. J. (1987). *J. Biol. Chem.* **262,** 4676–4682.

# Chapter 10

# Incorporation of Macromolecules into Living Cells

PAUL L. MCNEIL

*Department of Anatomy and Cellular Biology*
*Harvard Medical School*
*Boston, Massachusetts 02115*

## I. Introduction

Techniques for loading impermeant molecules into the cytoplasm of living cells have become essential cell and molecular biological tools. Indeed it is cell loading techniques that make possible transformation of cells with foreign DNA (Celis, 1984) and fluorescent analog cytochemistry (Taylor *et al.*, 1985; Wang *et al.*, 1982), to name just two of many possible examples. More and more frequently, it is becoming possible to replace the

153

test tube with the cell, and cell loading techniques have, in part, been responsible for this advance in experimental strategy.

The ideal technique for loading macromolecules into the cytoplasm of living cells would possess a formidable, perhaps an unattainable, combination of characteristics. It would be simple, inexpensive, rapid, and reproducible; applicable to diverse cell types; capable of uniformly loading large ($> 10^4$) numbers of cells with large ($> 10^6$) numbers of large ($> 10^5$ mol wt) macromolecules; and lastly, it would be completely without effect on short- and long-term cell function and viability.

It is perhaps no surprise that none of the now numerous (Table I) techniques available for loading cells meet all of these criteria. Fortunately, one or more of them will often suffice for a specific, if limited, purpose. But which one? This chapter, at a necessarily elementary and selective level, aims at providing a conceptual and practical basis for choosing, implementing, and evaluating a suitable cell loading technique. Because of the orientation of this volume, I have placed greatest emphasis on those cell loading techniques most suited to microscopic applications.

## II.  Strategies for Loading Living Cells

All loading methods use one of four general strategies for breaching cell plasma membrane. I call these the chemical, vehicle, mechanical, and electrical strategies (Table I). But these are by no means unambiguous categories. For example, there must be a mechanical element in any successful strategy, that is, an event, even if on the molecular scale in which membrane components are forced apart to allow entry of macromolecules.

First, then, in the "chemical" strategy cells are permeabilized simply by soaking them in the chemical agent, such as ATP or EGTA. The mechanisms by which such agents facilitate loading is unclear. A membrane receptor that opens a pore upon $ATP^{4-}$ binding is invoked to explain the effect of ATP (Steinberg et al., 1987), a generalized "weakening" of plasma membrane to explain the effect of $Ca^{2+}$-free medium. It is not clear how $CaPO_4$ and DEAE–dextran treatment lead to transfection with DNA; whether the complex of DNA with $CaPO_4$ or DEAE enters the cytoplasm directly, or is first pinocytosed, later to escape somehow into cytoplasm.

In the "vehicle" strategy, cells are loaded by causing them to fuse with a suitable membrane-delimited vehicle (usually a red blood cell ghost or liposome) which has previously been loaded itself with macromolecules (Doxsey et al., 1985; Godfrey et al., 1983; Pagano and Weinstein, 1978; Schlegel and Rechsteiner, 1975). Exceptionally efficient delivery by red

TABLE I

TECHNIQUES FOR LOADING MOLECULES INTO CYTOPLASM

| Mode of PM breach | Size of molecules loaded (mol wt)[a] | References[b] |
|---|---|---|
| **Chemical** | | |
| ATP | 1,000 | Rosengurt and Heppel (1975) |
| EDTA | 10,000 | Johnson et al. (1985) |
| CaPO₄ | DNA only? | Graham and van der Erb (1973) |
| DEAE–dextran | DNA only? | Vaheri and Pagano (1965) |
| **Vehicle** | | |
| RBC fusion | 300,000 | Schlegel and Rechsteiner (1975); Doxsey et al. (1985) |
| Vesicle fusion | No limit | Poste et al. (1976) |
| **Mechanical** | | |
| Microinjection | No limit | Graessmann and Graessmann (1971) |
| Hyposmotic shock | 10,000 | Borle and Snowdowne (1982) |
| Osmotic lysis of pinosomes | No limit | Okada and Rechsteiner (1982) |
| Scraping from substratum | 500,000, DNA | McNeil et al. (1984) |
| Agitation in cold | 500,000 | McNeil et al. (1984) |
| Sonication (mild) | 70,000, DNA | Fechheimer et al. (1986) |
| High-velocity microprojectiles | DNA | Klein et al. (1987) |
| Glass beads | 150,000 | McNeil and Warder (1987) |
| Scratching to wound culture | 150,000 | Swanson and McNeil (1987) |
| **Electrical** | | |
| Electric shock | DNA | Wong and Neumann (1982) |

[a] Largest molecules reported loaded. DNA is listed separately if it has been successfully introduced by a technique.

[b] Original reports or important variations.

blood cells is obtained when fusogenic viral hemagglutin is present on the target cell (Doxsey et al., 1985). This particular vehicle system possesses the potentially useful property (conferred by viral hemagglutin) that fusion, and therefore loading of adherent cells, can apparently be rapidly and synchronously triggered simply by reducing medium pH.

In the "electrical" strategy, cells are reversibly permeabilized to exogenous macromolecules by high voltage electric discharge, a technique used predominantly for transfection (Neumann et al., 1982). Polarization of membrane components by the electric field is proposed as the mechanism for "electroporation" (Neumann and Rosenheck, 1972).

There are several variations of the "mechanical" strategy for temporarily disrupting plasma membrane or its derivatives. Cells allowed to pinocytose macromolecules in hypertonic medium are loaded by osmotically lysing such pinosomes (Okada and Rechsteiner, 1982); or, more simply, cells are loaded by osmotically shocking them in the presence of macromolecules (Borle and Snowdowne, 1982). Cells penetrated by a glass microneedle are loaded by injecting (or microinjecting) a small volume of the macromolecules through the needle into cytoplasm (Graessman et al., 1974). Cells adherent to a substratum are loaded with exogenous macromolecules by scraping them off of that substratum (McNeil et al., 1984). Creating a wound in a confluent culture of cells will cause the loading of cells along the edge of the wound (Swanson and McNeil, 1987). Sonication permeabilizes suspended cells (Fechheimer et al., 1986). Finally, cultured cells are loaded with exogenous macromolecules by sprinkling glass beads onto them (McNeil and Warder, 1987).

Each of the four above strategies has advantages and limitations. ATP cannot load molecules larger than ~ 1000 mol wt, and is applicable only to transformed cells and a few other types. Vehicle-mediated loading introduces exogenous plasma membrane (of the vehicle) into that of the target cell; the vehicle can undergo phagocytosis rather than fusion; and, if virus mediated, vehicle loading is restricted to appropriately infected or transfected cells. Osmotic lysis of pinosomes appears to require actively pinocytotic cells, a very high concentration of the molecule to be loaded, and can only be used with macromolecules that do not adhere to the cell surface. Electric and osmotic shock of cells, as well as EGTA permeabilization, have not been characterized extensively enough for accurate evaluation as generally useful loading techniques. Mechanical strategies often result in loss of cells. Microinjection, for example, has been reported to cause an ~ 40% loss of injected Swiss/3T3 cells (Pastri et al., 1986). Surprisingly, such loss, as distinguished from viability of those cells remaining at various times after microinjection, seems seldom to have been assessed. Bead loading generally causes less cell loss than does scrape loading, especially in certain cell lines, such as the PtK-2 (McNeil and Warder, 1987).

## III. Selected Techniques for Loading Living Cells

In this section I provide a brief overview of several cell loading techniques and then describe them in sufficient detail for easy replication. I have necessarily been selective, choosing only those techniques with which I have personal experience and which I have found useful for fluorescence microscope applications. I have also provided a brief overview of microinjection. I have not indicated what concentrations of macromolecules are required for each loading technique, since this requirement is entirely dependent on the macromolecule used and the experiment undertaken.

## A. Bead Loading

Bead loading is an exceptionally rapid and simple new technique for loading large numbers of cultured cells with large macromolecules (McNeil and Warder, 1987). The culture medium of the cell monolayer is replaced by a small volume of the macromolecule to be loaded. Dry glass beads (75–100 $\mu$m diameter) are then sprinkled onto the cells, the cells are washed free of beads and exogenous macromolecules, and "bead loading" is completed (Fig. 1). The conditions for bead loading can readily be modified to accommodate cell type and loading objectives: for example, the amount of loading per cell increases if bead size is increased or if beads sprinkled onto the monolayer are subsequently agitated, but at the expense of increased cell loss. As many as 97% of a population of bovine aortic endothelial (BAE) cells can be loaded with 10,000 mol wt dextran; and 79% with a 150,000 mol wt dextran using bead loading. A variety of cell lines, including Swiss/3T3, J774.2, PtK-2, and MDCK, can be loaded using glass beads, and in many cases without disturbing culture confluency. Moreover, bead

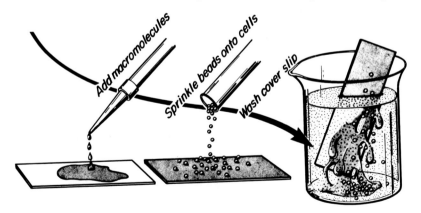

FIG. 1. A cartoon showing how to "bead load" cells. (1) Culture medium is replaced with a solution of macromolecule to be loaded. (2) Glass beads are carefully and evenly sprinkled onto the culturing surface. (3) Beads and unincorporated macromolecules are washed away from the culture. Many cells of the culture are now loaded, and, since they remain adherent, can be examined immediately in the microscope.

loading has the advantage of producing loaded cells that remain adherent and well spread, thus minimizing recovery time and allowing immediate microscopic examination (Fig. 2).

*Detailed Methods.* Glass beads (purchased from Sigma) of size ranges 450–500 $\mu$m or 75–150 $\mu$m are used for loading. Approximately 0.23 g (approximately 1000 beads) of the 450 $\mu$m or 0.12 g of the 75 $\mu$m beads are used for each cover slip (22 mm square) of cells. These weights were chosen because they are apparently sufficient to cover the cover slip with a single layer of beads.

Cells to be loaded are grown to any desired density on glass cover slips, slides, or petri dishes. Cultures growing on cover slips are rinsed three times with CMF ($Ca^{2+}$ and $Mg^{2+}$-free)–PBS (37°C) by dipping them in three successive baths of this medium. The culture is wicked free of excess CMF–PBS after the third rinse. We have found it convenient to then place the cover slip on a silicone–rubber sheet (North American Reiss Corp., Bellemead, New Jersey) supported by a glass slide, although the cover slip can also be held with forceps throughout the procedure. A (37°C) CMF–PBS solution (20–200 $\mu$l for a 22-mm square cover slip) of the macromolecule to be loaded is pipetted onto the culture, drawn off, and repipetted onto the cover slip in order to assure thorough mixing at the liquid interface with the cells. Beads are then carefully and evenly sprinkled onto the cover slip from a 6 × 50-mm culture tube (Kimble, Toledo, Ohio) held 1–3 cm above the horizontally oriented cover slip surface. If maximal loading is required, the beads are caused to roll around on top of the culture until evenly distributed over its surface by gently rocking the cover slip 3–6 times. The culture is rinsed free of glass beads and unincorporated macromolecules by dipping the cover slip in one or more PBS baths. Cultures can then be returned to the culture medium in their original dishes and allowed to recover at 37°C and 5% $CO_2$ for various intervals, or examined immediately in the microscope. The method can obviously be modified for cells growing in petri dishes or other culturing vessels.

Despite its simplicity, there are several important and readily controlled variables of bead loading which strongly influence the amount of loading and the yield of viable cells (McNeil and Warder, 1987). Perhaps most influential is the size of the beads employed and how much the beads resting on the cell monolayer are agitated. We measured, e.g., a 3- to 5-fold increase in the average extent of loading using 450-$\mu$m rather than 75-$\mu$m beads (see also Table II). Increased agitation of the 450-$\mu$m beads, caused by rocking of the cover slip to and fro six times, increased loading almost 3-fold over cultures onto which the 450-$\mu$m beads were sprinkled but not subsequently agitated. However, such increases in loading are traded for decreases in cell yield.

The surface properties of the beads moderately influence cell yield but

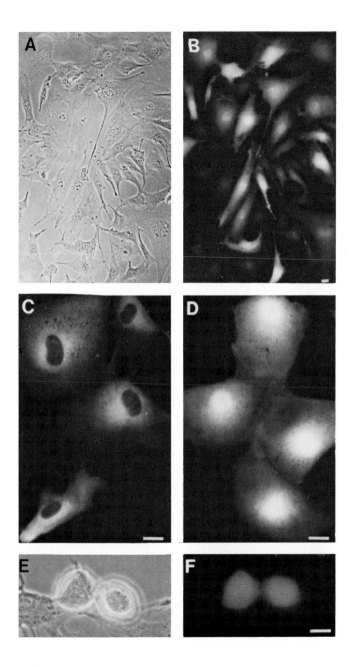

Fig. 2. Living cells of various types photographed after bead loading. (A and B) Bovine aortic endothelial (BAE) cells 18 hours after bead loading in FDx9 (10 mg/ml) using 75-$\mu$m beads. (C) BAE cells 4 hours after bead loading in FDx150 (20 mg/ml) using 75-$\mu$m beads. (D) PtK-2 cells 1 hour after bead loading in FDx9 using 450 $\mu$m alkali-washed beads. (E and F) Mitotic daughter BAE cells 18 hours after bead loading in FDx9 using 75 $\mu$m beads. Bar, 10 $\mu$m.

TABLE II

EFFECT OF BEAD SIZE AND THE VOLUME OF MACROMOLECULE
CARRIER SOLUTION ON BEAD LOADING

| Bead size | Volume carrier solution[a] | Yield cells[b] (% of control) | Loading index (pmol FDx per mg cell protein) |
|---|---|---|---|
| 75 $\mu$m | <20 | 69 | 74 |
|  | 50 | 93 | 67 |
|  | 200 | 98 | 41 |
| 450 $\mu$m | <20 | 94 | 42 |
|  | 50 | 94 | 52 |
|  | 200 | 88 | 160 |

[a] FDx 9 (200 $\mu$l) was added to the cultures of BAE cells growing on cover slips, drawn off, repipetted onto the culture, and then drawn off, leaving the indicated amount remaining on the culture. For experiments where <20 $\mu$l FDx were used, as much FDx was drawn off as possible without drying the culture. Beads were then sprinkled onto the culture without further agitation. Finally cultures were washed by dipping in PBS, and returned to culture medium for 1 hour before processing in the fluorometric assay (see text).

[b] Lowry protein assay. Control cultures received FDx 9 but no beads. Four replicate samples of each experiment are reported.

not extent of loading. This fact might be especially worthy of consideration for loosely adherent cell types (such as the NIH 3T3) or where the macromolecule to be loaded might adhere to the acid-washed bead obtained from the manufacturer. For example, glass beads (450 $\mu$m) washed in alkali (4 $M$ NaOH) rather than acid (5 $M$ HCl) increased cell yield almost 2-fold without diminishing the extent of loading. This effect is probably explained by the clearly diminished tendency of alkali-washed beads to adhere to cells (McNeil and Warder, 1987).

The minimal volume of macromolecule required for the 20-mm square cover slip is that which will prevent drying during the bead loading procedure, e.g., <20 $\mu$l (Table II). Smaller cover slips would obviously require even less volume.

## B. Scrape Loading

Scrape loading is another simple technique for loading cell populations with macromolecules (McNeil *et al.*, 1984). Adherent cells are scraped off of their substratum with a rubber policeman in the presence of the macromolecules to be loaded (Fig. 3). The percentage of cells loaded and numbers

## SCRAPE LOADING

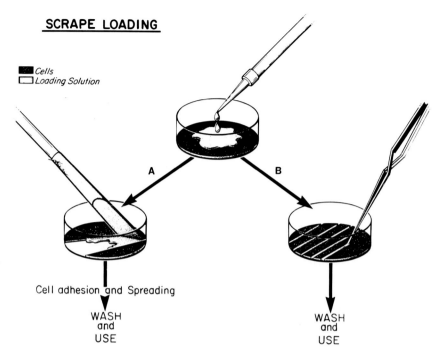

FIG. 3. A cartoon showing how to scrape load cells. Culture medium is replaced by a solution of the macromolecule to be loaded. Then *either* (A) all the cells of the dish are scraped off the substratum using a soft rubber policeman *or* (B) a small number only are removed by scratching the culture using a sharply pointed instrument. Cells scraped off of the substratum (A) are loaded with molecules present exogenously during scraping but must be replated and allowed to spread before microscopic examination or most other uses. Those lining the "wounds" of the scratched cultures (B) are loaded and they also remain adherent and well spread, and can therefore be used much sooner (<1 hour) for microscopy than cells from procedure A.

of macromolecules loaded per cell are comparable to those values measured for bead loading using 450-$\mu$m beads and agitation (see Section III,A). The main disadvantage of scrape loading is that loaded cells are obtained in suspension and must be repleted and allowed to spread, before microscopic examination. Scrape loading also appears to result in a lower yield of cells than bead loading (McNeil and Warder, 1987).

*Detailed Methods.* The scrape loading method described here is a modification of that previously characterized (McNeil *et al.*, 1984). Subconfluent cultures grown in 35-mm diameter dishes are rinsed three times with 3-ml aliquots of CMF–PBS (37°C). A 37°C solution of the molecule to be loaded (20–200 $\mu$l) is added to the dish, the dish is swirled to distribute that solution evenly, and then cells in the culture are scraped off of the dish with a rubber policeman. Culture medium (37°C, pH 7.4) is immediately added, and the cell suspension is plated on the culturing substratum, or cells can first be washed by centrifugation if immediate removal of unincorporated macromolecules is desirable. Cultures are placed in the incubator, washed with fresh culture medium 30 minutes later to remove unincorporated macromolecules, and then returned to the incubator until recovery of normal, well-spread morphology is complete. With appropriate adjustments in scale, dishes larger or smaller than the 35-mm example used above are perfectly suitable for scrape loading.

The major variable influencing loading and cell yield from the scrape loading procedure is the strength with which cells adhere to their substratum. Any means for loosening cells from their substratum will result in decreased loading and increased cell yield. For example, we (McNeil *et al.*, 1986) found that if macrophages were scraped from tissue culture plastic or untreated glass surfaces, cell loss was unacceptably high ($>90\%$). We therefore plated macrophages onto glass washed in alkaline detergent (to which they adhere less strongly), allowed them only 15–30 minutes after plating to adhere, and rinsed them thoroughly in CMF–PBS containing 10 $\mu$m EGTA before scraping from the substratum. In this way, by minimizing macrophage adherance to glass, we were able to obtain satisfactory loading *and* cell yield. Use of hypertonic saline (sucrose added to a concentration of 0.5 $M$) as the loading vehicle is another effective strategy for increasing cell yield from scrape loading, but also reduces the extent of loading achieved (unpublished results).

If scrape loading is performed in the cold (McNeil *et al.*, 1984) and/or after soaking of cells in hypotonic saline (Beckers *et al.*, 1987), resealing of disrupted plasma membrane, which occurs in seconds to minutes at 37°C, is prevented and cells are therefore killed and become irreversibly permeabilized. Such irreversibly permeabilized cells have been termed "semiintact" and the technique of scrape loading in the cold after cell swelling in hypotonic medium, or after forcing cells to adhere to a nitrocellulose filter, have been termed, respectively, the "hypotonic swelling" and "nitrocellulose" techniques (Beckers *et al.*, 1987).

## C. Scratch Loading

Scratch loading is simply a variation of the scrape loading technique (Swanson and McNeil, 1987). Zones of cells are denuded from a monolayer by scratching the culture substratum with a sharp instrument (Fig. 3). Most of those cells which line the edge of the wound thus created in the monolayer are loaded (Fig. 4). It is well known that cells lining such wounds migrate into the denuded zone where DNA synthesis and mitosis are then initiated (Todaro *et al.*, 1965; Dulbecco, 1970). We (Swanson and McNeil,

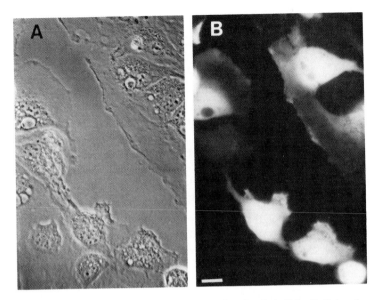

Fig. 4.   Phase (A) and fluorescence (B) micrographs of endothelial cells lining the mono-layer "wound" produced by scratching a cover slip with the sharp tip of jeweler's forceps in the presence of FDx9 (10 mg/ml). These cells were imaged 30 minutes after scratch loading and initiated locomotion into the denuded or "wounded" zone shown in this micrograph, and would later be expected to synthesize DNA and divide there. Bar, 10 $\mu$m.

1987) have used this technique to load fluoresceinated dextrans into cells that went on to divide 24 hours after scratch loading, and showed that these dextrans are excluded from the reforming nucleus: scratch loading is especially useful for the study of motile and mitotic cells in the microscope, since loading and the induction of these events are accomplished simultaneously and in the same cells. Moreover, it requires only a small volume of dissolved macromolecules (enough to prevent cover slip drying), is exceptionally rapid and easy to perform and produces cells which are clearly adherent, motile, and viable.

*Detailed Methods.*   Cover slips on which are growing subconfluent or confluent monolayers of cells are washed with CMF–PBS (37°C). A 37°C solution (10–100 $\mu$l for a 22-mm square cover slip) of the macromolecule to be loaded is pipetted onto the cells and cover slip, and then removed and repipetted onto the cover slip to ensure mixing at the liquid interface with the cells. The cover slip is then scratched with jeweler's forceps or any other equally sharply pointed instrument, creating many wounds in the monolayer. Culture medium is added gently back to the cells and, 30 minutes later, the cells are rinsed with further fresh culture medium to remove any remaining unincorporated macromolecules.

## D.  ATP

Soaking in ATP permeabilizes mast cells (Dahlquist and Diamaut, 1974; Gomperts, 1985), hepatocytes (Charest *et al.*, 1985), mononuclear and polymorphonuclear phagocytes (Becker and Henson, 1975), and several transformed lines (Rozengurt and Heppel, 1975). Clearly ATP sensitivity is not ubiquitous, but, given the unexplained nature of its distribution among cell types, it may be a property, if unknown, worthwhile testing on one's own cell line. Lucifer Yellow (mol wt 457) is an excellent indicator for successful ATP permeabilization (Fig. 5; Steinberg *et al.*, 1987). In general, any exogenous molecule < 1000 mol wt is rendered permeant (Rozengurt and Heppel, 1975; Gomperts, 1985) but in some cells it is only certain ions (Chahwala and Cantley, 1984; Dubyak and De Young, 1985).

*Detailed Methods.* To load sensitive cells with ATP one simply adds to their culture medium a volume of 100 m$M$ (pH 7.0) sodium ATP (Sigma or Boehringer-Mannheim) sufficient to give a concentration of 1–5 m$M$ (this concentration can vary somewhat depending on cell type) and a suitable concentration of the molecule to be loaded. Then, after 1–5 minutes at 37°C, the cells are washed free of ATP and unincorporated molecules, and can be used in experiments or allowed to recover for a suitable interval (1–4 hours).

## E.  Microinjection

Microinjection is capable of loading more molecules per cell at less expense per cell of those molecules than any other technique. The simple reason why is that a very small volume (picoliters?) is *forced* under pressure (e.g., injected) through a microneedle into cytoplasm. No other loading technique truly utilizes injection. Its disadvantages are that it requires considerable expertise and expensive apparatus, is tedious, and is capable of loading only a relatively small number of cells.

Microinjection is easily envisioned (after all, who hasn't received an injection!), but mastered only after considerable practice. The equipment used and cell type to be injected will dictate specific strategies and practice. Moreover, the technique has been reviewed in detail elsewhere (Graessman *et al.*, 1980). Therefore, I will give here only a brief overview.

First, then, the following equipment (commercially available) is essential: (1) a microneedle puller; (2) a micromanipulator and needle holder; (3) an upright, or preferably, an inverted microscope, equipped, again preferably, with a long working distance ×40 phase objective lens; (4) glass capillaries; and (5) a source of positive pressure (e.g., a syringe) and tubing for airtight

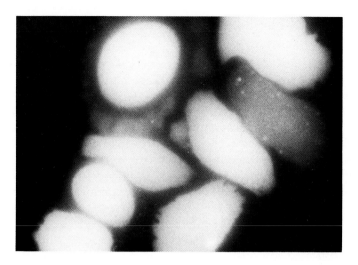

FIG. 5.   A fluorescence micrograph of J774 cells permeabilized by soaking them in 5 m*M* ATP for 5 minutes in the presence of 0.5 mg/ml Lucifer Yellow. [From Steinberg *et al.* (1987), with permission.]

coupling of syringe to needle or needle holder. A neurobiologist can often advise on the best selection of apparatus, or may share equipment.

A microneedle is pulled or drawn from a glass capillary tube using the microneedle puller. The diameter ($\sim 0.5$ $\mu$m) and shape of the needle tip are crucial to success but usually these variables are determined on the basis of one's experience with a particular puller, and sometimes also based on cell type. The best microneedle, conveniently viewed in a dissecting scope, is selected from among many pulled, and filled, either from the blunt end by capillary action (this requires specialized tubing productive of capillary action), or through the tip by application of negative pressure. The needle is assembled into the needle holder and micromanipulator, and coupled via tubing to the positive pressure source (e.g., syringe). The needle tip is micromanipulated while viewed through the microscope into position over a cell. The optimal needle position (usually adjacent to the nucleus in the thickest portion of the cell), the speed, and the angle for penetration, as well as many other details, are usually dictated by personal preference and long experience. Positive pressure is applied to the solution contained in the penetrating needle, delivering fluid into the cell and causing a slight expansion in cell volume (visible to the practiced eye!). The needle is removed. The cells are generally allowed to recover for several hours before experimentation.

# IV.  Evaluation and Analysis of Loaded Cells

In evaluating a new or variant loading technique, or, perhaps more likely, a novel application of an extant technique, one must rely on some form of qualitative or quantitative characterization. The depth of characterization undertaken will of course depend on one's needs. In the least demanding case, what may be desired is a simple and economical means for screening several loading techniques and choosing the optimal one for a particular cell line. Here fluorescein dextran of an appropriate size could be employed as a nonspecific indicator for loading and the microscope could be used for rapid screening of loading success. At the other extreme, one may wish to know how many molecules of, say, an antibody, are loaded on average into each cell, and to what extent the number loaded varies from cell to cell in the population. Here labeling of the antibody with a fluorophore and a more sophisticated technology—namely flow cytometry—would be required.

In this section, then, I will describe methods for answering the question, how much loading was obtained? Less attention has been paid to the question, how was cell viability (health) affected by loading? This is not because I underestimate its importance: success in studying any cell function by aid of a loading technique will ultimately depend on a demonstration that that function is uncompromised in the loaded cell itself. But the nature of such a demonstration will clearly differ for each particular cell line and/or cell function, and the appropriate tests are therefore best devised by each individual.

## A.  Macromolecular Indicators for Evaluating a Loading Technique

### 1.  Fluorescent Indicators

Fluorescent dextrans (see also Luby-Phelps, Chapter 4, this volume) are excellent indicators for characterizing a cell-loading technique. First, they process the inherent sensitivity of fluorescence. Second, dextrans can be purchased in a wide range of sizes (4,000–150,000 mol wt) labeled with fluorescein (from Sigma) or, for a smaller size range, with rhodamine and other fluorophores (from Molecular Probes). Therefore, using dextrans one can readily determine the size range of molecules that can be loaded by a particular technique. Third, dextrans are nontoxic, apparently indigestible, and are not rapidly segregated into lysosomes or other intracellular organelles. They are therefore especially well suited as a long-term (days) indicator of loading. Lastly, dextran is highly hydrophilic and possesses only the slight negative charge contributed by the fluorescein or rhodamine fluorophore. It therefore does not stick to cell surfaces and remains trapped in cytoplasm after loading only so long as the plasma membrane is intact. In this latter respect it is a vital dye, as well as an indicator of loading, but in the reverse sense of trypan blue.

For use in a loading technique, I generally prepare a 10–20 mg solution of fluoresceinated dextran (FDx) (purchased from Sigma) in a suitable loading buffer. I have not observed any untoward effect on the viability of cells as a consequence of cell loading with such FDx. However, if toxicity due to dextrans is suspected, then toxic contaminants should first be removed by dialysis against a suitable buffer or by gel filtration. Should commercially available size fractions be of insufficiently narrow range, gel filtration chromatography can be used for narrowing (McNeil *et al.*, 1984; Luby-Phelps *et al.*, 1986; see Luby-Phelps, Chapter 4, this volume). Should commercial preparations of fluorescently labeled dextrans be over- or underlabeled, there are published procedures for labeling dextran with fluorescein (de Belder and Granth, 1973; see also Gimlich and Braun, 1985) or rhodamine (Gimlich and Braun, 1985). Dextrans larger than 150,000 mol wt (up to 2,000,000 mol wt) can be purchased unlabeled (from Pharmacia).

Where one's interest in loading is confined to a specific protein, that protein can be labeled with a fluorophore and used as a direct indicator of loading success. Simple methods are available for labeling proteins with fluorescein and rhodamine (Nairn, 1969; Rinderknecht, 1962) as well as with many other fluorophores (cf. Haugland, Volume 30, this series).

## 2. RADIOLABELED TRACERS

Radiolabeled molecules can, of course, be used as tracers in quantitative measurements of amount of loading (see Section IV, C,2 below). Obviously, they are less useful for microscopic assays of loading, since such an assay would involve the tedious and difficult procedure of autoradiography. Methods for radiolabeling of proteins are well known and too various to cite here; radiolabeled dextrans can be purchased (from New England Nuclear).

## 3. HORSERADISH PEROXIDASE

Horseradish peroxidase (HRP) (I use the "enzymatic grade" from Boehringer-Mannheim, FRG) is an enzyme of 40,000 mol wt which can be caused to produce an opaque (light microscope) or electron dense (electron microscope) reaction product where present in cytoplasm (or any cell compartment). Since reaction product is generated after fixation of cells, HRP is clearly not an indicator for loading which can be observed in the living cell. Other drawbacks to HRP as an indicator are its expense (considerably greater than FDx); its poor sensitivity (10–50 mg/ml are, in my experience, required for EM observation of reaction product); and, lastly, its propensity to stick to the surface of some cells (Sung *et al.*, 1983). Its major advantage is the resolution conferred by visualization with EM. The method I use for developing the HRP reaction product is found in Adams (1977). HRP can be covalently coupled to proteins (Sell *et al.*, 1977), and can thereby serve as an indicator for loading of such proteins. This might be especially useful in conjunction with EM for high-resolution localization of a protein in cytoplasm.

## B.   Microscopy as a Qualitative Tool for Evaluating Loading

Phase contrast microscopy provides a rapid way of visually assessing cell health and fluorescence microscopy of qualitatively assessing the extent to which individual cells and the population as a whole are loaded (assuming the cells have been loaded with a fluorophore). Since one is viewing living cells, a cell chamber providing a healthy environment and also good optical characteristics is needed. I construct a simple chamber for viewing living cells by supporting on fragments of #1 cover slips, the cover slip on which the loaded cell population has been cultured. After drying the uppermost, cell-free surface of the cover slip in preparation for receiving immersion oil, the narrow cell chamber created between slide and cover slip is filled with saline or culture medium. I then seal the chamber with "valap" (a 1:1:1 mixture of paraffin, vaseline, and anhydrous lanolin; see also Swanson, Chapter 9, this volume), melting it over a flame in the tip of a Pasteur pipette, and then applying the melted valap with the pipette around the cover slip's circumference. The sealed preparation can be used in either an inverted or upright microscope, and cells can be kept in an apparently healthy state for many hours.

The phase contrast morphology of the loaded, living cell (which, e.g., can be identified by its fluorescence after loading with FDx) is a rapid and reasonable first indication of how well a cell line has tolerated a loading technique (Fig. 6). Its morphology should closely resemble that of cells on the same cover slip but which were not loaded (most techniques do uniformly load all of the cells in a population); and it should resemble the morphology of cells on a separate cover slip not subject to loading conditions. Trypan blue (filtered 0.1% wt/vol in PBS) can be used to score microscopically numbers of viable cells in the population.

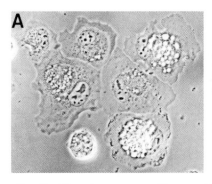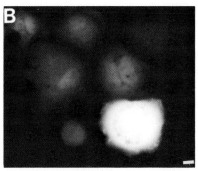

FIG. 6. The microscope provides a rapid means for assessing cell health and loading success. Phase contrast (A) and fluorescence (B) micrographs of BAE cells 24 hours after "electroporation" in the presence of 10 mg/ml FDx9. Cells suspended in PBS (without $Ca^{2+}$ or $Mg^{2+}$) were electrically shocked in a Bio-Rad Gene Pulser (Bio-Rad Laboratories, Richmond, California) at settings of 1000 V and 25 $\mu$F (resulting time constant = 0.4 msec). Cell morphology is seen to be abnormal in A (compare with Fig. 4), and loading of FDx9 unevenly distributed among cells in B. Bar, 2 $\mu$m.

One can also use the microscope to quickly estimate the extent of loading of a fluorophore. In particular, fluorescence microscopy rapidly shows how uniform was the loading achieved in the cell population (Fig. 6). But judging the extent of loading visually, as brightness of cellular fluorescence, is a highly subjective procedure and where critical distinctions must be made the quantitative techniques described below are recommended. If two different loading techniques are to be compared by fluorescence microscopy, then some objectivity can be gained by determining the lowest excitation (lamp) level required to see loaded cells in each population.

## C. Quantitative Evaluation of Loading

### 1. FLUOROMETRIC

Our fluorometer-based assay is an adaptation of Swanson et al.'s (1985) method for quantitating pinocytosis. It measures average picomoles of a macromolecule loaded per unit cell protein, and requires that the cells have been loaded with a fluorescently labeled macromolecule. We prefer FDx as an indicator for this sensitive assay, for reasons given above, but the following assay can readily be adapted for any fluorophore and conjugate macromolecule.

One hour (or any suitable interval) after completion of the loading procedure, the culture of cells adherent to a cover slip (22-mm square) is thoroughly washed free of exogenous fluorescence by dipping in six beakers containing PBS. Excess PBS is wicked off of the cover slip between each rinse. Cell protein and loaded fluorescein dextran are then extracted by soaking cover slips in 0.5 ml of 0.05% triton X-100 and 0.02% $NaN_3$ in water. After harvesting extracted cell remainders from the cover slips with a rubber policeman, 0.2 ml of the resultant lysate are used for protein determination in a Lowry (Lowry *et al.*, 1951) assay, or equivalent, and the remaining 0.3 ml are diluted with 1.7 ml of 0.05% triton X-100, 0.02% $NaN_3$, 1 mg/ml BSA, 10 m$M$ Tris (pH 8.0). The fluorescence (excitation = 490 nm; emission = 520 nm) intensity of the latter solution is measured using a spectrofluorometer. Relative fluorescence values thus measured are converted to molar dextran concentrations using known or measured molar dye to dextran ratios and using as concentration standards solutions of FDx calibrated spectrophotometrically assuming an extinction coefficient for fluorescein at pH 8.0 of $6.8 \times 10^4$ U/mol. A "loading index" is then calculated as picomoles dextran loaded per milligram cell protein. Control cultures receive FDx, but are otherwise subject to the conditions required for loading. Controls are used as a blank for fluorescence. A representative data set obtained using this assay is given in Table II.

## 2. RADIOMETRIC

By slight modification of the above fluorometric procedure, one can measure loading with a radiolabeled macromolecule. Thus, after loading with a radiolabeled indicator, completion of the above washing and extraction procedures, and suitable sample preparation, e.g., addition of scintillant, the cell extract would be counted radiometrically (rather than fluorometrically).

## 3. FLOW CYTOMETRIC ANALYSIS OF SINGLE CELLS

Flow cytofluorometry can rapidly (<3 minutes) measure the fluorescence associated with each of 20,000 (or more) cells in a population subject to loading with a fluorescent molecule (Fig. 7). From this detailed record of how the fluorescence intensities are distributed in a loaded population one can, by calibration, calculate the average number of molecules loaded per cell, and determine the degree to which loading varies from cell to cell (i.e., whether loading is uniform or heterogeneous throughout the population).

Cell viability can also be measured by flow cytofluorometry and this can be done simultaneously (in a dual laser beam instrument) with the fluorescence measurement of loading. Forward scatter is lower for dead cells than for living and this parameter is readily measured by flow (Fig. 7). The

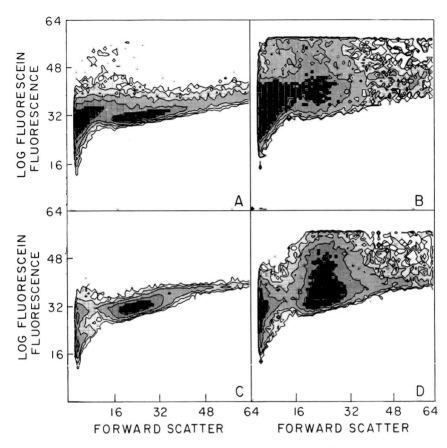

FIG. 7.   Flow cytofluorometry can be used for rapid and quantitative cell-by-cell measurement of extent of individual cell loading and viability. Flow cytometric analysis of the fluorescence and light scattering properties of each 25,000 fibroblasts loaded by scraping in medium lacking FDx 70 (A and C) and in medium containing solution of FDx 70 (B and D). Contour lines are drawn connecting those histogram bins containing 95, 90, 75, 50, and 25% of the 25,000 cells measured. Clearly, scrape loading introduces a highly variable amount of FDx70 from cell to cell in the population. This is evident in the two log spread of fluorescence intensities of panels B and D. Also obvious are a low-scattering population of dead cells (to the left of each histogram) immediately after scrape loading (A and B) but much reduced (<5% of total population) 24 hours later (C and D). The arbitrary values of fluorescence measured by cytofluorometry can readily be calibrated in terms of numbers of molecules of FDx loaded (see the text). (Reproduced from *The Journal of Cell Biology,* 1984, Vol. 98, pp. 1556–1564 by copyright permission of The Rockefeller University Press.)

difference, however, between the forward scatter of living and dead cells is not always unambiguous, although this method has the advantage of requiring no additional sample preparation. A clearer flow cytometric distinction between living and dead cells can be made by cell staining prior to flow cytometry with propidium iodide (PI), which permeates and stains the DNA of dead cells only.

Our method for preparing cells for flow cytofluorometry follows (McNeil and Warder, 1987); techniques for operating a flow cytometer and analysis of data are beyond the scope of this chapter and, in any case, these functions are usually carried out by a skilled technician.

Cultures that have been loaded with a fluorescent molecule are rinsed six times in CMF–PBS as described above and trypsinized as usual for the cell line. Calf serum is added to a final concentration of 10%. Suspensions are centrifuged at 4°C and 1000 $g$ for 3 minutes. The pellet ($> 10^5$ cells) is resuspended in 200–500 $\mu$l PBS supplemented with 10% calf serum. If viability is to be assayed as PI exclusion, then 10 minutes before the "run" PI is added to 100 $\mu$g/ml. Finally, fluorescence from macromolecules within loaded cells, forward angle light scatter, and/or PI fluorescence are then recorded simultaneously using the flow cytometer from each of 20,000 or more cells. For ease of subsequent data analysis, data recording is generally best done in the "list" mode. For simultaneous measurement of fluorescein and PI fluorescence, a 50% beam splitter can be used: half the light being passed through a standard fluorescein filter set and half through a 620-nm long pass filter (McNeil *et al.*, 1984). Relative fluorescence values measured by flow cytometry from cells are calibrated in terms of absolute numbers of fluorescein molecules per cell using as standards microbeads labeled with known numbers of fluorescein equivalents (Quantitative Fluorescein Microbead Standards Kit, Flow Cytometry Standards Corporation, North Carolina). Numbers of dextrans loaded per cell are then calculated from the known or measured molar fluorescein-to-dextran ratio.

## REFERENCES

Adams, J. C. (1977). *J. Histochem. Cytochem.* **29**, 775–780.
Becker, E. L., and Henson, P. M. (1975). *Inflammation* **1**, 71–84.
Beckers, C. J. M., Keller, D. S., and Balch, W. E. (1987). *Cell* **50**, 523–534.
Borle, A. B., and Snowdowne, K. W. (1982). *Science* **217**, 252–254.
Chahwala, S. B., and Cantley, L. C. (1984). *J. Biol. Chem.* **259**, 13717–13722.
Charest, R., Blackmore, P. F., and Exton, J. H. (1985). *J. Biol. Chem.* **260**, 15789–15794.
Dahlquist, R., and Diamant, B. (1974). *Acta Pharmacol. Toxicol.* **34**, 368–384.
de Belder, A. N., and Granth, K. (1973). *Carbohydr. Res.* **30**, 375–378.
Diacumakos, E. G. (1978). *Methods Cell Biol.* **7**, 288–311.
Doxsey, S. J., Sambrook, J., Helenius, A., and White, J. (1985). *J. Cell Biol.* **101**, 19–27.
Dubyak, G. R., and De Young, M. B. (1985). *J. Biol. Chem.* **260**, 10653–10661.

Dulbecco, R. (1970). *Nature (London)* **227**, 802–806.

Fechheimer, M., Denny, C., Murphy, R. F., and Taylor, D. L. (1986). *Eur. J. Cell Biol.* **40**, 242–247.

Fechheimer, M., Boylan, J. F., Parker, S., Sisken, J. E., Patel, G. L., and Zimmer, S. G. (1987). *Proc. Natl. Acad. Sci. U.S.A.* **84**, 8463–8467.

Gimlich, R. L., and Braun, I. (1985). *Dev. Biol.* **109**, 509–514.

Godfrey, W., Doe, B., and Wop, L. (1983). *Proc. Natl. Acad. Sci. U.S.A.* **80**, 2267–2271.

Gomperts, B. D. (1985). *In* "Secretory Processes" (R. T. Dean and P. Stahl, eds.), pp. 18–37. Butterworths, London.

Graessman, A., and Graessman, M. (1971). *Hoppe. Seilers 2, Physiol. Chem.* **352**, 527.

Graessman, A., Graessman, M., Hoffmann, H., Niebel, J., Brandler, G., and Nueller, N. (1974). *Fed. Eur. Biol. Soc. Lett.* **39**, 249–251.

Graessman, A., Graessman, M., and Mueller, C. (1980). *In* "Methods in Enzymology" (L. Grossman and K. Moldave, eds.), Vol. 65, pp. 816–825. Academic Press, New York.

Graham, F. L., and van der Erb, A. J. (1973). *Virology* **52**, 456–467.

Johnson P. C., Ware, J. A., Clivedon, P. B., Smith, M., Quorak, A. M., and Salzman, E. J. (1985). *J. Biol. Chem.* **260**, 2069–2076.

Klein, T. M., Wolf, E. D., Wu, R., and Sanford, J. C. (1987). *Nature (London)* **327**, 70–73.

Lowry, O. H., Rosenbrough, N., Farr, A. L., and Randall, R. J. (1951). *J. Biol. Chem.* **193**, 265–275.

Luby-Phelps, K., Taylor, D. L., and Lanni, F. (1986). *J. Cell Biol.* **102**, 2015–2022.

McNeil, P. L., and Warder, E. (1987). *J. Cell Sci.,* **88**, 669–678.

McNeil, P. L., Murphy, R. F., Lanni, F., and Taylor, D. L. (1984). *J. Cell Biol.* **98**, 1556–1564.

McNeil, P. L., Swanson, J. A., Wright, S. D., Silverstein, S. C., and Taylor, D. L. (1986). *J. Cell Biol.* **102**, 1586–1592.

Nairn, R. C., ed. (1969). "Fluorescent Protein Tracing." Livingstone, Edinburgh.

Neumann, E., and Rosenheck, K. (1972). *J. Membr. Biol.* **10**, 279–290.

Neumann, E., Schager-lidder, M., Wang, Y., and Hofschneider, P. H. (1982). *EMBO J.* **1**, 841–845.

Okada, D. Y., and Rechsteiner, M. (1982). *Cell* **29**, 33–41.

Pagano, R. E., and Weinstein, J. N. (1978). *Annu. Rev. Biophys. Bioeng.* **7**, 435–468.

Pastri, G., Lacal, J.-C., Warren, B. S., Aaronson, S. A., and Blumberg, P. M. (1986). *Nature (London)* **324**, 375–377.

Poste, G., Papahadjopoulos, D., and Vail, W. J. (1976). *Methods Cell Biol* **14**, 33.

Rinderknecht, H. (1962). *Nature (London)* **193**, 167–168.

Rozengurt, E., and Heppel, L. A. (1975). *Biochem. Biophys. Res. Commun.* **67**, 1581–1588.

Schlegel, R. A., and Rechsteiner, M. C. (1975). *Cell* **5**, 371–379.

Sell, S., Linthicum, D. S., Bass, D., Bahu, R., Wilson, B., and Nakane, P. (1977). *In* "Advances in Pathobiology: Differentiation and Carcinogenesis" (C. Borek, Ferroglio, C. M., and King, D. W., eds.) Vol. IV, pp. 272–305. Stratton, New York.

Steinberg, T. H., Neuman, A. S., Swanson, J. A., and Silverstein, S. C. (1987). *J. Biochem.* **262**, 8884–8888.

Sung, S.-S. J., Nelson, R. S., and Silverstein, S. C. (1983). *J. Cell. Physiol.* **116**, 21–25.

Swanson, J. A., and McNeil, P. L. (1987). *Science,* **238**, 548–550.

Swanson, J. A., Yirrec, B. D., and Silverstein, S. C. (1985). *J. Cell Biol.* **100**, 851–859.

Taylor, D. L., Amato, P. A., Luby-Phelps, K., and McNeil, P. L. (1985). *Trends Biochem. Sci.* **9**, 88–91.

Todaro, G. J., Lazar, G., and Green, H. (1965). *J. Cell. Comp. Physiol.* **66**, 325–334.

Vaheri, A., and Pagano, J. S. (1965). *Virology* **27**, 435–436.

Wang, Y.-L., Heiple, J., and Taylor, D. L. (1982). *Methods Cell Biol.* **24**, 1–11.

Wong, T.-K., and Neumann, E. (1982). *Biochem. Biophys. Res. Commun.* **107**, 584–587.

# Chapter 11

# Hapten-Mediated Immunocytochemistry: The Use of Fluorescent and Nonfluorescent Haptens for the Study of Cytoskeletal Dynamics in Living Cells

GARY J. GORBSKY AND GARY G. BORISY

*Laboratory of Molecular Biology*
*University of Wisconsin*
*Madison, Wisconsin 53706*

## I. Introduction

Hapten-mediated immunocytochemistry (HMI) is an alternative and a complement to the direct imaging of fluorescent analogs in living cells.

175

HMI has evolved from the use of antifluorescein antibodies to investigate hapten–antibody interactions (Lopatin and Voss, 1971), to block the activity of fluorescein-labeled ligands (Sklar *et al.*, 1981), and to construct affinity matrices for labeled proteins (Luna *et al.*, 1982). HMI can increase the signal-to-noise ratio to improve resolution and obviate photodamage in studying the localization and dynamics of both fluorescent and nonfluorescent analogs. Briefly, the procedure is as follows. An analog is first produced by derivatization of the native macromolecule with a fluorescent or nonfluorescent haptenic group. The analog is then incorporated into living cells and, after a time, the cells are fixed. To reveal the distribution of the analog, the cells are labeled with antibodies to the hapten and in turn with secondary antibodies conjugated to markers visible by light or electron microscopy (see Luby-Phelps *et al.*, 1984; Spiegel *et al.*, 1984; Gorbsky *et al.*, 1987).

HMI can be used for two general purposes: to study the equilibrium distribution of analogs introduced into living cells and to analyze their dynamics. While the technique is applicable to analogs of many cellular constituents, we will illustrate its usefulness with examples from the study of microtubule biology.

For equilibrium localization, cells are allowed to incorporate the analog for a period of time prior to fixation and immunolabeling. For analyzing cellular dynamics two time points must be chosen. The final time point is always the moment of fixation. The initial time point can be selected in a number of ways. One method is to choose the initial time as the moment the analog is introduced into the cell, e.g., the time of microinjection. Alternatively, the analog can be allowed to equilibrate within the cell and the initial time chosen at the moment of some cell stimulus such as the application of a hormone. A more precise method for choosing an initial time point in studying cellular dynamics when a fluorescent analog is used is through fluorescence photobleaching. Antibodies prepared against native fluorescein show little binding to bleached fluorescein (Gorbsky *et al.*, 1987). Thus, after indirect immunolabeling, a photobleached region is reflected in the final specimen as an area of reduced secondary label.

## II.   Advantages of Hapten-Mediated Immunocytochemistry

While many cellular processes can be studied by direct observation of fluorescent analogs in living cells (see Wang, Chapter 1, this volume), images of relatively low signal-to-noise ratio are often obtained due to background fluorescence. For example, microinjection of cytoskeletal analogs generally results in a diffuse fluorescent background due to the normal cellular equilibrium between polymerized and unpolymerized protein.

A second problem with direct fluorescent observation is that microinjection of a fluorophore and illumination of living cells for fluorescence can lead to photodamage (see Taylor and Salmon, Chapter 13, this volume). In addition, the illumination used in direct imaging can cause bleaching of the analog resulting in attentuation of the signal. For these reasons video systems of very high sensitivity are required for direct imaging. These systems generally produce fairly noisy images that must be improved through digital processing.

HMI is an approach that is complementary to the direct imaging of fluorescent analogs in living cells and includes a number of distinct advantages. Chief among these is the ability to analyze cellular dynamics with the electron microscope. However, HMI also holds some significant advantages in light microscopy. HMI is capable of detecting both fluorescent and nonfluorescent analogs. For some cellular components such as the cytoskeleton where structures of interest are detergent-insoluble, background due to denatured and unincorporated analog can be greatly reduced by permeabilization and extraction of cells prior to fixation. In many instances, this preextraction step greatly improves visualization of the distribution of a fluorescent analog in detergent-insoluble structures even when subsequent antibody labeling is not used. However, significant amplification of the signal from a fluorescent analog is achieved with immunolabeling. This amplification occurs because each primary antibody directed against the fluorophore can bind several secondary antibodies. Tertiary complexes involving an additional layer of antibodies, Protein A, or biotin–avidin reagents can produce yet stronger signals. Fluorescence signals from fixed specimens may also be optimized by including antibleaching agents and adjusting the pH of the mounting medium.

Another advantage of HMI is that photodamage can often be completely avoided because cells need not be illuminated for fluorescence while alive. Unlike direct fluorescence observation of living cells in which the sensitive video systems and digital image processors are generally necessary, HMI requires only the use of standard fluorescence or electron microscopes. Finally, direct fluorescence imaging and HMI are not mutually exclusive methods. Cells that have been imaged by fluorescence in the living state can be subsequently fixed and immunolabeled for more detailed analysis by light or electron microscopy (Luby-Phelps *et al.*, 1984; Amato and Taylor, 1985; Gorbsky *et al.*, 1987).

Naturally, the advantages of HMI are not realized without some cost. First, it should be noted that fixation, permeabilization, and the processing of cells for microscopy may cause alterations in cell shape or could possibly affect the distribution of the structures of interest. Second, the direct fluorescent observation of living cells generally allows the recording of multiple images from the same cell, thus revealing changes with time in the distribution of the analog. Unless combined with direct fluorescent imaging, HMI generally relies upon phase contrast or differential interference contrast to provide information about the cell in the living state. Only after the cell is fixed is the distribution of the analog known. Thus, to establish a sequence of changes in the distribution of an analog it is necessary to examine a series of cells fixed at varying times.

# III. Preparation of Reagents

## A. Choice of Hapten

Of fluorescent haptens studied, the most well known is fluorescein. Numerous reactive fluorescein derivatives are available commercially (Molecular Probes, Eugene, Oregon; Research Organics, Cleveland, Ohio). Fluorescein is strongly antigenic, and antibodies of high affinity are easily produced (Kranz and Voss, 1984). Anti-fluorescein antibodies are also available commercially (Chemicon, El Segundo, California; Reagents International, Windham, Maine). As noted previously, fluorescein antibodies have low affinity for bleached fluorescein, a feature of great usefulness in photobleaching studies. Antibodies, including monoclonal antibodies, to fluorophores other than fluorescein have been produced (Voss, 1984; Haaijman et al., 1986) and have been used for HMI (Spiegel et al., 1984).

If a fluorescent hapten is not required, a wider choice is available. Basically any small group to which antibodies may be produced or purchased is suitable if it can be conjugated to the macromolecule of interest for incorporation into the living cells. The most commonly used molecule of this class is biotin. Biotin can be gently linked to proteins (Bonnard et al., 1984; Schulze and Kirschner, 1986) and incorporated into nucleotides (Manuelidis et al., 1982). Forms containing spacer arms are available (Enzo Biochem, Inc., New York, New York; Pierce Chemical Co., Rockford, Illinois) that permit greater accessibility to affinity probes. Biotin can be localized through the use of antibodies, with avidin from egg white or with streptavidin from Streptomyces avidini. These reagents are available commercially from a wide variety of suppliers.

## B. Antibody Preparation

Anti-hapten antibodies are prepared by coupling the hapten to a carrier macromolecule, generally a protein, and injecting the hapten-carrier conjugate into suitable animals for the production of immune serum. Carrier proteins will in general be recognized as foreign by the immune system of the host. Because the serum will contain antibodies to the carrier any proteins that might be present in the cells to be studied or in their medium (e.g., bovine serum albumin) should be avoided. Keyhole limpet hemocyanin has proven to be a useful carrier for many types of hapten.

The reaction used to link the hapten to the carrier is of great importance for the successful production of antibodies. A wide variety of strategies are available (for review, see Makela and Seppala, 1986). Many haptens containing reactive groups are commercially available. These need only to be added to the carrier under the proper conditions. Generally, derivatization of the carrier is performed under conditions designed to maximize the hapten-to-carrier ratio. (> 20 : 1). Some linkers appear to result in conjugates that are more immunogenic than others even when the identical hapten is coupled. In the case of fluorescein it was found that injection of rabbits with conjugates prepared with the isothiocyanate resulted in high affinity antibodies while the injection of conjugates made with dichlorotriazinylamino fluorescein failed to elicit antibodies (Voss, 1984). The reason for the failure of the latter conjugate to elicit antibody production is unknown. However, the antibodies prepared from the isothiocyanate conjugate did bind to test proteins made with dichlorotriazinlyamino fluorescein.

A possible complication arises if the linker used in the construction of the hapten-carrier antigen is itself immunogenic. If the identical linker is used in the preparation of the analog that is to be introduced into the cells, antibodies may bind the linker rather than the hapten (Briand et al., 1985). In many cases the extra binding will not cause a problem and may in fact increase the signal. However, if the experiments involve the destruction of a photolabile antigen such as the photobleaching of fluorescein, antibodies to the linker would be detrimental. In this situation, it is preferable to use different linking reactions for the preparation of the hapten-carrier conjugate and for the analog that is to be introduced into the cells.

Preimmune and immune sera collected from animals are tested for their ability to bind to hapten conjugated to a test protein. The test protein should be different from the carrier, bovine serum albumin being an acceptable example. Again it may be preferable to conjugate the hapten to the test protein by an alternative linking chemistry from that used to prepare the hapten carrier. Sera may be assayed in a number of ways, e.g., Ouchterlony double diffusion, ELISA in microtiter plates, or dot blotting on nitrocellulose. Antibodies to fluorophores may also be tested by their ability to alter the fluorescence of the fluorophore in solution (Lopatin and Voss, 1971).

In many cases, whole sera appropriately diluted and preabsorbed can be used directly. If further purification of antibodies is desired, immunoglobulin fractions may be isolated in the conventional manner by salt fractionation and DEAE chromatography. Affinity purification of anti-hapten antibodies may be performed on columns prepared by binding the hapten via a spacer arm to agarose beads (Kranz and Voss, 1984). Affinity purification is of value when it is necessary to prepare labeled primary antibodies, but it should be noted that the highest affinity antibodies in a serum often fail to elute from affinity columns and are thus lost (Hudson and Hay, 1980).

## C.   Secondary Labels

Three general classes of secondary reagents are in common use in immunohistochemistry. These are antiimmunoglobulin antibodies, Protein A which binds the Fc region of some but not all classes of IgG from several species, and avidin or streptavidin which bind to biotin. While all these reagents have potential use in hapten-mediated immunocytochemistry, most studies have relied upon secondary antibodies (Luby-Phelps et al., 1984; Spiegel et al., 1984; Soltys and Borisy, 1985; Gorbsky et al., 1987; Sammak et al., 1987) even when biotin was the hapten (Mitchison et al., 1986; Schulze and Kirschner, 1986). Affinity-purified antibodies made in a variety of host species are commercially available from a large number of sources.

The reporter groups, the indicators to which the secondary label is coupled to render it visible, are chosen based on the level of resolution desired (light or electron microscopic), and the sensitivity necessary to detect the introduced analog. Generally, the haptenized analog is introduced into the cells in tracer amounts compared to the endogenous level of the macromolecule. For this reason, the most sensitive detection methods are often required. For light microscopy, immunofluorescence appears to provide the clearest signal. High-quality fluorescent secondary antibodies are readily available from commercial sources and their use is straightforward.

At the electron microscopic level, studies involving hapten-mediated immunocytochemistry have been done with secondary antibodies coupled to horseradish peroxidase (Luby-Phelps et al., 1984) and colloidal gold (Mitchison et al., 1986). Peroxidase has the advantage of generating signals detectable by both light and electron microscopy. On the other hand, at the electron microscopic level, the reaction product is often diffuse and possesses no characteristic morphology that allows it to be unambiguously distinguished from heavy metal staining of the cytoplasm. From the standpoint of an unequivocal ultrastructural marker, colloidal gold is ideal. It also functions well at the light microscopic level when intensified by photochemical silver reaction (Holgate *et al.*, 1983; Springall *et al.*, 1984). Recently, kits that permit the silver enhancement to be done under room lights have become commercially available (E. Y. Laboratories, San Mateo, California; Janssen Pharmaceutica, Piscataway, New Jersey). The most significant disadvantage of colloidal gold is that, in some instances, penetration of the probes into aldehyde cross-linked cytoplasm is incomplete (Wolosewick *et al.*, 1983). Colloidal gold-secondary antibodies can be purchased but are expensive. With some practice they can be routinely prepared in varying sizes (DeMey, 1983; Slot and Geuze, 1985). The most useful sizes for hapten-mediated immunocytochemistry with the transmission electron microscope are from 4 to 20 nm.

## IV. Application of Hapten-Mediated Immunocytochemistry

### A. Derivatization of Analogs and Incorporation into Cells

A flow chart depicting the steps in processing samples for HMI is shown in Fig. 1. The derivatization of fluorescent analogs and the incorporation of analogs into living cells are topics covered in detail elsewhere in this volume. With HMI, use of nonfluorescent analogs is possible. For derivatization of proteins, biotin has many advantages. Proteins may be labeled in a mild reaction at neutral pH with the *N*-hydroxysuccinimide form. A photoactivatible reagent is also available for derivatizing nucleic acids and proteins (Pierce Chemical Co., Rockford, Illinois).

In addition to haptenic groups that can be chemically coupled to macromolecules, nucleic acid analogs can be prepared through the introduction of modified nucleotides during *in vitro* polymerization (Manuelidis *et al.*, 1982) or by metabolic incorporation in living cells (Gratzner, 1982; Ito and McGhee, 1987). Antibodies specific for the modified nucleotides are then used for immunolabeling.

### B. Fixation

The goal of fixation in HMI is to preserve cell structure, maintain antigenicity of the hapten, and permit access of immunological probes. Fixation should rapidly immobilize the analog without introducing artifacts. To some extent these objectives are mutually incompatible and a reasonable compromise must be achieved. The degree of structural preservation required will be determined by the level of resolution (light or electron microscopy) and by the biological question being addressed. The

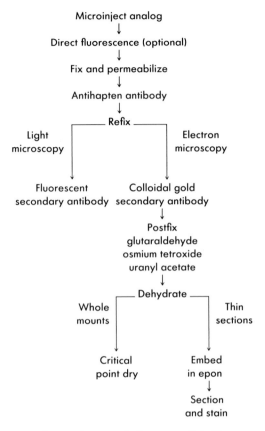

Microinject analog
↓
Direct fluorescence (optional)
↓
Fix and permeabilize
↓
Antihapten antibody
↓
Refix

Light                                    Electron
microscopy                               microscopy

Fluorescent              Colloidal gold
secondary antibody   secondary antibody
↓
Postfix
glutaraldehyde
osmium tetroxide
uranyl acetate
↓
Dehydrate

Whole                                    Thin
mounts                                   sections

Critical                  Embed
point dry                 in epon
↓
Section
and stain

FIG. 1.    Flow chart of processing steps for hapten-mediated immunocytochemistry.

fixation protocol may also be designed to remove certain amounts of material from the living cell. For example, in studies of the cytoskeleton, it is often desirable to eliminate unincorporated analog molecules by extraction of cells with detergent prior to fixation.

Protein fixatives routinely used for immunocytochemistry of cell cultures can be expected to be useful for HMI. Naturally if other cellular constituents, e.g., nucleic acids or lipids, are under investigation, appropriate fixation methods must be developed. The protein fixatives are of two general classes: precipitants and covalent cross-linkers. The most commonly used precipitants are methanol and acetone applied to cell cultures at $-20°C$. These fixations are simple, appear to have little effect upon antigenicity, but lead to unacceptable ultrastructural preservation.

The most commonly used covalent cross-linkers are formaldehyde and glutaraldehyde and combinations thereof. Formaldehyde, freshly prepared from paraformaldehyde, is generally applied at a concentration of 1 to 4%. Formaldehyde usually produces somewhat better structural preservation than the precipitants although ultrastructural preservation is still suboptimal. High-quality glutaraldehyde (electron microscopic grade) at 0.5–2% produces the best ultrastructural preservation but may have adverse effects on antigenicity. However, with HMI, the antigen is completely defined, and tests on the effects of glutaraldehyde or other fixatives on antibody binding are easy to perform by use of test proteins conjugated to hapten. Glutaraldehyde also induces a high level of background fluorescence. This background can be reduced although not completely eliminated by treatment of samples with sodium borohydride (Osborn and Weber, 1982).

While it is beyond the scope of this chapter to describe the various alternative fixatives that have been employed for immunohistochemistry and thus may be useful for HMI, other fixations may be suitable for particular cellular components. For example, microtubules are notoriously difficult to preserve. Glutaraldehyde results in excellent preservation but has the disadvantage of inducing considerable background fluorescence. We discovered that, for light microscopy, fixation of cytoskeletons with ethylene glycolbis (succinic acid N-hydroxysuccinimide ester) (Sigma Chemical Co., St. Louis, Missouri) produced excellent preservation of the microtubule network (Gorbsky et al., 1987).

Another major consideration in designing a fixation protocol for HMI is optimizing the penetration of the applied antibodies. Apart from target molecules exposed on the upper surfaces of cultured cells, permeabilization of the plasma membrane is generally required. Methanol or acetone fixations generally result in permeabilization through extraction of membrane lipids. Other protocols require the use of detergents applied before, during, or after fixation. Extraction of cells prior to fixation leaves behind only the detergent-insoluble cytoskeleton. This method also results in the most thorough permeabilization. The use of detergents during or after fixation generally permits adequate penetration of fluorescent and enzyme-linked antibodies, but the complete penetration of colloidal gold-coupled antibodies is less certain.

## C. Immunolabeling

The goal in immunolabeling is to maximize specific binding of primary and secondary antibodies and to minimize any nonspecific binding that contributes to background. For this reason antibodies of the highest titer that can be used at the greatest dilutions are desirable. For HMI it is easy to prepare large amounts of pure antigen containing a highly immunogenic hapten and thus obtain high titer antibodies. In addition, for many haptens, quality antibodies can be purchased. Because immunolabeling requires the use of relatively little antibody, commercial sources are frequently the most efficient means of obtaining both primary and secondary antibodies.

Another means for decreasing the background binding of both primary and secondary antibodies is preadsorption. Because the hapten is not a normal cellular component, monolayers or pellets of the cells used for study provide an ideal material for preadsorption. Antibodies may also be treated with acetone powders prepared from normal tissue such as liver.

Regardless of whether preadsorptions are carried out, one of the most essential ingredients for reduction of background labeling is the use of a blocking protein. The alternatives include bovine serum albumin, ovalbumin, gelatin, nonfat dehydrated milk, or normal serum from the animal in which the secondary antibody was produced. These agents, at concentrations of from 0.1 to 10% are used to dilute antibodies and are sometimes included in the washing solutions. Background is further reduced if cells are first incubated in the blocking protein alone for a short time to saturate nonspecific protein binding sites prior to the application of the primary antibodies.

One problem we have encountered when using antifluorescein followed by a fluoresceinated secondary antibody is the tendency for antibody complexes to artifactually clump during the immunolabeling. For example, in cells injected with fluorescein-tubulin, microtubules labeled with antitubulin appeared as long continuous fibers. In contrast, labeling with antifluorescein and fluoresceinated secondary antibody resulted in a punctate pattern. However, when a brief secondary fixation was used just after application of the primary antifluorescein, the labeling was continuous, presumably because rearrangement of primary antibody was avoided. A second fixation step is also often included after application of colloidal gold antibodies to prevent their loss from the cells during subsequent processing for electron microscopy.

One of the most important features of HMI is the ability to double-label specimens for the hapten and for the endogenous macromolecule. For example, upon equilibration an analog should be distributed in all areas where the endogenous molecules are found. To test this equilibration, a double-label experiment is performed. The easiest method for double labeling is the use of primary antibodies prepared in different host animals. Secondary antibodies with different fluorophores or coupled to different-sized gold particles are then applied. Because the hapten is generally present in lesser amounts than the endogenous protein, we normally apply the antihapten primary and secondary before labeling for the endogenous molecule. In this manner we optimize binding of the antihapten antibody, the one apt to generate the weaker signal, and we avoid steric blocking of its binding sites that might occur if both antihapten and antiendogenous molecule were applied together.

## D.   Final Sample Preparation

For light microscopy (Fig. 2), final sample preparation consists of mounting cover slips containing cells in a medium that optimizes the signal from the fluorescent antibodies. This is generally accomplished for fluorescein by adjusting the pH to about 8.5–9.0 and adding antibleaching agents. Some mounting formulations result in superior phase-contrast images. The formula we use is one based on the use of polyvinyl alcohol (Sammak *et al.*, 1987). When dried at room temperature overnight this material hardens at the edges of the cover slip to immobilize it to the slide. Cover slips so prepared are preserved for weeks if kept at −20°C. Eventually the anti-

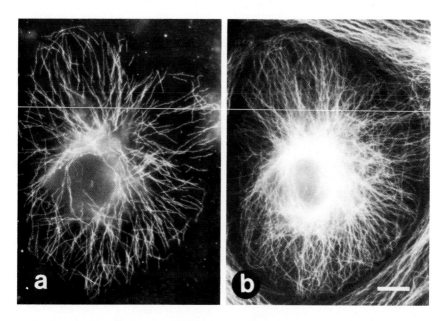

Fig. 2.   Visualization of microinjected biotin–tubulin by light microscopy. A TC-7 African green monkey kidney cell was injected with biotin–tubulin at 22°C. Approximately 2 minutes later the cell was lysed with detergent in microtubule-stabilizing buffer and fixed with ethylene glycol bis-(succinic acid *N*-hydroxy succinimide ester). To reveal the distribution of the injected biotin–tubulin, cells were labeld with Texas Red–streptavidin (a). To reveal the cellular microtubules, a rat monoclonal antitubulin antibody and fluorescein anti-rat IgG were applied (b). The biotin–tubulin have incorporated onto the ends of preexisting microtubules. ×900. Bar, 10 μm. (Micrographs courtesy of Dan Webster.)

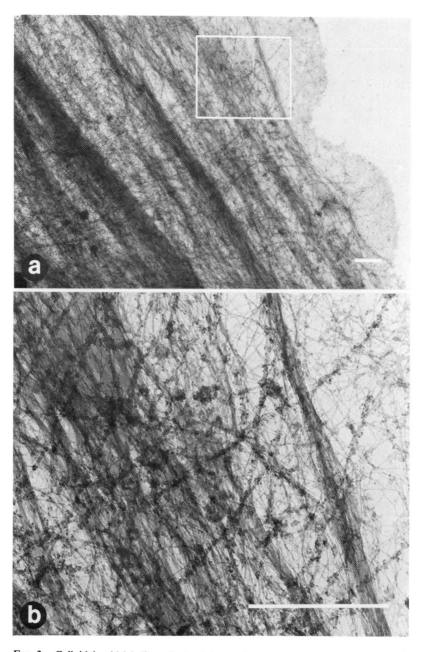

FIG. 3. Colloidal gold labeling of microinjected fluorescein–tubulin. An LLC-PK cell, grown on a formvar- and carbon-coated grid, was injected with fluorescein–tubulin, allowed to equilibrate, lysed with detergent, and fixed with glutaraldehyde. The cell was then labeled with rabbit antifluorescein followed by anti-rabbit IgG adsorbed to 8-nm colloidal gold particles. After postfixation, the specimen was dehydrated and preserved by critical point drying. (a) Low magnification electron micrograph of area near the cell edge. (b) Enlargement of boxed region of a showing colloidal gold labeling of microtubules but not other, thinner filaments. Viewed at 1000 kV with AEI-7 high voltage electron microscope. Bars, 1.0 μm.

bleaching agent may begin to develop an unacceptable background fluorescence. The sample can be rescued by floating off the cover slip in a dish of warm water, washing extensively, and remounting in fresh mounting medium.

The preparation of samples for HMI with the electron microscope is more demanding. If cells have been grown on formvar and carbon-coated grids they may be prepared as whole mounts (Fig. 3). After immunolabeling the cells are postfixed with glutaraldehyde and then treated with osmium tetroxide and uranyl acetate. They are quickly dehydrated through a graded series of ethanol to 100% ethanol that has been stored over molecular sieve. The samples are then critical point dried, taking care to avoid the introduction of moisture during the process (Ris, 1985). The cell may be lightly coated with carbon to stabilize structure, and the grids are stored in a dessicator until observation.

Cells to be sectioned are grown on carbon-coated cover slips. Identification of injected cells is greatly facilitated if a 400-mesh locator grid (Ted Pella, Inc., Tustin, California) is placed on the cover slip while coating with carbon. When the grid is removed a carbon impression of the locator pattern remains on the cover slip. The cover slips are baked (180°C overnight) to adhere the carbon. After immunolabeling, cells are postfixed with glutaraldehyde, treated with osmium and uranyl acetate, and dehydrated. The cells are then infiltrated and embedded in epon. Inversion of a Beem capsule, partially filled with epon over the locator-pattern region of the cover slip, facilitates subsequent removal of the cover slip and allows the pattern to be examined for trimming with an inverted microscope. Freeing the cover slip from the epon is accomplished by dipping the sample in liquid nitrogen. The samples are then trimmed, sectioned, and stained by conventional means.

## V. Conclusion and Future Outlook

HMI is at present a useful complement to the direct imaging of fluorescent analogs. It provides a bridge integrating the powerful tools of immunocytochemistry and electron microscopy with the study of dynamic processes within the living cell. Currently, the use of the photolabile hapten, fluorescein, provides a powerful means for marking cellular macromolecules at a specific time point in order to analyze their movements. An example of the use of fluorescein-tubulin to study microtubule movements in mitosis is provided in Figs. 4 and 5.

Future developments should enhance the utility of HMI. For example, attempts are being made to develop photoactivable fluorophores (Ware *et*

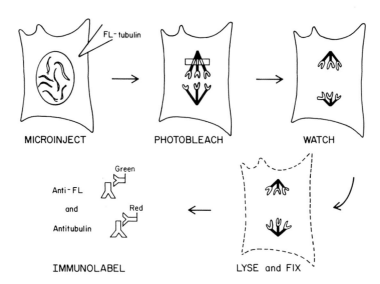

FIG. 4. Design of a photobleaching experiment with the use of fluorescein (FL)-tubulin and HMI. Cells are microinjected with fluorescein–tubulin during prophase and photobleached with a bar-shaped laser beam just after the onset of anaphase. Cells are observed with phase contrast microscopy until late anaphase when they are lysed with detergent and fixed. The cells are then labeled with rabbit antifluorescein and fluoresceinated anti-rabbit antibody. This is followed by the application of rat antitubulin and Texas Red anti-rat antibody.

*al.*, 1986). Similarly, the introduction of activable hapten derivatives would be a significant advance. These haptens would become capable of binding antibody only after a specific stimulus, for example, the photolysis of a particular bond brought about by laser illumination. Unlike the photobleaching experiments with fluorescein, photoactivating a hapten would result in a positive signal on a negative background and thus be more clearly discernible. Other improvements in HMI will be forthcoming with advances in conventional immunocytochemistry and in the preparation and use of fluorescent and nonfluorescent analogs.


# VI.   Technical Appendix

## A.   Preparation of Antifluorescein in Rabbits

### 1.   PREPARATION OF IMMUNOGEN

Dissolve 50 mg of keyhole limpet hemocyanin (Calbiochem, San Diego, California) in 5 ml of 0.25 $M$ carbonate buffer, pH 9.0. While vortexing,

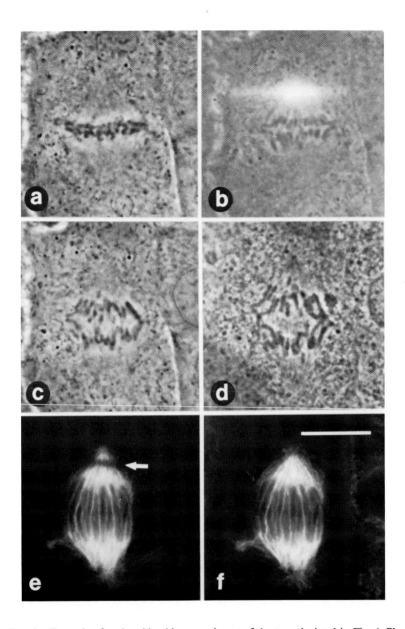

FIG. 5. Example of a photobleaching experiment of the type depicted in Fig. 4. Phase contrast micrograph of living LLC-PK cell at the onset of anaphase (a). This cell had previously been injected with fluorescein–tubulin. A short time later, the laser was used to bleach a band across the upper half spindle. The light emitted from the fluorescein during the bleaching is superimposed on a phase image of the cell (b). The living cell continued in anaphase and is shown 30 seconds after photobleaching (c). At 60 seconds, the cell was lysed in detergent and fixed (d). The cell was then processed for indirect immunofluorescence with antifluorescein antibodies (e) and antitubulin antibodies (f). Arrow in e indicates the bleached domain of the microtubule bundles. The results of such experiments showed that the microtubule bundles did not move toward the poles as the chromosomes advanced in anaphase. ×1914. Bar, 10 μm.

add 20 mg of fluorescein isothiocyanate, isomer 1 (Molecular Probes, Eugene, Oregon). Incubate overnight at room temperature. Add 1 ml of 1.0 $M$ Tris buffer, pH 8.0, and spin at 48,000 $g$ for 15 minutes to remove large aggregates. Chromatograph the sample on a 20-ml column of Sephadex G-25 (Pharmacia, Piscataway, New Jersey) equilibrated with 140 m$M$ NaCl, 3 m$M$ KCl, 10 m$M$ PO$_4$, pH 7.3 (PBS). A clear separation should be seen between the leading yellow peak that consists of the fluorescein protein conjugate and the trailing peak of free fluorescein. If necessary, rechromatograph the conjugate.

### 2.    INJECTION OF IMMUNOGEN INTO RABBITS AND COLLECTION OF SERA

Test preimmune sera at a 1 : 10 dilution by immunofluorescence on the cells to be studied. Those animals whose sera exhibit unacceptably high background levels of staining are discarded. Prior to injection, adjust a portion of the hapten carrier to a protein concentration of 2 mg/ml and then mix at a 40 : 60 ratio Freund's complete adjuvant to form an emulsion. This may be accomplished by extensive vortexing at high speed or by use of two syringes connected by a double Luer lock barrel. Using a total of 0.8 mg hapten-carrier conjugate, inject rabbits intradermally at multiple sites along the back and once intramuscularly in the thigh of a hind leg. Boost at 3- to 4-week intervals with injections of 0.5 mg of protein emulsified with Freund's incomplete adjuvant. After the second boost, collect sera and test for reaction with fluorescein-conjugated bovine serum albumin by Ouchterlony double diffusion and by dot blotting. Bleed those animals exhibiting the best responses several times to accumulate sera. Aliquot and store sera at $-70°$C.

### 3.    PURIFICATION OF IMMUNOGLOBULIN FRACTION

Precipitate immunoglobulins from 50 ml of crude serum by adding an equal volume of saturated ammonium sulfate at room temperature. After 30 minutes centrifuge the precipitate at 3000 $g$ for 30 minutes at room temperature. Resuspend the precipitate in 25 ml of PBS and reprecipitate by adding 20 ml of saturated ammonium sulfate. After centrifugation, resuspend the precipitate in 12 ml of 0.08 $M$ phosphate buffer, pH 8.0, and dialyze extensively against the same buffer. Equilibrate 50 ml, settled wet volume of DEAE-cellulose (Whatman DE-52) with the 0.08 $M$ phosphate buffer, and pack the ion exchanger into a column. Apply the sample and collect the initial flow through peak. This material is the IgG fraction and is relatively free of contaminants. Dialyze into PBS, aliquot, and store at $-70°$C.

## B.  Hapten-Mediated Immunocytochemistry with Fluorescein–Tubulin

### 1.  REAGENTS

Secondary antibodies, unlabeled and conjugated to fluorophores are obtained from commercial sources. Colloidal gold-coupled secondary antibodies are prepared by the method of Slot and Geuze (1985).

### 2.  PREADSORPTION OF PRIMARY AND FLUOROPHORE-CONJUGATED SECONDARY ANTIBODIES

Dilute immune serum or antibody $1:20$ with 10% normal goat serum. Add 250 mg/ml goat liver acetone powder (Sigma Chemical Co., St. Louis, Missouri). Incubate 1 hour on rocker table at room temperature. Spin 18,000 rpm in Sorvall at $4°C$ for 1 hour. Remove supernatant. Spin again or filter through 0.22 $\mu$m filter. Aliquot, freeze, and store at $-70°C$.

### 3.  EXTRACTION AND FIXATION OF CELLS

Rinse cells on cover slips twice in 60 m$M$ PIPES, 25 m$M$ HEPES, 10 m$M$ EGTA, 2 m$M$ MgCl$_2$ (PHEM) at room temperature and lyse for 90 seconds with a solution consisting of 0.5% Triton X-100 in PHEM. For light microscopy, fix cells for 20 minutes in 5 m$M$ ethyleneglycolbis-(succinic acid $N$-hydroxy succinimide ester) (Sigma Chemical Co., St. Louis, Missouri) in PHEM. The fixative is prepared as a 100-m$M$ stock in DMSO and diluted into PHEM just before application to the cells. For electron microscopy, fix cells in 1.0% glutaraldehyde in PHEM for 15 minutes. After three rinses with PBS, glutaraldehyde-fixed cells are treated with 2 mg/ml NaBH$_4$ in water for 15 minutes. Transfer cover slips to PBS where they may be stored before immunolabeling.

### 4.  IMMUNOLABELING

Invert cover slips containing cells onto drops of 25% goat serum in PBS on parafilm for 30 minutes at $37°C$. Rinse in PBS and incubate cover slips inverted on 40 $\mu$l of antifluorescein serum (our preparation at a final total dilution of $1:500$ to $1:1000$ in 10% goat serum). Wash in PBS for 30 minutes. Refix with 2.5 m$M$ ethyleneglycolbis (succinic acid $N$-hydroxy succinimide ester) in PBS for 10 minutes and wash again for 10 minutes in PBS. For light microscopy, incubate in fluorophore-conjugated secondary antibody for 45 minutes at $37°C$. Fluorophore-labeled secondaries are used at 20 $\mu$g/ml in 10% goat serum. For electron microscopy, colloidal gold secondaries, generally diluted to a medium pink color with 10% normal goat serum, are incubated about 2 hours at 37°C. Wash again in PBS. If desired, apply antitubulin primary and appropriate secondary antibodies for double labeling. For light microscopy, mount cover slips in 10% polyvinyl alcohol with antibleaching agent (for formula, see Sammak *et al.*, 1987). For electron microscopy, postfix cells with 2% glutaraldehyde, 0.2% tannic acid in PBS, rinse several times with water, and treat with 0.2% osmium tetroxide, 1.0% uranyl acetate in water for 30 minutes. After dehydration through 50, 70, and 95% and three changes of dry 100% ethanol, critical point dry or embed samples for sectioning.

## ACKNOWLEDGMENTS

We thank Dan Webster, Laboratory of Molecular Biology, University of Wisconsin, for contributing original micrographs. We also thank the staff of the Integrated Microscopy Resource for Biomedical Research, University of Wisconsin, for assistance with the high voltage electron microscopy and Steve Limbach and Leslie Rabbas for technical illustration.

This work was supported by National Institutes of Health grants GM25062 to G. G. Borisy and RR00570 to the Integrated Microscopy Resource for Biomedical Research.

## REFERENCES

Amato, P. A., and Taylor, D. L. (1985). *J. Cell Biol.* **102,** 1074–1084.

Bonnard, C., Papermaster, D. S., and Kraehenbuhl, J.-P. (1984). *In* "Immunolabeling for Electron Microscopy" (J. M. Polak and I. M. Varndell, eds.), pp. 95–110. Elsevier, Amsterdam.

Briand, J. P., Muller, S., and Van Regenmortel, M. H. V. (1985). *J. Immunol. Methods* **78,** 59–69.

DeMey, J. R. (1983). *In* "Immunohistochemistry" (A. C. Cuello, ed.), pp. 347–372. Wiley, New York.

Gorbsky, G. J., Sammak, P. J., and Borisy, G. G. (1987). *J. Cell Biol.* **104,** 9–18.

Gratzner, H. G. (1982). *Science* **218,** 474–475.

Haaijman J. J., Coolen, J., Krose, C. J. M., Pronk, G. J., and Ming, Z. F. (1986). *Histochemistry* **84,** 363–370.

Holgate, C. S., Jackson, P., Cowen, P. N., and Bird, C. C. (1983). *J. Histochem. Cytochem.* **31,** 938–944.

Hudson, L., and Hay, F. C. (1980). "Practical Immunology." Blackwell, Oxford.

Ito, K., and McGhee, J. D. (1987). *Cell* **49,** 329–336.

Kranz, D. M., and Voss, E. W. (1984). *In* "Fluorescein Hapten: An Immunological Probe" (E. W. Voss, ed.), pp. 15–22. CRC Press, Boca Raton, Florida.

Lopatin, D. E., and Voss, E. W. (1971). *Biochemistry* **10,** 208–213.

Luby-Phelps, K., Amato, P. A., and Taylor, D. L. (1984). *Cell Motil.* **4,** 137–149.

Luna, E. J., Wang, Y. L., Voss, E. W., Branton, D., and Taylor, D. L. (1982). *J. Biol. Chem.* **257,** 13095–13100.

Makela, O., and Seppala, I. J. T. (1986). *In* "Handbook of Experimental Immunology" (D. M. Weir, and L. A. Herzenberg, eds.), pp. 3.1–3.13. Blackwell, Oxford.

Manuelidis, L., Langer-Safer, P. R., and Ward, D. C. (1982). *J. Cell Biol.* **95**, 619–625.

Mitchison, T., Evans, L., Schulze, E., and Kirschner, M. (1986). *Cell* **45**, 515–527.

Osborn, M., and Weber, K. (1982). *Methods Cell Biol.* **24**, 97–132.

Ris, H. (1985). *J. Cell Biol.* **100**, 1474–1487.

Sammak, P. J., Gorbsky, G. J., and Borisy, G. G. (1987). *J. Cell Biol.* **104**, 395–405.

Schulze, E., and Kirschner, M. (1986). *J. Cell Biol.* **102**, 1020–1031.

Sklar, L. A., Oades, Z. G., Jesaitis, A. J., Painter, R. G., and Cockran, C. G. (1981). *Proc. Natl. Acad. Sci. U.S.A.* **78**, 7540–7544.

Slot, J. W., and Geuze, H. J. (1985). *Eur. J. Cell Biol.* **38**, 87–93.

Soltys, B. J., and Borisy, G. G. (1985). *J. Cell Biol.* **100**, 1682–1689.

Spiegel, S., Schlessinger, J., and Fishman, P. H. (1984). *J. Cell Biol.* **99**, 699–704.

Springall, D. R., Hacker, G. W., Grimelius, L., and Polak, J. M. (1984). *Histochemistry* **81**, 603–608.

Voss, E. W. (1984). *In* "Fluorescein Hapten: An Immunological Probe" (E. W. Voss, ed.), pp. 3–14. CRC Press, Boca Raton, Florida.

Ware, B. R., Brvenik, L. J., Cummings, R. T., Furukawa, R. H., and Kraft, G. A. (1986). *In* "Applications of Fluorescence in the Biomedical Sciences" (D. L. Taylor, A. S. Waggoner, R. F. Murphy, F. Lanni, and R. R. Birge, eds.), pp. 141–157. Liss, New York.

Wolosewick, J. J., DeMey, J., and Meininger, V. (1983). *Biol. Cell* **49**, 219–226.

# Chapter 12

# Culturing Cells on the Microscope Stage

NANCY M. MCKENNA AND YU-LI WANG

*Worcester Foundation for Experimental Biology*
*Shrewsbury, Massachusetts 01545*

## I. Introduction

Keeping live, healthy cells on the microscope is crucial for many types of experiments, in particular, fluorescence microscopy of dynamic processes. Although cultured cells may be observed for a short period of time without any special device, a culture system is usually necessary for observations which last for more than 10 minutes. Since the development of tissue culture techniques, many different approaches for this purpose have been taken. In this chapter, we will review the requirements which govern the design of microscope culture systems and discuss in detail several different systems, including one designed and used successfully in this laboratory. We will, however, not describe simple chambers for short-term maintenance of living cells (see, e.g., McNeil, Chapter 10; Chen, Chapter 7; and Swanson, Chapter 9, this volume, for several designs of simple observation chambers).

195

## II.  Requirements of the Microscope Culture System

### A.  Requirements for Maintaining the Cell

The major requirements for maintaining the cell are temperature, pH, osmolarity, and nutrients. Cultured mammalian cells should generally be maintained at a temperature close to 37°C. However, with the exception of temperature-sensitive mutants, most cells can tolerate a slightly lower temperature (e.g., 32–34°C), as well as slight variations in the temperature (±1°C).

Generally, pH of the medium should be maintained between 7.2 and 7.4. Although cells may be kept in a nonbicarbonate-based buffer system for a limited period of time, the bicarbonate system is still the ideal choice for long-term cell culture. However, its use requires either sealing the chamber or providing the chamber with $CO_2$ in order to maintain the proper pH. Sometimes HEPES is added to the medium to improve the buffering capacity. It slows down and decreases the extent of the change in pH, but does not eliminate entirely the requirement of $CO_2$ (Freshney, 1983).

The osmolarity of the medium must also be maintained at a physiological level. Drift in osmolarity is usually caused by evaporation. Cells are also sensitive to rapid changes in osmolarity, which may be induced, for example, by a sudden replacement of evaporated medium with fresh medium.

Finally, nutrients and minor components (e.g., growth factors) must be supplied to the cell by perfusion or periodic replacement of the culture medium. Although normally each change of medium can last for 24 to 72 hours, more frequent replacement may be required for microscope culture chambers with a very small volume.

The required performance of a microscope culture system depends on the period of observation, the nature of the experiment, and the property of the cell. On the one hand, for short-term observations, the best design may be simply a sealed cover slip. On the other hand, for long-term experiments of highly sensitive cells, precise monitoring of the environment coupled to automatic perfusion of solutions may be necessary. For most purposes, the optimal solution is likely to be a compromise between the two extremes.

### B.  Practical Requirements

The culture system should also fulfill various requirements of the experiments. First, any culturing system should be easily sterilized. Once sterilized, there should be little or no manipulation required on the substrate for cell culture. The container should be easily covered or sealed during observation to minimize the exposure to sources of contamination. However, it should also offer good accessibility if the experiment involves microinjections or physical manipulations of the cell. Access to the medium is also important for the addition of external agents in some experiments. If the experiment involves cell cloning or preparation for immunofluorescence after microscopic observations, the culture system should also allow such tasks to be easily performed.

The dimensions of the culture chamber, including the overall size, the surface area for cell culture, and the depth of the chamber are important considerations for many experiments. The culture container should fit easily onto the microscope stage, yet large enough to accommodate a sufficient number of cells in easily accessible and observable areas. The depth should be shallow enough to ensure good optical quality for transmitted light and adequate accessibility of cells, but large enough to provide enough medium around cells.

Good optical quality is essential for microscopic observations. Plastic tissue culture dishes are excellent in many respects, but they cannot be used with high resolution objectives. Not only is their thickness incompatible with the short working distance of most objectives for fluorescence optics, the strong fluorescence emitted by the plastic material makes fluorescence microscopy essentially impossible. In order to obtain high-quality images, glass cover slips must be used. They should be washed thoroughly and treated with acid–alcohol or extracellular matrix proteins, such as collagen or fibronectin, to promote the attachment of cells. Although most objectives are designed for No. 1 cover slips, No. 2 cover slips can often be used here to provide a higher mechanical strength of the chamber without seriously compromising the quality of images.

Finally, different culture systems may vary widely in terms of ease of setting up, the versatility, the sturdiness and reliability, and the cost. While a highly sophisticated and expensive setup may be ideal for certain experimental protocols, frequently it is more productive to have a large number of simple culture dishes. It should also be clear that there is no perfect system for all purposes, and the strength and weakness of each system must be evaluated in relation to the specific experiments. In the following sections, we will discuss a small number of representative culture systems.

## III.   Cell Culture Chambers

The most sophisticated chambers incorporate mechanisms for the control of various parameters, without the need of a separate microscope incubator. The Leiden Culture System (Fig. 1; Ince *et at.,* 1983, 1985; commercially available from Medical Systems Corp., Greenvale, New York) is such an example. A round glass cover slip of 24-mm diameter is assembled with a Teflon ring to form a dish, which may be placed in a regular $CO_2$ incubator during the initial plating of cells. The dish is inserted into a microincubator before setting up on the microscope. The microincubator maintains the temperature by activating a heating coil and by directing warm $CO_2$ across the surface. With the temperature probe immersed directly into the medium, the temperature can be controlled precisely (see Section IV, B). Evaporation is controlled by layering mineral oil across the surface, an effective technique used by electrophysiologists for many years. Otherwise, the chamber remains open and allows access of microelectrodes, although the actual area accessible is limited by the hindrance of the rim of the chamber and the small diameter of the cover slip. Direct replacement of medium is also difficult due to the presence of oil. In addition, the culture chamber is connected to a number of tubings and cables, which may be a nuisance. Finally, the system is designed primarily for the inverted micro-

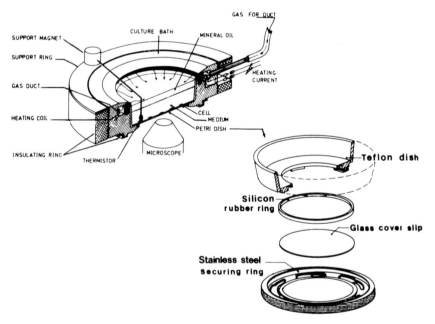

FIG. 1.   Diagram of the Leiden Culture System (kindly supplied by Medical Systems Corp.).

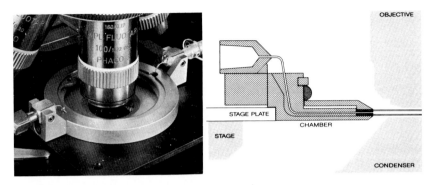

FIG. 2.   Diagram and photograph of the Dvorak-Stotler Chamber (kindly supplied by Nicholson Precision Instruments).

scope; both the oil and the rim interfere with the use of objectives in upright microscopes.

The Dvorak-Stotler Chamber (Fig. 2; Dvorak and Stotler, 1971; commercially available from Nicholson Precision Instruments, Inc., Gaithersburg, Maryland) represents one of the sealed chambers with ports for perfusion. Unlike the Leiden Culture System, both the top and bottom of the chamber consist of cover slips. The two cover slips are separated by a very short distance, 1.2 mm. The most important advantage of the shallow depth is the ideal optical qualities for transmitted illumination in both upright and inverted microscopes. It allows the use of condensers of very short working distances, such as those for high-resolution Nomarski optics. However, cells are inaccessible for physical manipulations and are subject to fluid shear unless the rate of perfusion is very slow. Setting up the chamber without introducing air bubbles may also require practice. The condition of the medium is maintained by the sealed chamber and by a continuous perfusion of medium. Since the chamber has no temperature control system, it must be heated with a separate incubator (such as an air curtain incubator, discussed later). The relatively high price prohibits the acquisition of multiple systems. Therefore, when more than one cover slip is involved in the experiment, they must be placed in plastic culture dishes during initial cell culture, removed before observation, and mounted into the chamber: a tedious process with risks of contamination.

Less complicated culture chambers are also available. The Gabridge Chamber (also called Chamber/Dish; Fig. 3; Gabridge, 1981; commercially available from Bionique Laboratories, Inc., Saranac Lake, New York) is basically just a cover slip held tightly in place against a Teflon plate with the help of a metal pressure plate and an O-ring. The chamber is assembled and

FIG. 3. Diagram of the Gabridge Chamber (kindly supplied by Bionique Laboratories).

autoclaved before plating cells. It is then covered with the top of a regular 35-mm petri dish or with a layer of mineral oil as for the Leiden Culture System. The former allows direct replacement of the medium and avoids any possible effect of mineral oil. However, it is not as effective as the oil in preventing evaporation and sometimes collects condensation on its lower surface. Because of the open nature, the chamber allows cells in the center and distal areas to be manipulated, although a significant fraction of cells hide under the edge of the Teflon core and are inaccessible to microelectrodes. Like the Leiden System, Gabridge chambers are designed primarily for the inverted microscope in conjunction with condensers of relatively long working distances. The chamber also requires the use of a separate device for the maintenance of temperature and pH. However, one major advantage of the chamber is the low cost, which makes it practical to acquire multiple sets in a regular laboratory.

The culture chamber used in our laboratory represents a simplification of the Gabridge Chamber (Fig. 4). It consists of a 70 × 50 × 6 mm plexiglass or Teflon plate with a 35-mm diameter annulus drilled through the center. These dimensions can be varied to suit individual circumstances. One side

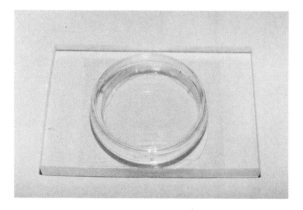

FIG. 4. Photograph of the culture chamber used in our laboratory.

of the annulus is covered by attaching a $45 \times 50$ mm cover slip with a continuous strip of Dow-Corning high-vacuum grease around the annulus. The assembled chamber is autoclaved and used much like a Gabridge Chamber. Because of the similarity to the Gabridge Chamber, most comments about the Gabridge chamber also apply here.

## IV.   Microscope Incubators

For most culture chambers, it is necessary to maintain the culture condition, as discussed previously, with separate devices. However, few microscope incubators are commercially available, and those available often require the addition of extra components to cover all the requirements. Therefore, we will discuss pH, osmolarity, and temperature controls separately.

### A.   Maintenance of pH and Osmolarity

For a sealed culture chamber with perfusion ports, these parameters are normally maintained by continuous perfusion. A syringe pump is usually used to maintain a constant flow.

For open chambers, even with an oil overlay, the pH of the bicarbonate buffer must be maintained by a continuous supply of $CO_2$. Although the actual requirement of $CO_2$ is about 5%, we found a gentle stream of pure $CO_2$ more effective due to the leakiness of the microscope incubation system in general. A simple system may be constructed with a gas flow regulator, which controls the flow of $CO_2$ up to 1000 ml/minute, and a tubing held near the culture chamber. The flow is adjusted to maintain the pH in an optimal range, as judged by the color of phenol red in the medium. The required flow of $CO_2$ varies with the design and the manipulation involved. A higher rate is required if the microscope is not enclosed or if the enclosure is open for manipulations.

For a more precise regulation of $CO_2$ concentration, an enclosure (see next section) with a $CO_2$ sensor may be built. A solenoid control valve is then regulated based on the $CO_2$ concentration detected inside the enclosure. Such devices are available from manufacturers of incubators. However, in a leaky enclosure the triggering of the control solenoid may be frequent and annoying. In addition, since it is difficult to obtain a uniform distribution of $CO_2$ inside the enclosure, the location of the probe may be critical.

Maintaining the osmolarity in an open dish is difficult, especially if hot air is used for temperature regulation (see section below). The problem may be minimized by covering or sealing the chamber as much as possible, by making a special cover with slots matching the requirement of accessibility (Albrecht-Buhler, 1987), by placing a dish of water at the outlet of hot air to humidify the air (see next section), by periodic replacement of the medium, and by the oil overlay technique as mentioned above.

## B.   Temperature Regulation

Essential components for temperature regulation are a temperature sensor, a heater, and a controller which regulates the heater according to the signal from the sensor.

The most direct and precise way for monitoring temperature is immersing a sensor into the medium (Bright *et al.,* 1987). Its disadvantage is also clear: risks of contamination and/or extra efforts in setting up. Placing the sensor anywhere else introduces the possibility that the temperature of the medium may be significantly different from the set temperature. For example, since the culture chamber is in contact with both the air and the microscope stage, unless the two are identical in temperature, neither is ideal for placing the probe. In addition, unless air temperature is regulated both above and below the culture chamber, their differences in temperature will both make the medium temperature unpredictable and cause warping of cover slips, which leads to drifts in focusing.

One possibility to solve these problems is to enclose the stage and the entire air space around the chamber within an insulated incubator (see below). The probe may then be located on the stage while still allowing a precise control of the temperature. In a more open system, consistent results may also be obtained if (1) enough time is allowed for the system to reach a steady state before use; (2) the medium temperature is calibrated as a function of the set temperature; (3) a constant amount of medium is used in all experiments; and (4) a constant temperature is maintained in the laboratory.

The controller for the heater must be of the proportional type. At steady state, it will maintain the heater at a steady and low power. Inexpensive on–off types alternate only between the fully on or off states, and the fluctuation in temperature may be very large.

There are several types of heaters. Those incorporated into the microscope stage are most convenient, but available only for a limited type of microscopes. In addition, unless the cover slip is directly in contact with the stage, the heating may not be efficient. A second type, referred to as "air stream" or "air curtain" incubators, simply blows warm air at the specimen (Orion Research, Cambridge, Massachusetts; and Nicholson Precision Instruments; it may be built simply by combining an appropriate hair dryer with a proportional controller and a probe). It can be positioned conveniently with little interference to the use of the microscope; however, the temperature of the medium may vary with the positioning and continual warming and cooling can strain high-quality objectives. In addition, if the warm air is directed at an open dish, evaporation and loss of $CO_2$ may be rapid. A variation of the warm air blower uses an infrared source (Opti-Quip, Highland Mills, New York), thus reducing the problem of air stream. However, the infrared radiation is easily picked up by most electronic detectors, such as low light level video cameras.

The warm air blower may be used in conjunction with an enclosure incubator for the microscope, so that the air in the enclosure is recirculated continuously through the heater. This has several advantages: the enclosed air mass serves as a temperature buffer and reduces the air flow required for maintaining a steady state; the $CO_2$ concentration is better maintained; and warm air does not have to be directed at the culture chamber. The disadvantage, physical hindrance caused by the enclosure, may be minimized with a proper design incorporating one or more sliding or hinged doors.

While enclosure incubators are difficult to design for the upright microscope, they are commercially available for Nikon and Leitz inverted microscopes. Such enclosure incubators, however, appear overpriced and may not suit the particular experimental needs. Therefore, we designed and built our own enclosures with plexiglass.

Our incubator is built to rest on the edges of the gliding stage of a Zeiss IM35 microscope (Fig. 5). With a mechanical stage, a simple frame has to be added under the incubator to avoid interference with the moving sample carrier. The sides of the enclosure are insulated with foam pads. Access to the chamber is provided by sliding doors on its left- and right-hand sides. Two round plexiglass ports at the rear wall permit connection to the inlet and outlet of a warm air blower, which is a compact, adjustable hair dryer set at a power of 600 W. Flexible aluminum pipes (2-in. Volkswagon preheating duct) are used to bring warm air from the blower into the enclosure, in order to isolate any vibration. Humidification of the air is achieved by placing a dish of water at the outlet of warm air (where it enters the enclosure). The hair dryer is plugged into a proportional temperature controller (Yellow Springs Instruments, Yellow Springs, Ohio; Model 72), which receives signals from a surface probe (YSI Mode 421) attached to the stage with silicone glue.

Due to the dissipation of heat around the objective, the temperature of the medium is usually $1-2°C$ lower than the set temperature. The drift in focusing may also be severe if a large difference in temperature exists between air masses inside and outside the enclosure. For experiments where

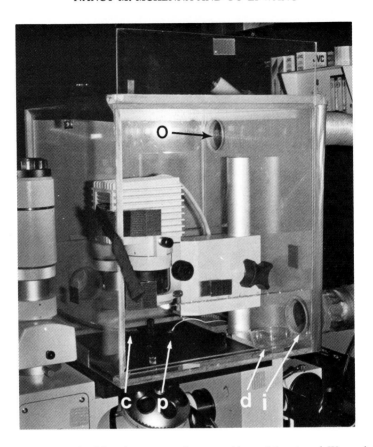

FIG. 5. Photograph of the microscope enclosure used in our laboratory. i, Warm air inlet; o, air outlet; d, dish with water; p, temperature probe; c, $CO_2$ inlet. Insulations for the door and the front side are not installed.

a precise control and a high stability are required, the area under the sample and around the microscope nosepiece (lower chamber) may be enclosed with plates attached to the sides of the microscope. In this case, warm air is pumped into one side of the lower chamber and directed to the main body of the incubator through a piece of flexible, insulated (using a wool sock with the tip cut off) pipe attached to the opposite side. This configuration is capable of holding the medium temperature to within 0.1 °C of set temperature. Through a combination of temperature regulation, $CO_2$ infusion, and medium replacement, we have succeeded in culturing cells for more than 3 days in our microscope incubators.

## ACKNOWLEDGMENTS

The authors wish to thank Dr. M. Gabridge of Bionique Laboratories, Dr. K. Randolph of Medical Systems Corp., and Nicholson Precision Instruments, Inc., for supplying photographs. Our research was supported by grants from National Institutes of Health (GM-32476), National Science Foundation (DCB-8796359), and the Muscular Dystrophy Association.

## REFERENCES

Albrecht-Buchler, G. (1987). *Cell Motil. Cytoskeleton* **7,** 54–67.

Bright, G. R., Fisher, G. W., Rodowska, J., and Taylor, D. L. (1987). *J. Cell Biol.* **104,** 1019–1033.

Dvorak, J. A., and Stotler, W. F. (1971). *Exp. Cell Res.* **68,** 144–148.

Freshney, R. I. (1983). "Culture of Animal Cells." Liss, New York.

Gabridge, M. G. (1981). *In vitro* **17,** 91–97.

Ince, C., Ypey, D. L., Diesselhoff-Den Dulk, M. C. D., Visser, J. A. M., de Vos, A. M., and van Furth, R. (1983). *J. Immunol. Methods* **60,** 269–275.

Ince, C., van Dissel, J. T., and Diesselhoff, M. M. C. (1985). *Pfluegers Arch.* **403,** 240–244.

# Chapter 13

## Basic Fluorescence Microscopy

### D. LANSING TAYLOR

*Department of Biological Sciences*
*Center for Fluorescence Research in Biomedical Sciences*
*Carnegie-Mellon University*
*Pittsburgh, Pennsylvania 15213*

### E. D. SALMON

*Department of Biology*
*University of North Carolina, Chapel Hill*
*Chapel Hill, North Carolina 27514*

METHODS IN CELL BIOLOGY, VOL. 29

# I.  Introduction

## A.  Attributes of Fluorescence Microscopy

There are five major attributes of fluorescence as a tool in microscopy:

1. *Specificity.* Fluorescent molecules absorb and emit light at character-istic wavelengths. Therefore, fluorescent probes can be selectively excited and detected in a complex mixture of molecular species.

2. *Sensitivity.* It is possible to detect a small number of fluorescent molecules. Approximately 50 molecules can be detected in a cubic micro-meter volume of a cell with the fluorescence microscope. This number should decrease with further advances in probe chemistry, experimental methods, and photodetectors.

3. *Spectroscopy.* Fluorescent molecules can be designed to be extremely sensitive to the immediate physical–chemical environment. A variety of spectroscopic parameters that are discussed in detail in Volume 30 of this series can be employed to measure chemical and molecular properties such as pH, free calcium ion concentration, membrane potential, hydrophobic-ity, charge distribution, microviscosity, molecular distances, diffusion coef-ficients, and molecular orientations.

4. *Temporal Resolution.* Fluorescence measurements are limited in temporal resolution to those events that occur with a frequency equal to or greater than the inverse of the time between absorption and emission of light. Therefore, those biological processes occurring at a rate on the order of $\sim 10^8$ seconds$^{-1}$ or slower can be detected and measured. This time domain includes many chemical and molecular changes occurring in living cells.

5. *Spatial Resolution.* Fluorescence signals can be measured from cellu-lar domains as small as single molecules if the molecules contain a sufficient number of fluorophores. The resolution of structures is still limited to the resolving power of the light microscope, which is a function of the numeri-cal aperture of the objective and the wavelength of emission. However, within the limiting resolution, molecular distances can be determined by resonance energy transfer (see Herman, Volume 30, this series).

Immunofluorescence has been the most common application of fluorescence microscopy in cell biology. It combines the specificity, sensitivity, and spatial resolution of fluorescence microscopy with the selective binding of antibodies to restricted regions of antigen molecules termed epitopes. Multiple epitopes can be localized in the same cell, on the same or on different molecules, by choosing fluorophores with different fluorescence colors (see Waggoner et al., Volume 30, this series). The same concept has recently been used to perform fluorescence-based in situ hybridization.

Fluorescence microscopy is also an important biophysical tool for studies in living cells and in reconstituted preparations in vitro. Fluorescent molecules can be incorporated into living cells to measure local physiological changes in cystolic pH (see Tsien and Poenie, as well as Bright et al., Volume 30, this series), $Ca^{2+}$ (see Tsien and Poenie, as well as Bright et al., Volume 30), membrane potential (see London et al., 1986), cell structure (see Luby-Phelps, Chapter 4, this volume), as well as the dynamics of fluorescent analogs of specific biological macromolecules (see Wang, Maxfield, Angelides, and Pagano, this volume), lateral mobility of lipids and proteins in membranes (Wolf, Volume 30, this series), and rotational diffusion (Axelrod, Volume 30, this series). Dimensional changes in the association of molecules on the order of 5.0 nm can also be detected by resonance energy transfer measurements (Herman, Volume 30, this series).

Recent advances in video image recording methods and digital image processing techniques now provide a means of obtaining multispectral, two- and three-dimensional, time-resolved measurements of the distribution of fluorescent probes in cells (see Waggoner et al., and Agard and Sedat, Volume 30, this series).

This chapter presents the basic principles of fluorescence microscopy and the fundamental practical considerations required for fluorescence microscopy in cell biology. A variety of biological applications of fluorescence microscopy are provided elsewhere in Volumes 29 and 30, this series. Some basic information concerning general fluorescence technology has been summarized recently (Taylor et al., 1986).

## B. Nature of Fluorescence

Fluorescence is a type of luminescence where light is emitted from molecules for a short period of time following the absorption of light. When the delay between absorption and emission is on the order of $10^{-8}$ seconds or less, the emitted light is termed fluorescence. If the delay is $\sim 10^{-6}$ seconds it is termed delayed fluorescence, while a delay of greater than $\sim 10^{-6}$ seconds results in phosphorescence.

When light interacts with matter, it may be either scattered (diffracted light) or absorbed. Light absorption occurs in discrete amounts termed quanta. The energy in a quantum is given by:

$$E = \hbar v = \hbar c / \lambda$$

where $\hbar$ is Planck's constant, $c$ is the velocity of light in a vacuum, $\lambda$ is the wavelength of light, and $v$ is the frequency of the vibration of light.

As an example, the energy in ergs of a photon of wavelength equal to 200 nm is:

$$v = c / \lambda$$

$$v = \frac{3.0 \times 10^{10} \text{cm/seconds}}{2.0 \times 10^{-5} \text{ cm}}$$

$$v = 1.5 \times 10^{15} \text{ seconds}^{-1}$$

$$E = h v$$

$$E = (6.6 \times 10^{-27} \text{ erg seconds})(1.5 \times 10^{15} \text{ seconds}^{-1})$$

$$E = 9.9 \times 10^{-12} \text{ ergs/photon}$$

This amount of energy is imparted to an atom or molecule when a single photon of this wavelength is absorbed. There is less energy in photons of longer wavelengths.

The power of light in watts is defined as the amount of energy produced per unit of time or the rate of energy production:

$$1 \text{ W} = 10^7 \text{ ergs/second}$$

Microscopy is concerned with the power of light per unit of cross-section since the light is spread out over some area of the specimen. The power per unit cross-section is defined as intensity ($I$) or irradiance, such that

$$I = \text{W/cm}^2$$

When a quantum of light is absorbed by a molecule, a valence electron is boosted up into a higher energy orbit forming an excited state. When this electron returns back down to its original lower energy orbit, termed the ground state, a quantum of light may be emitted.

The absorption spectrum of a molecule depends on the number of energy levels possible for the electronic state of the molecule. Valence electrons can exist only in discrete energy levels (Fig. 1). Absorption occurs only at wavelengths of light whose quantum energy is equivalent to the difference in energy between the ground electronic state and the excited state. For an atom, there are few possible energy levels, and light absorption occurs only at discrete wavelengths (lines) of equivalent light energy (Fig. 2). The absorption spectrum is much broader for a molecule (Fig. 2), since nuclei

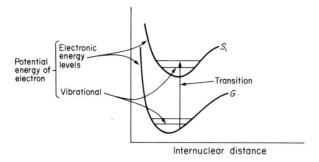

FIG. 1. Diagram of the potential energy of electrons in the ground ($G$) and the lowest singlet excited state ($S_1$) of a molecule.

rotate and vibrate relative to the center of molecular mass. Discrete electronic configurations exist, but both the ground state and the excited state are further subdivided into substates of vibrational and rotational energy. This increases the number of possible energy levels and broadens the absorption spectrum in comparison to that exhibited by a single atom.

Fluorescence has been described as a relaxation process starting with the absorption of light and ending with the emission of light. It is the emission of light produced by deexcitation from the lowest singlet excited state to the ground state (path b in Fig. 3). The wavelength of the emitted fluorescence is usually longer than the wavelength of absorbed light (termed Stokes' Law) as shown in Fig. 4. Exceptions occur only when collisions between molecules impart extra energy to the electron in the excited state. Usually such molecular collisions are infrequent and some energy of the excited electron is lost before deexcitation. As a consequence, the quantal energy of emitted light is usually less than the absorbed light. The emission spectrum is shifted to correspondingly longer wavelengths in comparison to the absorption spectrum (Fig. 4). As with the absorption spectra, the emission spectra of molecules exhibits a broad range of wavelengths in comparison to the narrow range of emission spectra exhibited by atoms.

## C. Pathways of Deexcitation

Fluorescence is one of several possible pathways of the deexcitation process by which an excited state electron gives up energy and returns to the ground state (Fig. 3). Fluorescence lifetime is the average time that a molecule remains in the excited state. The time required for the process of absorption (about $10^{-15}$ seconds) is instantaneous compared to the typical lifetime of fluorescence (about $10^{-8}$ seconds). Fluorescence is delayed (about $10^{-6}$ seconds) if the excited electron moves to a forbidden triplet state before moving back to the lowest singlet excited state and emitting

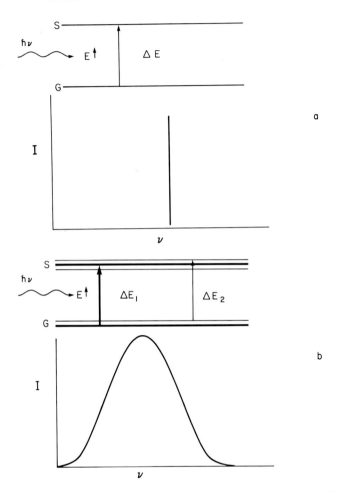

FIG. 2. Diagram of the energy difference ($\Delta E$) between the ground state *(G)* and the excited state *(S)* of an atom after absorbing a photon *(ħv)*. A single frequency (*v*) matches the energy difference between the ground and excited states (a). Diagram of the energy differences between ground states *(G)* and excited states *(S)* of a molecule after absorbing a photon *(ħv)*. A distribution of frequencies matches possible energy differences between the ground and excited states (b).

light (d in Fig. 3). Phosphorescence (e in Fig. 3) begins with a transition to the forbidden triplet state. In contrast to delayed fluorescence, emission occurs by a transition of the excited electron from this lower energy level to the ground state. Phosphorescence has much longer lifetime of the excited state (up to several seconds) and the emission spectra is red-shifted in comparison to fluorescence.

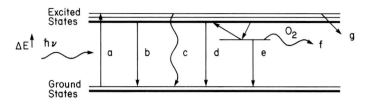

FIG. 3.   Diagram depicting the possible pathways of deexcitation from an excited state to the ground state of a molecule. (a) Absorption, (b) fluorescence, (c) radiationless loss of energy to the medium, (d) delayed fluorescence, (e) phosphorescence, (f) photobleaching in the presence of $O_2$, and (g) chemical reactions.

There are a variety of nonradiative pathways of deexcitation (c in Fig. 3). Energy of the excited electron can be dissipated by interactions with the solvent or other molecules in the sample. Energy absorbed by one molecule can be passed to another molecule for use in driving chemical reactions such as occurs in the light reactions of photosynthesis in plants. Finally, some fluorescent molecules can be destroyed by the excitation process in the presence of molecular oxygen, a phenomena termed photobleaching. Photobleaching can be a serious technical problem in fluorescence photomicroscopy (see Section III, D), but it can also be used as a powerful tool to measure the mobility of molecules (see Wolf, Volume 30, this series).

## D.   Fluorescent Probes

Autofluorescence is the fluorescence of naturally occurring molecules in cells. Most of the autofluorescence in mammalian cells excited in the near

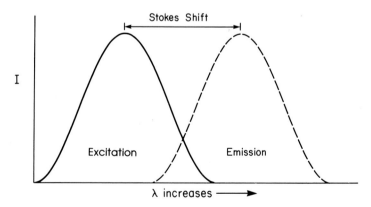

FIG. 4.   Generalized excitation and emission spectra of a fluorescent molecule showing the loss of energy between absorption and emission.

UV and blue region of the spectrum is due to NADH, riboflavin, and flavin coenzymes. Changes in autofluorescence have been used to measure molecular reactions of mitochondrial electron transport (see Kohen et al., 1981). However, naturally occurring fluorescence has had limited value in cell biology in comparison to the application of exogenous fluorescent probes. In fact, autofluorescence is a great source of noise in many experiments (see Section III, A, below).

Fluorophores or fluorochromes are fluorescent dyes or probes that are added to cells. Fluorophores are chosen or synthesized for particular applications based on several criteria. These criteria include absorption and emission spectra, extinction coefficient, quantum yield, environmental effects, and chemical reactivity. Some of the most common fluorophores used in cell biology are listed in Table I of Waggoner et al., Volume 30, this series.

## E. Parameters of Fluorescent Probes

The extinction coefficient, $\epsilon$, and quantum yield, $\Phi$, in addition to fluorescence lifetime, $\tau$, are the three fundamental parameters of fluorescence. The extinction coefficient is a measure of the probability of absorption. A large extinction coefficient indicates a high probability of absorption. Large extinction coefficients range from ~40,000 to 250,000 $M^{-1}$ $cm^{-1}$. If there is a high probability of absorption for fluorescent molecules, then there is also a high probability of emission. Therefore, the intrinsic lifetime of the excited state must be short. In fact, the intrinsic lifetime is inversely proportional to the probability of absorption (see below). In solutions that obey Beer's law in the spectrophotometer, the extinction coefficient is given from:

$$\text{optical density (OD)} = -\log I/I_0 = \epsilon c l$$

and

$$E = OD/cl$$

where $I$ is the light intensity after passing a distance $l$ through the sample, $I_0$ is the incident intensity, $\epsilon$ is the extinction coefficient, and $c$ is the concentration of absorber. The common unit of $\epsilon$ is $M^{-1}$ $cm^{-1}$.

The quantum yield, $\Phi$, is a measure of the efficiency of fluorescence relative to all the possible pathways of deexitation. Quantum yield can be expressed as the ratio of the number of quanta emitted divided by the number of quanta absorbed. Fluorescent molecules usually studied in biomedical samples have quantum yields less than 1, but useful fluorescent molecules usually have quantum yields greater than 0.1. The fluorescence intensity is also a function of the quantum yield. Therefore, the product of $\Phi$ and $\epsilon$ will determine the fluorescence intensity of a probe:

$$I_{\text{fluorescence}} = I_0 \Phi \epsilon c l$$

Fluorescence lifetime, $\tau$, is the average time that a molecule remains in the excited state. The intrinsic lifetime, $\tau_0$, is the maximum possible average lifetime, $\tau$, and this occurs when the $\Phi$ is a maximum. This requires that all of the pathways of deexcitation other than fluorescence be eliminated. Most useful fluorescent probes have lifetimes on the order of $1 - 100 \times 10^{-9}$ /seconds.

## II. Microscope Design

### A. Basic Concepts

The fundamental principle in the design of a fluorescence microscope is to maximize the collection of fluorescent light while minimizing the collection of excitation light. This is achieved by optimizing the optical configuration and components of the fluorescence microscope.

A major constraint in the design of a fluorescence microscope is the fact that the intensity of the fluorescence from the specimen is usually several orders of magnitude less intense than the intensity of illumination. Fluorescence images must be recorded using low light level video cameras or cooled CCD imagers and digital image processing to minimize photobleaching damage. Image contrast depends critically on the ability of the microscope to pass fluorescent light to the detector while substantially blocking the excitation light.

The sensitivity and contrast of fluorescence microscopy depends on the optical configuration of the illumination and imaging paths in the microscope and the optical performance of the microscope components. We will first describe the various optical arrangements used in fluorescence microscopy and then discuss the parameters to be considered to obtain optimal performance of the optical components.

A fluorescence microscope is a compound microscope. The optical paths for specimen image formation are similar to that of a standard bright field microscope. There are three basic types of illumination methods used in fluorescence microscopy: (1) full aperture transmitted light illumination (Fig. 5); (2) darkfield (or darkground) transmitted light illumination (Fig. 6); and incident light illumination (Fig. 7).

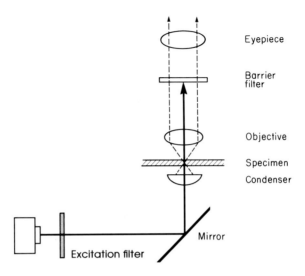

Fig. 5.  Diagram of a fluorescence microscope assembled by the addition of an excitation filter and a barrier filter in a standard bright field microscope. Solid line, excitation light; broken line, emitted light.

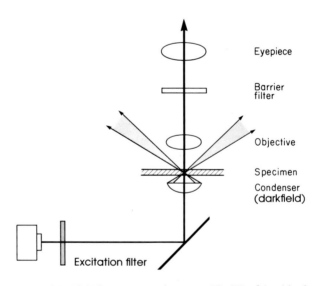

Fig. 6.  Diagram of darkfield fluorescence microscope. The NA of the objective is designed to miss the direct rays from the darkfield condenser (thin lines). Only emitted fluorescence can enter the objective (thick line).

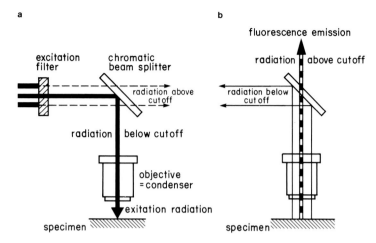

FIG. 7. Diagram of an incident or epiillumination fluorescence microscope. (a) The excitation of the spectrum using the objective as a condenser. The chromatic beam splitter or dichroic mirror has the characteristics that light below a designed cutoff wavelength is reflected through the objective (condenser) and excites the specimen (heavy black line). (b) The emission of light from the specimen. Emitted light (heavy dashed line) that enters the objective above the designed cutoff is not reflected, but passes to the detector. Wavelengths below the designed cutoff are reflected away from the detector.

## B. Full Aperture Transmitted Light Illumination

Fluorescence microscopy can be performed with a conventional bright field light microscope by inserting excitation filters between the illuminator and the condenser, as well as the insertion of emission filters (also called barrier or suppression filters) between the objective and the recording device as shown in Fig. 5. Excitation filters are band-pass filters chosen to pass light at the absorption spectra of the fluorophore, while blocking the longer wavelength light of the fluorescence spectrum. In contrast, emission filters are chosen to pass the light in the emission spectrum while blocking the light of the excitation spectrum. As seen in Fig. 8, the absorption and emission spectra of a typical fluorophore usually overlap and their peaks may be separated by only 20 nm as shown by the spectra for fluorescein. Since the excitation light is usually much higher in intensity than the fluorescent light, band-pass filters with very sharp cut-offs must be used to achieve usable image contrast. Practically, this has proven difficult to achieve, and other illumination methods have been developed to prevent light from the illumination beam from entering the objective. The advantages of full aperture transmitted light illumination are the simple modification of a bright field microscope required to perform fluorescence and the

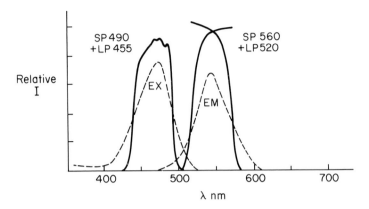

FIG. 8. Excitation and emission spectra of fluorescein with common filter sets employed to separate the excitation and emission light. SP, Short pass filter; LP, long pass filter.

ability to use high numerical aperture (NA) objectives. The disadvantages include the possible presence of inner filter effects in a thick sample (see Section II, C, below) and the difficulty in aligning the condenser and the objective. However, recent improvements in the production of interference filters makes this approach potentially very useful.

## C.   Darkfield Illumination

Until about 1970, the problem of removing excitation light was solved primarily by using darkfield illumination configurations as shown in Fig. 6. A darkfield condenser produces an annular cone of illumination whose aperture is greater than the aperture of the objective. Unscattered light from the illumination beam does not enter the objective. Image contrast was far superior than with full objective aperture illumination using the excitation and emission filters available in the early 1970s.

The major disadvantage of this design is that the efficiency of collecting the fluorescence emission by the objective is greatly reduced. Weakly fluorescent specimens are difficult to detect, since the ability of the objective to collect emitted light from the specimen depends on the NA of the objective to the second power. In order for the exciting light in a dark-field microscope to miss the objective, the objective aperture must be less than the illumination aperture. Objectives with aperture diaphrams must then be used. The working aperture is typically no greater than NA = 0.7, where good quality objectives can have NA = 1.4. The reduction in potential fluorescent light intensity for equivalent illumination and magnification is

$(0.7/1.4)^2 = 0.25$. This reduction is a major disadvantage for the detection and measurement of weak fluorescence and the prevention of photobleaching. Darkfield fluorescence is no longer a viable choice in most applications in cell biology.

## D. Incident or Epiillumination

The illumination and contrast problems were optimally solved in the late 1960s by modifying conventional vertical illuminators with the addition of newly developed interference filters (Ploem, 1967). As seen in Fig. 7, incident or epiillumination occurs through the objective. The objective is both the condenser lens and the objective lens of the system. The novel component of the system is the chromatic beam splitter (also called a dichroic mirror). The chromatic beam splitter is constructed with an interference coating that has a high reflectance at 45° for the wavelengths transmitted by the excitation filter and not the emission filter. Conversely, it transmits the wavelengths passed by the emission filter, but not the wavelengths passed by the excitation filter.

In combination with excitation and emission interference filters, excellent contrast of weakly fluorescent specimens can be achieved by epiillumination with the full aperture of the objective. Since the objective serves as the condenser, alignment of the instrument is simple. In addition, the fluorescence for thick specimens is brighter than in transmitted light excitation schemes, since the illuminating beam does not have to pass through the specimen.

A significant advantage of the incident or epiillumination scheme is that it can be combined with conventional transillumination methods such as phase contrast, polarization, and differential interference contrast. It is often valuable to compare the distribution of fluorescence in a specimen with the structure of the specimen which can be seen by the above transmitted light contrast methods.

Sensitivity and compatability with transmitted light contrast methods usually make incident illumination the preferred optical configuration for fluorescence microscopy. As a consequence, incident illuminators are standard equipment for almost all types of research microscopes manufactured today for the biological sciences.

## E. Koehler Illumination

The illumination path for an incident light illuminator is designed according to the principles of Koehler as shown in Fig. 9. In Koehler illumina-

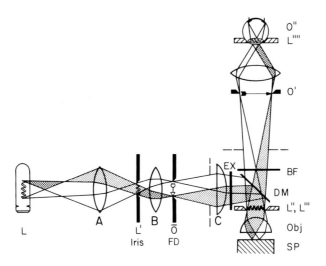

Fig. 9.   Diagram of the illumination pathway and imaging pathway of light in an incident or epiillumination fluorescence microscope. Conjugate planes for the specimen or object plane are labeled (O, SP, O', and O") and for the lamp (L, L', L", L''', L''''). Everywhere in the optical system where the specimen plane is in focus the plane of the lamp is out of focus. The field diaphragm (FD) is conjugate with the specimen, while the image of the lamp filament or arc (L) is conjugate with the back focal plane of the objective (L", L'''). This relationship is called Koehler illumination. Lenses A, B, and C, the excitation filter (EX), the dichroic mirror (DM), and the barrier filter (BF), make up the key optical components of the epiillumination system (see text).

tion, the image of the light source (L) is focused at the iris diaphragm (L') which is also conjugate to the entrance pupil (or back aperture, L"L''') of the objective lens (Obj), while the image of the field diaphragm (FD) is focused at the specimen plane (SP). When the field diaphragm is in focus at the specimen plane, the image of the light source is out of focus. This arrangement produces even illumination of the specimen field in spite of the uneven illumination intensity typical of most light sources. The field diaphragm controls the size of the illuminated area of the specimen without affecting the intensity of the illuminated area. The iris diaphragm (Iris) controls the size of the light source projected into the objective entrance pupil. Opening and closing the iris diaphragm increases and decreases the intensity of illumination of the specimen without affecting the size of the field illuminated. If the iris diaphragm is not included in the illuminator design, then the intensity of illumination of the specimen is usually adjusted by inserting neutral density filters into the illumination path at a convenient position where the image of the filter surfaces will be out of focus at the specimen plane.

## F.   Illuminators

Shutters are critical in fluorescence microscopy. Continuous illumination of the specimen can greatly reduce specimen fluorescence due to photobleaching and continuous illumination is not healthy for living cells. Furthermore, fluorescence may not be desirable while the specimen is being viewed or photographed by transillumination methods. Usually a mechanical shutter is placed in the illumination path to provide on–off control of illumination. Electronic shutters (i.e., Vincent Associates, Rochester, New York) are not expensive and they can provide rapid (10 msec), remote control of illumination (see Wampler and Kutz, this volume).

Modern fluorescence microscopy requires a range of light sources to meet the demands of a variety of applications from ultralow light irradiance over a broad range of wavelengths from the UV to the infrared to intense irradiance at specific wavelengths. Several types of white light lamps and lasers have emerged as standard sources of illumination.

Four major characteristics of lamps must be considered when choosing a white light source: (1) the spectral concentration of radiance from the lamp; (2) the size of the filament or arc compared to the area of the back focal plane of the objective (condenser in epiillumination); (3) the stability of the light source over time and space within the illuminated field; and (4) the uniformity of illumination of the field.

Lamps can be characterized in radiometric terms by the spectral concentration or spectral density of radiance $(B_\lambda)$. This is the proper way to define the light output of a lamp since it is a measure of the radiant intensity per area, a physical measurement independent of the human visual sensitivity. This characteristic is specific for an individual lamp. A portion of the area of the luminous surface of a lamp emits light within a solid angle with a particular spectral range:

$$B_\lambda = \phi/Fw\Delta\lambda$$

where $\phi$ is the radiant flux in watts (W), $F$ is the area of luminous surface in cm², $w$ is the solid angle of emitted cone of light in steradians (sr), and $\Delta\lambda$ is the spectral bandwidth selected in nanometers (nm).

The light output of the lamp is expressed in the units: $W\ cm^{-2}\ sr^{-1}\ nm^{-1}$. This value represents the radiant flux per unit area, unit solid angle, and unit spectral bandwidth. The absolute value will depend on how the microscope is adjusted. The radiant flux is determined by the size of the luminous surface selected by the iris diaphragm of the epiilluminator. The solid angle of the emitted cone of light is determined by the NA of the collecting lens of the illuminator. The spectral bandwidth is determined by the filters and/or monochrometer employed.

The light housing is usually fitted with a well-corrected collector lens that is positioned so that the filament or arc is near its principal focal point. In Koehler illumination the lamp collector lens serves as an enlarged secondary light source. An image of the filament or the arc must then be projected onto the back focal plane of the objective which serves as the condenser during excitation in epiillumination. The filament or arc should fill the back aperture of the objective to both maximize the radiance and to ensure an even illumination of the field by Koehler illumination (see above). A ground glass is sometimes inserted in front of the lamp to maximize the uniform illumination, especially with arc lamps that tend to produce "hot spots" of illumination.

The illumination of the specimen must be constant over time and space across the field of view. The stability of the light source is also critical for ensuring the precision of measurements. Instability over time reflects the temporal fluctuations of radiance of the lamp and is derived primarily from the variations in the electrical supply to the lamp. Instability of the lamp in space is often detected in arc lamps and is called "flicker." Flicker occurs when different surface elements of the arc lamp exhibit different fluctuations arising from the migration of plasma across the electrodes. These spatial variations can be caused by small variations in the resistance of the electrodes, fluctuations of the power supply, and/or mechanical vibrations. The filament-based light sources are very stable, especially when they are operated under constant current with a power supply. In contrast, the arc lamps are inherently less stable. DC-powered arc lamps have improved the stability considerably, but they must be tested. It has been demonstrated that the smaller the distance between electrodes the more stable the arc lamp (lower wattage lamps). Xenon arcs are more stable than mercury arcs. The addition of a small percentage of xenon to a mercury arc lamp can increase the stability of the arc (Oriel, Strafford, Connecticut). All types of lamps exhibit aging which occurs by the gradual deposition of metal on the surface of the bulb. Aging can be detected by direct observation of the bulb.

Choosing a light source and collector lens that causes the back focal plane of the objective to be filled with an image of the filament or arc is the first step. Taking precautions to stabilize the lamp radiance over time and space is the second step. Finally, the microscope should be adjusted for Koehler illumination. Careful attention to these characteristics of lamps will optimize the use of fluorescence microscopy as a quantitative tool.

## 1. Specific White Light Sources

Based on the characteristics of lamps described above, several different white light sources have become standards in fluorescence microscopy. The choice depends on the spectral radiance of excitation over the range of wavelengths required and the stability of radiance required. The major lamps include: tungsten; quartz–halogen; 50 and 100 W mercury arcs; and 75 W xenon arc (Fig. 10).

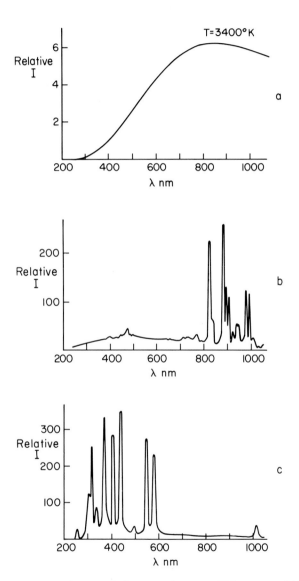

FIG. 10. Diagrams showing the relative spectral output of (a) tungsten, (b) 75 W xenon-arc (XBO-75), and (c) 100 W mercury arc (HBO-100) lamps. Note the large output of near infrared light by the xenon arc.

## 2. Lasers

Lasers provide stable, intense monochromatic light. There are two major intensity peaks in an argon ion laser, one at 488 nm and one at 514 nm. The laser emission can contain on or both laser lines by tuning the resonant frequency of the laser. The 488 line is ideal for exciting fluorescein isothiocyanate (FITC). The 514 line is marginal for exciting rhodamine isothiocyanate (RITC).

In general, a different type of laser is required for different spectral emissions. High power, stable laser light sources are expensive (5,000–$20,000) and difficult to maintain. Because of their cost and limited spectral output, lasers have thus far been used mainly by biophysicists in applications such as measurement of fluorescence redistribution after photobleaching to analyze the mobility of molecules in living cells and in reconstituted preparations *in vitro* (see Wolf, Volume 30, this series). In the near future, technological developments in lasers and laser diodes should bring them closer to the ideal light source required for the majority of applications in fluorescence microscopy, particularly in conjunction with laser-based confocal scanning microscopy (see Brakenhoff *et al.*, Volume 30, this series).

## G.  Filters

There are four different types of filter construction: neutral density, glass filters, colored glass filters, gelatin filters, and interference filters. Neutral density filters attenuate all colors or the spectrum uniformly. They are frequently made by depositing a thin film of metal on a flat glass surface. This surface is sealed by covering the metal surface with another flat glass plate.

Colored glass filters transmit light in limited regions of the spectrum because of light absorption, which is usually due to the metal composition of the glass. Gelatin filters consist of a gelatin layer containing organic dyes. The gelatin layer is usually mounted between two flat glass plates.

Interference filters consist of many layers of films with different refractive indices sequentially deposited upon a flat glass surface and sealed by another flat glass surface. The transmission characteristics of the filter are produced by interference of light reflected at the surfaces of the different film layers. It is important to recognize that light not transmitted by an interference filter is reflected. Interference filters designed for selecting spectral regions in part by reflection are called dichroic beam splitting mirrors. Interference filters can be designed to transmit or reflect light in discrete bandwidths from the UV to the infrared regions of the spectrum.

Filters are characterized by their transmission and reflection characteristics. Band-pass filters are identified by their peak wavelength of transmission, the percent of incident light at that wavelength transmitted, and the half-bandwidth of transmitted wavelengths. High-quality band-pass filters have sharp cutoff transmission characteristics. High-quality short-pass filters and long-pass filters also have sharp cutoffs between the range of wavelengths transmitted and those absorbed or reflected. Colored glass filters and gelatin filters usually do not have sharp cutoffs, but they are inexpensive. Interference filters provide the best performance, but they are expensive (typically $100 or more).

The bandwidths and cutoff wavelengths of filters are chosen to maximize energy transmission without spillover of fluorescence between different filter sets. Figure 11 shows the practical filter sets recommended for Hoechst, FITC, and RITC. They are designed to isolate the excitation and emission spectra for each fluorophore as can be seen by comparing the transmission characteristics of these filter sets with the absorption and emission spectra of the probes. The excitation and emission spectra of Hoechst and FITC are well separated so that broad band filters for excitation and emission of Hoechst can be used to maximize light intensity. The excitation and emission spectra of FITC and RITC overlap to a small degree and narrow band-pass filters must be used. Be careful in using the filter sets supplied with many microscopes. The manufacturers often sacrifice spectral separation for fluorescence through-put. Also, be careful to use blocking filters for the infrared. Many detectors are very sensitive in the infrared and can detect radiation that is invisible to the human eye.

The excitation filter set, dichroic mirror (DM), and emission or blocking filter set (BF) are usually mounted in a mechanical cube positioned above the objective. In most commercially available illuminators, a slider or rotatable filter holder is used to select between two to four different sets. This permits rapid switching of filter sets to provide views of fluorescence from multiple fluorescent probes in the same specimen. Lenses (A, B, and C in Fig. 9) make the light propagate parallel to the optical path, perpendicular to the emission filter surface and at 45° to the dichroic mirror. For all the filter sets in an epiilluminator, the excitation and emission filters should be oriented perpendicular and the dichroic mirror oriented at 45° to the optic axis of the microscope. Otherwise, the image of the specimen will be displaced laterally in the image plane to different degrees for the different filter sets.

A partially reflecting mirror, which is available from most manufacturers, can be used in place of the dichroic mirror in an incident illuminator. This configuration does not have the selectivity and energy transmission capabilities of a dichroic mirror, but it can be used for fluorescent

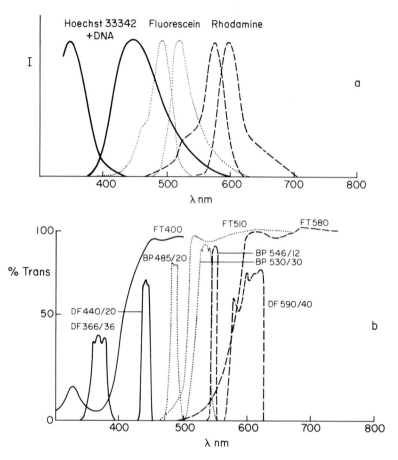

FIG. 11.   Excitation and emission spectra of Hoechst 33342, fluorescein, and rhodamine (a). The filter sets that we use to image Hoechst 33342, fluorescein, and rhodamine in the same cells are shown in (b).

probes for which no dichroic mirror is commercially available. The excitation and emission wavelengths are isolated by appropriate choice of interference filters for excitation and emission wavelengths.

## H.   Objective Lenses

The brightness of the fluorescence image varies with the objective employed. The relative brightness of the image can be predicted if we ignore the effects of the transmission and reflectivity of the lens elements. The intensity of light gathered by an objective varies as the square of the numerical aperture. The NA is a measure of the light gathering power of a lens and is defined by:

$$NA = n \sin \theta$$

where $n$ is the refractive index between the specimen and the objective and $\theta$ is the half-angle of the cone of rays entering the objective. In addition, the brightness of the image also varies as the inverse of the square of the objective magnification. In transmitted light fluorescence the condenser usually remains fixed while the objective is changed. When a higher magnification objective is used with a fixed condenser, the illuminating light from the condenser is spread out over a field larger than that viewed by the objective. Therefore, some of the illuminating light is lost. In summary, the brightness of the image in transmitted light fluorescence is related to:

$$\text{image brightness} \propto \frac{\text{objective } NA^2 \times \text{condenser } NA^2}{\text{magnification}^2}$$

In epiillumination the objective also serves as the condenser. When the objective magnification is increased, the same light intensity is focused on a smaller field. The result is that the image brightness depends on:

$$\text{image brightness} \propto \frac{\text{objective } NA^4}{\text{magnification}^2}$$

A practical rule is to maximize the NA of the objective and to minimize the total magnification for electronic imaging. At very low light levels, when light is limiting, it is often more important to increase the signal-to-noise ratio at the expense of magnification. Ploem has described the use of a demagnifying lens in combination with a high NA objective to maximize image brightness (Ploem, 1986). The need to minimize the irradiance of excitation light in order to minimize photobleaching damage must also be considered when balancing the selection of optical components. It is important to define the irradiance at the specimen plane and this can be performed simply using a power meter (see Section IV, B).

Fluorescent specimens act as self-luminous objects. Point sources in the self-luminous specimen behave as independent sources, like two distant stars observed with a telescope. Therefore, there is no coherent relationship between the phases of the emitted light. Light from each point source is diffracted by the objective aperture, which produces an airy disk image of the point source. The light from two point sources are incoherent, so that Raleigh's criterion for resolving power can be applied. The resolution of two adjacent point sources, $\gamma$, is given by:

$$\gamma = \frac{0.61\lambda}{NA}$$

This consideration of resolution is operable for the transverse ($XY$ axes) resolution only. Axial resolution ($Z$ axis) involves consideration of the depth of field.

The depth of field of an optical system is the physical distance along the $Z$ axis through the specimen that is in focus at the image plane. Another way of defining depth of field is the distance that a single point object can be moved in the $Z$ axis before the image changes. This distance is determined primarily by wave optical considerations at high NA, while geometrical optical considerations are significant at low NA. We will consider high NA objectives in our treatment. The depth of field is determined by the refractive index, $n$, of the medium, the NA of the objective, and the wavelength of the illuminating light. The depth of field can be defined mathematically by:

$$D = \frac{\lambda}{4\,n\,\sin^2\,(\theta/2)}$$

Calculating the depth of field for one common objective will illustrate the considerations. We will use a 63 × objective (water immersion, $n = 1.33$) with an NA = 1.2, assuming a wavelength of 500 nm (0.5 $\mu$m). Since the NA = $n \sin \theta$, we can solve for $\sin^2 (\theta/2)$ in the equation for depth of field by substitution:

$$NA = n \sin \theta$$

$$\sin \theta = \frac{NA}{n}$$

$$\sin \theta = \frac{1.2}{1.33} = 0.9$$

$$\theta = \sin^{-1} 0.9 = 64.5$$

$$\sin^2 \frac{\theta}{2} = 0.28$$

therefore,

$$D = \frac{0.5}{4(1.33)(0.28)}$$

$$D = 0.34 \;\mu\text{m}$$

This objective will therefore maintain a sharp image of a point source for a distance of 0.34 $\mu$m in axial movement of the specimen. It must be emphasized that light emanating from other point sources outside of the distance equal to the depth of field of the objective will be distributed into the plane of focus. The out-of-focus light from regions outside of the depth of field is noise that must be removed or avoided to obtain the best information from the fluorescence microscope. The chapters by Agard and Sedat and Brakenhof *et al.* in Volume 30, this series, deal with this important issue in two different ways (see also Lanni, 1986).

Improvements in the design of objectives for fluorescence microscopy have been extensive during the past 5 years. Objectives have been developed with a maximum or high NA for both low and high magnification and with a range of working distances. In addition, immersion objectives for use with different oils, water, and glycerol have been designed for use, with or without cover glasses. The new objectives have been constructed with minimum fluorescence from optical components including cements. Fluorite and quasifluorite optical lenses have optimized transmission properties in the wavelengths down to ~350 nm, while quartz lenses are available for wavelengths below ~350 nm. It is our opinion that quartz lenses are not required for most probes excited around 350 nm including Fura-2, if a light source with a high output in this spectral region is employed.

## I. Detectors

See chapters by Spring and Lowy, Wampler and Kutz, and Aikens *et al.*, in this volume.

## III. Practical Issues of Epiillumination Fluorescence Microscopy

## A. Autofluorescence

Autofluorescence is the fluorescence of naturally occurring molecules in cells. Most cells will exhibit some degree of autofluorescence when excited with the proper wavelength at high intensity of excitation. Recent investigations indicate that most of the autofluorescence from mammalian cells excited in the near UV and visible region of the spectrum is due to NADH, riboflavin, and flavin coenzymes (Benson *et al.*, 1979; Aubin, 1979). These molecules have a broad excitation spectrum ranging from ~350 to 500 nm. The level of autofluorescence in cells is also a function of the stage of the cell cycle and physiological state. Proteins fluoresce when excited at 250–280 nm due to the presence of tryptophan, tyrosine, and phenylalanine.

All studies with the fluorescence microscope must start with the examination of autofluorescence of the cells under the same conditions that will be employed with the exogenous fluorescent probes. These conditions include the objective, lamp, irradiance at the specimen plane, and detector. The signal-to-noise ratio will be determined in part by the relative strength of the autofluorescence compared to the fluorescence of the exogenous probe. It may be necessary to correct for the autofluorescence signal or to avoid the autofluorescent region of the spectrum by shifting the exogenous probe to the longer wavelengths (see Waggoner *et al.*, Volume 30, this series). Fluorescence and phosphorescence is often detected in slides, cover slips, immersion oil, and culture media. All components that will be placed in the optical path must be checked separately.

## B.  Detectability

A fluorescence microscope is able to collect a higher percentage of emitted fluorescence from a sample than a standard fluorometer due to the high NA of the microscope objective. Several studies have demonstrated the extreme sensitivity of fluorescence microscopy. Barak and Webb were able to detect single low-density lipoproteins (LDL) particles labeled with ~50 fluorophores (Barak and Webb, 1981). The lower limit of detectability has not been reached and even single fluorophore detection is feasible under the optimal conditions (Mathies and Stryer, 1986). Hirschfeld originally demonstrated the ability to detect single biological molecules by labeling polyethyleneamine with multiple fluorophores and then coupling the labeled "kite tail" to single IgG molecules (Hirschfeld, 1976a,b). This is a technically challenging area in the field of fluorescence microscopy.

## C.  Quantitation of Fluorescence Intensity

The fluorescence intensity of a solution is directly proportional to its concentration only in highly dilute solutions that follow Beer's law (Fig. 12); see Section I,E. At higher concentrations of fluorescent probes the fluorescence intensity can actually decrease. There are three major reasons that can account for this phenomenon. The inner filter effect occurs when the excitation light does not penetrate through the sample due to the light absorbance of the sample. Therefore, all of the fluorophores do not absorb to the same extent. This can occur in very thick and/or concentrated microscope samples. Epiillumination is particularly valuable under these conditions since this configuration is similar to "front surface" fluorescence performed in fluorometers. This is the only way to study the fluorescence of this type of sample. The fluorescence can also decrease at higher fluorophore concentrations due to the formation of probe dimers produced in the ground state or dimers formed only in the excited state, which are called

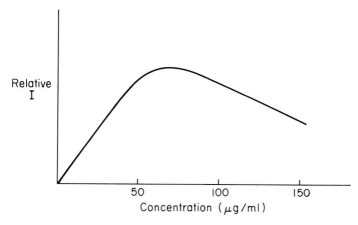

Fig. 12. Diagram of the effect of fluorophore concentration on the relative intensity of fluorescence (see the text).

excimers. These complexes usually cause quenching of the fluorescence. In addition, trivial reabsorption of the emitted fluorescence can occur since the emission spectrum of many probes overlaps the excitation spectrum.

The presence of a concentration effect can be tested by increasing and decreasing the concentration of the fluorescent probe used in an experiment. This is a dose response fluorescence control. A sure sign of a concentration artifact is a measured increase in fluorescence intensity when the fluorophore concentration is decreased. These concentration problems are usually not a problem with thin mammalian cells labeled with standard probes. However, each new probe and experimental situation must be evaluated.

## D. Bleaching

Bleaching is defined as the permanent destruction of fluorescence by a light-induced conversion of the fluorophore to a chemically nonfluorescent compound. This process requires light and molecular oxygen for most of the commonly used fluorophores. However, the photochemistry is very complex and no absolute mechanism has been defined. Removing oxygen is not a viable option when studying living cells, so the total dose of light must be regulated. This is the main reason for using a shutter on the light source. The sample should be illuminated only during the time of recording an image or making a measurement. Furthermore, the total dose of light must be kept to a minimum, where dose is defined as the intensity of excitation multiplied by the time of illumination (see Bright *et al.*, 1987).

Excitations at low irradiances over a relatively long time interval are required with image intensifiers, since they typically have very low intrascene dynamic ranges. In contrast, high irradiances for very short times can be obtained with cooled CCD cameras, due to the very large intrascene dynamic range (see Bright *et al.*, 1987). The same total dose can be delivered by both of these approaches, yet the temporal resolution is optimized in the latter.

The generation of toxic compounds including free radicals during photobleaching is an additional problem. The loss of fluorescence signal, as well as the loss of physiological significance of the experiment, are two important reasons to control the bleaching of probes introduced into living cells. Solutions to the problem usually involve a combination of choosing a more photostable probe such as rhodamine instead of fluorescein; decreasing the total dose of illumination by decreasing the number and duration of excitation periods; and increasing the sensitivity of the detector. Better probes and approaches to imaging should improve this most serious limitation to fluorescence microscopy.

## E.   Light Scattering

The scattered light in an epiillumination fluorescence microscope is primarily forward light scattering which passes through the specimen and does not enter the objective. The presence of an optimal dichroic mirror and barrier filter removes any back-scattered light that does enter the objective. However, excessive light scattering due to the content of the specimen could affect spectral measurements and must be corrected. Microspectrofluorometric analyses of the excitation and emission spectra should be performed. It is encouraging that the total light scattering of mammalian cells in culture is very small due to the small specimen pathlength.

## F.   Environmental Effects

The major power of fluorescence spectroscopy is the environmental sensitivity of many fluorescent probes. However, this same sensitivity must be well understood and controlled in order to make the proper interpretations of the fluorescence signals. It is easy to obtain a fluorescence signal, but it is very difficult to interpret it properly. In practice, the behavior of fluorescent probes under different environmental conditions *in vitro* must be well understood before attempting interpretations in living cells (see Waggoner, 1986). A good example of the environmental sensitivity is the effect of solvent polarity on the quantum yield and excitation, as well as emission spectrum of some probes. Some of the membrane potential sensitive probes (see Waggoner, 1986) exhibit both an increase in quantum yield and a red shift in the excitation and emission spectra when the probe enters the more hydrophobic environment of the membrane from the cytoplasm. A distribution of this class of probe into different environments would yield a complex fluorescence signal. Microspectrofluorometric measurements of the probe *in vitro* and in different regions of cells can identify environmental changes in probes.

Fluorescence quenching is the loss of fluorescence due to the interaction of the fluorophores with other molecules in the environment. Static quenching is the decrease in fluorescence due to the interactions of fluorophores in the ground state with other molecules. This can include self-association called ground state dimerization. Static quenching causes a decrease in the fluorescence intensity since the quenched fluorescent molecules do not fluoresce with the same quantum yield and/or with the same spectral properties. Fluorescent probes may be more quenched in cells compared to buffer solutions because of the presence of quenching activity in the local cell environment.

Dynamic quenching is the decrease in fluorescence intensity at defined wavelengths due to the interaction of the fluorophore in the excited state with quenching molecules in the environment. Therefore, the excitation spectrum is not altered, but the emission spectrum is changed. Dynamic quenching forms a new molecular species with distinct fluorescence properties. This type of quenching is dependent on the rate of diffusion in solution, so that it is affected by temperature and viscosity. Determination of the quantum yield and spectral properties of the probe in the living cell compared with solution measurements *in vitro* is the only way to characterize this problem (see Wampler, 1986; Kohen *et al.*, 1981).

## G. Standards

Standards for quantitative fluorescence microscopy are important to ensure the reproducible performance of the system from day to day and to compare results between laboratories (see Sisken, Volume 30, this series). A variety of standards for fluorescence intensity, quantum yield, detector sensitivity, and detector corrections must be developed. This is an important, yet underdeveloped area in fluorescence microscopy.

## H. Spectral Corrections

Excitation and emission spectra must be corrected for instrumental variables including optical components, quality of monochromators or filters, sensitivity of the detector, spectral output of the lamp, etc. Uncorrected spectra cannot be reproduced in other laboratories with different equipment. There are a variety of methods for correcting both excitation and emission spectra but the simplest are standard solutions that have been calibrated under control conditions. Standardized, "corrected" spectra are published in a number of books (see Argauer and White, 1964; Berlman, 1971). The essence of the correction scheme is to determine the correction factor at each wavelength which converts the experimental spectrum of a standard solution, normalized to a peak height of 1.0, to the published corrected spectrum of the same solution also normalized to a peak height of 1.0. This conversion factor at each wavelength can then be used to correct a spectrum of any other solution measured under the same conditions. This issue only becomes a problem when accurate spectral measurements are attempted (see Herman, Volume 30, this series).

# IV.   Basic Setup of an Epiillumination Fluorescence Microscope

## A.   Alignment of Lamp and Microscope

1. A fluorescent bead slide can be prepared by mounting commercially available beads on a standard microscope slide. Beads with a diameter of $\sim 5-10\ \mu$m and labeled with a variety of fluorophores are diluted and dried on a microscope slide (sources for the beads include Polysciences, Inc., Paul Valley Industrial Park, Warrington, Pennsylvania). A drop of UV curing cement is added, a cover slip is applied, and the slide is irradiated for $\sim 1$ hour. This standard slide can then be used to set up the microscope from day to day.

2. Select an objective that will be used in the experiment. Align the microscope for Koehler illumination for transmitted light microscopy before setting up epiillumination. This is performed as follows: (a) bring the specimen into focus; (b) close the field diaphragm (FD) and focus the condenser until the edge of the FD is in focus with the specimen; (c) center the condenser with the adjustment screws until the image of the FD is in the center of the field; (d) use a focusing telescope to view and to center the iris diaphragm; and (e) focus an image of the lamp on the plane of the iris diaphragm. This can be accomplished by closing down the FD and adjusting the screws on the lamp housing.

3. Now place a thin piece of white paper on the microscope stage in place of a specimen. Select the lamp for epiillumination. Rotate the chosen objective out of place so that a blank objective slot allows the image of the filament or arc to be visible on the paper by epiillumination. The image of the arc is then centered and focused using the adjustment screws on the lamp. Close down the iris diaphragm if your microscope has one in the incident light path. The image of the filament or arc should be coincident with the image of the iris diaphragm to attain Koehler illumination. The focused and centered filament or arc should now illuminate the microscope field uniformly when the objective is swung back into place. The optimal situation is when the filament or arc just fills the back focal plane of the objective (circular aperture observed on the white paper in the absence of the objective).

4. Place the bead slide on the microscope stage and bring one bead into focus. Check the uniformity of illumination by moving the bead into different quadrants of the field. A properly aligned microscope will produce an even illumination of the field so that the test bead will exhibit uniform fluorescence. Repeat the steps above if you do not achieve uniform illumination.

5. This procedure can be quantified by using a photomultiplier (PMT). The output of the PMT should be constant if the illumination system is optimal. The camera will add the problem of shading due to the characteristics of the detector (see Spring and Lowy, Chapter 15, and Aikens et al., Young, and Jericevic et al., Volumes 29 and 30, this series.

6. The field diaphragm in the epiillumination pathway can now be adjusted to limit the area of the field that is illuminated.

7. The iris diaphragm can be adjusted over a small range to control the irradiance at the specimen plane (see Section IV, B, below). The irradiance can be controlled in microscopes not possessing an iris in the epiillumination pathway by inserting neutral density filters between the light source and the objective. The voltage on filament lamps should never be used to control irradiance, since the color temperature of these lamps changes significantly with operating voltage.

8. At this point you should use some type of low light level camera to view the experimental samples. Remember, you want to minimize the total dose of illumination (irradiance) to the specimen.

9. Check to make sure that the camera is viewing the center of the field.

10. If your microscope has two lamps, one for epiillumination and the other for transmitted light illumination, you can observe the specimen either sequentially or simultaneously by both transmitted and fluorescence microscopy.

## B. Measurement of Irradiance at the Specimen Plane

Characterization of the irradiance at the specimen plane is an important parameter in fluorescence microscopy. Publication of this value along with the time of illumination should optimize the ability of other investigators to reproduce experiments. We use an NRC (Newport Research Corporation, Fountain Valley, California) laser power meter, model 820 with a detector area of 1.0 cm², to measure the irradiance.

1. Select the objective and fluorescence filter set to be used and set the irradiance at the specimen plane that will be used in the experiment by using neutral density filters and/or the incident light iris diaphragm.

2. Adjust the field diaphragm to a specific diameter. This can be accomplished with an eyepiece reticle and a stage micrometer. Determine the area illuminated by this setting of the field diaphragm.

3. Place the surface of the detector photodiode on the stage in place of the specimen. Measure the uncorrected power output at the selected excitation wavelength with the laser power meter.

4. Correct power reading for spectral sensitivity using the table supplied with the power meter. Divide the uncorrected value in microwatts by the correction factor.

5. Divide the corrected power by the area of illumination measured above. This will yield the intensity or irradiance in microwatts per squared centimeter.

### REFERENCES

Agard, D., Hiraoka, Y., Shaw, P., and Sedat, J. (1989). *Methods Cell Biol.* **30,** 353–377.
Aikens, R., Agard, D., and Sedat, J. (1989). *Methods Cell Biol.* **29,** 291–313.
Angelides, K. (1989). *Methods Cell Biol.* **29,** 29–58.
Argauer, R., and White, C. (1964). *Anal. Chem.* **36,** 368–371.
Aubin, J. (1979). *J. Histochem. Cytochem.* **27,** 36–43.
Axelrod, D. (1989). *Methods Cell Biol.* **30,** 245–270, 333–352.
Barak, L., and Webb, W. (1981). *J. Cell Biol.* **90,** 595–604.
Benson, R., Meyer, R., Zaruba, M., and McKhann, G. (1979). *J. Histochem. Cytochem.* **27,** 44–48.
Berlman, I. (1971). "Handbook of Fluorescence Spectra of Aromatic Molecules," 2nd Ed. Academic Press, New York.
Brakenhoff, G., Van Spronsen, E., Van der Voort, H., and Nanninga, N. (1989). *Methods Cell Biol.* **30,** 379–398.
Bright, G., Fisher, G., Rogowska, J., and Taylor, D. (1987). *J. Cell Biol.* **104,** 1019–1033.
Bright, G., Fisher, G., Rogowska, J., and Taylor, D. (1989). *Methods Cell Biol.* **30,** 157–192.
Herman, B. (1989). *Methods Cell Biol.* **30,** 219–243.
Hirschfeld, T. (1976a). *Appl. Opt.* **15,** 2965–3135.
Hirschfeld, T. (1976b). *Appl. Opt.* **15,** 3135.
Jeričević, Ž., Wiese, B., Bryan, J., and Smith, L. (1989). *Methods Cell Biol.* **30,** 47–83.

Kohen, E., Thorell, B., Woiters, J., Kohen, C., Bartick, P., Salmon, J.-M., Viallet, P., Schahtschabel, D., Rabinovitch, A., Minty, D., Meda, P., Westerhoff, H., Nestor, J., and Ploem, J. (1981). *In* "Modern Fluorescence Spectroscopy" (E. L. Wehry, ed.), Vol. 3, pp. 295–346. Plenum, New York.

Lanni, F. (1986). *In* "Applications of Fluorescence in the Biomedical Sciences" (D. L. Taylor, A. Waggoner, R. Murphy, F. Lanni, and R. Birge, eds.), pp. 505–521. Liss, New York.

London, J., Zecevic, D., Loew, L., Orbach, H., and Cohen, L. (1986). *In* "Applications of Fluorescence in the Biomedical Sciences" (D. L. Taylor, A. Waggoner, R. Murphy, F. Lanni, and R. Birge, eds.), pp. 423–447. Liss, New York.

Luby-Phelps, K. (1989). *Methods Cell Biol.* **29,** 59–73.

Mathies, R., and Stryer, L. (1986). *In* "Applications of Fluorescence in the Biomedical Sciences" (D. L. Taylor, A. Waggoner, R. Murphy, F. Lanni, and R. Birge, eds.), pp. 129–140. Liss, New York.

Maxfield, F. (1989). *Methods Cell Biol.* **29,** 13–28.

Pagano, R. (1989). *Methods Cell Biol.* **29,** 75–85.

Ploem, J. (1967). *Z. Wiss. Mikrosk.* **68,** 129–142.

Ploem, J. (1986). *In* "Applications of Fluorescence in the Biomedical Sciences" (D. L. Taylor, A. Waggoner, R. Murphy, F. Lanni, and R. Birge, eds.). pp. 289–300. Liss, New York.

Spring, K., and Lowry, R.J. (1989). *Methods Cell Biol.* **29,** 269–289.

Taylor, D., Waggoner, A., Murphy, R., Lanni, F., and Birge, R. (1986). "Applications of Fluorescence in the Biomedical Sciences." Liss, New York.

Tsien, R. (1989). *Methods Cell Biol.* **30,** 127–156.

Waggoner, A. (1986). *In* "Applications of Fluorescence in the Biomedical Sciences" (D. L. Taylor, A. Waggoner, R. Murphy, F. Lanni, and R. Birge, eds.), pp. 3–28. Liss, New York.

Waggoner, A., DeBiasio, R., Conrad, P., Bright, G., Ernst, L., Ryan, K., Nederlot, M., and Taylor, D. (1989). *Methods Cell Biol.* **30,** 449–478.

Wampler, J. (1986). *In* "Applications of Fluorescence in the Biomedical Sciences" (D. L. Taylor, A. Waggoner, R. Murphy, F. Lanni, and R. Birge, eds.), pp. 301–319. Liss, New York.

Wampler, J., and Kutz, K. (1989). *Methods Cell Biol.* **29,** 239–267.

Wang, Y.-L. (1989). *Methods Cell Biol.* **29,** 1–12.

Wolf, D. (1989). *Methods Cell Biol.* **30,** 271–306.

Young, T. (1989). *Methods Cell Biol.* **30,** 1–45.

# Chapter 14

# Quantitative Fluorescence Microscopy Using Photomultiplier Tubes and Imaging Detectors

JOHN E. WAMPLER

*Department of Biochemistry*
*University of Georgia*
*Athens, Georgia 30602*

KARL KUTZ

*Georgia Instruments, Inc.*
*Atlanta, Georgia 30341*

---

## I.  Introduction

The sensitivity and selectivity of fluorescence detection has excited the interest of microscopists since the 1950s. For a detailed analysis of the fundamentals and history of fluorescence microscopy, the reader is referred

239

to the recent Ploem and Tanke monograph (1987) and Chapter 13 by Taylor and Salmon, this volume. Since the advent of epifluorescence illumination geometries (Brumberg, 1959), there has been steady growth in the use of fluorescence to visualize and localize subcellular components. Two natural outgrowths of these efforts were inevitable: to quantitate the fluorescence signal using photometric attachments (Thaer, 1966), and to capture the microscopic image using video detector systems (Reynolds, 1968). Early efforts were primarily photometric, i.e., measurements of total fluorescence flux. Advances in this area are marked by increases in the precision of the measurement and spatial resolution (for review, see Piller, 1977). While some role has been found from such analyses with a quantitative biological interpretation placed on the size of the total fluorescence signal, this approach is limited because of the multitude of artifacts which can influence the signal level. The emerging techniques in quantitative fluorescence microscopy revolve around spectrally sensitive measurements where the information is less corrupted by optical and physical artifacts. Typically, the factors influencing a fluorescent probe are evaluated by analyzing the difference or ratio of signals at two wavelengths, or in the more sophisticated approaches, complete spectral analysis of excitation and/or emission spectra (Kohen *et al.,* 1981; Wampler, 1986; Bright *et al.,* 1987a,b).

With modern technical advances — developments of laser light sources, computer automation and control, and digital imaging — the two major approaches to quantitation have pursued parallel goals. Photometric systems coupled with laser excitation beam scanning or computer controlled stage positioning quantitate spatially resolved fluorescence signals with high-signal measurement precision. Digital video based microspectrofluorometers analyze entire images for spectral properties with less precision in the measurement, but with more impressive temporal and spatial response. Thus, the focus of this chapter is on detector systems for quantitative micro*spectro*fluorometry and the special demands in hardware and the optical configuration of these systems.

In the limit of the ideal, the detector for a microspectrofluorometric system for studies of biological systems should deliver performance sufficient to allow the following:

1. Spatial resolution at the limit of the optical microscopy (i.e., submicrometer).
2. Spectral resolution sufficient for biological fluorophores (typically ~ 10 nm).
3. Temporal resolution sufficient for electrophysiological measurements (micro- to milliseconds).

4. Sufficient sensitivity to allow excitation irradiance levels low enough to avoid irradiation damage ($<50 \ \mu W/cm^2$).
5. Detectability equivalent to micromolar concentrations of fluorophores.

These needs translate very simply. The detector system must have single photon sensitivity, for it is this sensitivity that one trades off for the other aspects of performance listed. With enough sensitivity wavelength selection can be more rigorous and excitation irradiance levels can be lowered. Sensitivity also translates into lower detection limits, since the system can utilize narrower wavelength selection and the concomitant reduction in stray light (the major limit to most fluorescence analyses).

The need for this level of sensitivity is not obvious. Excitation irradiance levels of 50 $\mu W/cm^2$ of green light, for example, are equivalent to around $1.3 \times 10^{14}$ photons/second/cm$^2$. This is, of course, a large number, too large even for photon counting by a detector system irradiated at this level. However, this irradiance is not delivered to the detector! A typical 100-$\mu m^2$ sample area on a microscope stage would receive only $1.3 \times 10^{10}$/second of these photons. Temporal resolution in the millisecond range would further reduce the exposure to $1 \times 10^7 \hbar v$ per measurement interval. With video detectors, one video time frame (1/30th second) would give $4 \times 10^8 \hbar v$/ frame time. If the sample were 10-$\mu m$ thick and the fluorophore had an extinction coefficient of $5 \times 10^4$, a 10-$\mu M$ concentration would absorb about 0.1% of this light. With a quantum yield of 50%, the entire fluorescence emission band from such a sample would be represented by only about $10^5$ photons. With narrow, selective spectral isolation, even with relatively wide bandwidths, only a small portion of these photons would be delivered in the detection bandwidth ($<10\%$ at the emission maximum with a typical biological fluorophore and 20-nm band pass). Even with a high collection angle objective lens, the collection efficiency and transmittance of the optical system would further degrade this signal. Thus, the detector would see only a few thousand photons and the best photocathode convert only 10% or so of these to an electronic signal. The bright macroscopic irradiance is thus represented by a quantum limited fluorescence image on the microscopic scale. Not only must the detector be single photon sensitive but some form of averaging or counting is essential. Obviously, spatial resolution to the micron scale would mean a further loss (1/10,000). In fact, the above figures tend to be optimistic in several cases of widely used fluorescent probes. Thus, single photon detectors are an absolute necessity for state-of-the-art performance!

There have been a number of instrumental approaches to obtaining spectrally resolved information in fluorescence microscopy. Much of the current interest is in instrument systems which take the ratio of images obtained at two different exciting wavelengths (Tanasugarn *et al.,* 1984; Tsien and Poenie, 1986; Bright *et al.,* 1987a,b). However, there has also been a considerable interest covering over three decades of literature in instrument systems which can scan emission or excitation spectra from microscopic sources (Rousseau, 1957; Olson, 1960; Pearse and Rost, 1968; Rost, 1973; Ploem *et al.,* 1974; Jotz *et al.,* 1976; Kohen and Kohen, 1977; Hirschberg *et al.,* 1979; Rich and Wampler, 1981; Wampler, 1986; Kurtz *et al.,* 1987). While most of these systems employ photomultiplier tubes for quantitative analysis, the potential for measurements using imaging detectors has also long been recognized. The additional advantage of spatial information has been thought to outweigh the disadvantage of losses in sensitivity over the photomultiplier detector. However, to date only a few scanning spectroscopic imaging systems have been built (Rich and Wampler, 1981; Wampler, 1986; Kurtz *et al.,* 1987). Most quantitative systems rely on discrete filters for wavelength selection and are thus limited to measurements at only a few wavelengths at a time (Tsien and Poenie, 1986; Bright *et al.,* 1987a,b).

The first applications of sensitive video based detector systems to biological systems were by Reynolds (1964, 1968, 1972). Recent advances in instrumentation design, lowered costs, and the advent of new fluorescence probes have continued to expand these applications. Several recent reviews discuss the growth in hardware and applications (Reynolds and Taylor, 1980; Barrows *et al.,* 1984; Arndt-Jovin *et al.,* 1985; Benson *et al.,* 1985; Wampler, 1985a,b; Bright *et al.,* 1987a,b).

Of course, the ideals summarized above are not reached with a video based detector. For instance, standard video frame rates (U.S. standard = 30 frames/second) limit temporal resolution and the low signal-to-noise ratios of such detectors imposes further time limits because multisample averaging must be used to obtain precise quantitative measurements.

The basics of fluorescence microscopy are discussed by Taylor and Salmon, Chapter 13, this volume. The following overview (Sections I,A and I,B) summarizes the most typical geometry and the changes needed to make quantitative, spectrally resolved measurements.

## A.  Overview of Illuminating Light Path for Fluorescence Microscopy

The most typical geometry for fluorescence microscopy is the, so-called, epiillumination geometry, where the exciting light is delivered to the sample through the objective lens. The cross-section diagram of Fig. 1 illustrates the illuminating light path of such instruments. The component parts of the

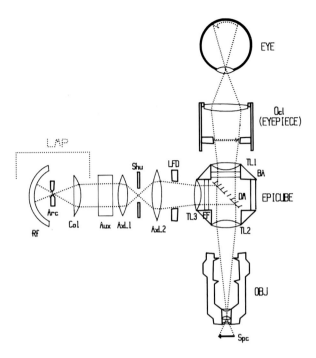

FIG. 1. The optical geometry of a typical fluorescence microscope. The optical components of an epiillumination fluorescence microscope geometry are indicated as discussed in the text. Arc, arc lamp; Aux, auxiliary components (e.g., filters); AxL1 and AxL2, auxiliary lens; BA, barrier filter; Col, source collector lens; EPICUBE, epiilluminator optical components for wavelength selection and beam splitting; DM, dichroic mirror; EF, excitation filter; LFD, luminous field diaphragm; LMP, arc lamp source; OBJ, objective lens; Ocl, ocular (eyepiece); Rf, lamp reflector; Shu, mechanical shutter; Spc, specimen; TL1, TL2, and TL3, telan lenses.

illuminating light train are the source wavelength selecting components, intervening optics, and the objective lens used as a condenser. The main purpose and design criteria for such systems are to uniformly irradiate the sample, through the objective, with a cone of wavelength selected light which fills the field of view (Koehler illumination) (see Taylor and Salmon, Chapter 13, this volume). Inhomogeneity in either the irradiance or wavelength over the field are major problems for quantitative measurements of subsequent fluorescence signals. A secondary goal is to control the irradiation intensity either to increase the fluorescence signal by delivering high irradiance levels or to control photodegradation of fluorophores by using low irradiance.

The illuminating light path of the fluorescence microscope begins at the illumination source (LMP in Fig. 1). The source for epifluorescence is typically a lamp selected either for brightness at selected excitation wavelengths or for broad spectral output or uniformity, depending on the application.

The main role of lenses in the illumination light path are to deliver the light to the limiting aperture and control its uniformity. Light rays emitted from the source are collected by a focusable lens or group of lenses called the collector lens (Col in Fig. 1). Typically, the lamp housing is also backed by a reflector (Rf). The collector lens/reflector system is designed to efficiently gather the light emanating from the source and should be selected on the basis of the arc size to be imaged, the aperture to be filled by the arc image, and a glass composition that allows efficient transmission of light at the desired wavelengths. A proper collector system should provide even and intense illumination at the specimen. A focusable collector lens is preferential to accommodate a variety of illumination sources and lamp positions. In the more complex systems, the collector lens is used as a collimator, i.e., the light rays leave it traveling nearly parallel.

As the light rays leave the collector lens they pass through auxiliary components (Aux) such as heat filters and, in the case of microspectrofluorometers, wavelength selecting components. They are then focused (by AxL1) forming a light source image plane. At this light source image plane the rays are bundled closely together. This is a good location for a shutter (Shu) and for stray light control via a limiting aperture. Rays leaving this image plane are collected by an auxiliary lens (AxL2) which matches the illumination beam to the collection angle of the objective and focuses an image of the source at the back focal plane of the objective.

As with all optical systems and very particularly with fluorescence systems, stray light must be minimized. Apertures and diaphragms are placed on the excitation light path to reduce off angle light, spurious reflections, etc. The limiting aperture (luminous field diaphragm, LFD) must be uniformly filled with light so that the sample is uniformly illuminated. To this end some manufacturers place a diffuser of some type at or just before the LFD or at the source. Of course, this may increase stray light and decreases illumination.

In commercial fluorescence microscopes, the wavelength selection components generally rest in the microscope body along with any beam splitter, prism, or lenses used to combine the illumination and imaging light paths. The entire group of components (epicube in Fig. 1) may be installed with a removable core, slider, or turret in order to allow rapid substitution of the wavelength selecting components for different fluorophores. In a conventional epifluorescence microscope, the reflector/beam splitter plays a role in wavelength selection. Typically, a dichroic mirror (DM) is used where coatings on the glass surface are designed to be reflective over one range of wavelengths and transmissive over other. Such devices are often also called cold mirrors or edge interference filters in the optical literature. It's important that the coated surface of such a beam splitter be on the side toward the source and objective, not the eyepiece side. Otherwise refractive effects on illumination passing through the glass substrate can cause image position shifts that are wavelength dependent.

For fluorescence microscopy, the edge where the transition from reflection to transmission occurs must be sharp (Fig. 2) and selected to match the fluorescence properties of the probe as indicated. If this edge is not sharp

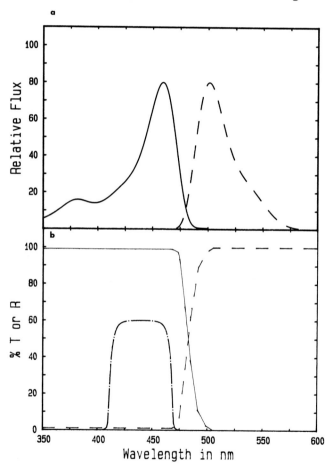

FIG. 2. Idealized description of the match between excitation and emission filters and the spectra of a fluorophore. (a) Excitation (solid line) and emission (dashed line) spectra showing the Stokes shift and the partial overlap of excitation and emission spectra. (b) Reflection spectrum (solid line) of dichroic filter used as a beam splitter. Transmission spectrum (dashed line) of same filter (idealized). Excitation band (dashed line with dots) which would give less stray light and restrict autofluorescence.

there is significant mixing of stray exciting light with the fluorescence and this degrades sensitivity. Typically, the epiilluminator "cube" will also contain other filters in addition to the beam splitter. The ideal excitation band, as shown in Fig. 2, allows full excitation of the fluorophore, but cuts off on the blue side as well as the red to limit fluorescence from other biological materials (autofluorescence), and transmission of near-infrared and infrared light from the lamp (heat filter).

This selection of exciting wavelengths should occur with the light rays parallel so as not to introduce a shift in the image when examining the same specimen at different wavelengths. Such shifts are dependent on the angle of the dichroic reflector or other reflecting element, its thickness, and its refractive index. This shift will manifest itself as a spatial displacement of an image or a parasitic image. Manufacturers tend to overcome this problem by adding telan lenses (TL1, TL2, and TL3 in Fig. 1) in their fluorescence illuminators, keeping the light rays parallel in the vicinity of the beam splitter–reflector. However, each additional air interface is a source of lowered transmission efficiency.

The wavelength-selected light travels through the objective lens (OBJ in Fig. 1) and is condensed by the objective onto the specimen (Spc). The objective should be selected on the basis of glass compound, number of glass elements, degree of correction, numerical aperture, and maximum transmittance. Generally, fluorite lenses are preferred over crown- or flint-type glass lenses. However, new technologies in lens fabrication now make it possible to use more highly corrected lenses (planapochromatic) in fluorescence application over a broad wavelength range. The old rule that planapochromatic lenses should not be used at excitation ranges below 400 nm does not apply. High numerical apertures are now being offered by a variety of manufacturers with high transmittance. Objectives must also be chosen on the basis of optical correction to limit biased information. Plan or flat field optics are desirable, if an investigator wishes to sample information off the optical axis. If a variety of wavelengths are sampled then an objective which is sufficiently corrected over the range of wavelengths is necessary to retain spatial register of the information in the image.

In the epifluorescence microscope, the objective functions as a condenser. This relationship yields a gain in brightness as a function of the fourth power of the numerical aperture of the objective (Inoue, 1986; Taylor and Salmon, Chapter 13, this volume). This is one of several reasons that high numerical aperture (NA) objectives are often used in fluorescence microscopy. However, as discussed below (Section I,B,2) there are several negative aspects in the use of high NA objectives.

The light rays emanating from the objective irradiate the specimen. The specimen should be prepared with attention to details such as the light transmission of mounting medium, cleanliness, transmission and thickness of cover slips, and sample thickness.

For quantitative measurements many of these "typical" microscope components are not acceptable or are difficult to adapt. About the only quantitative measurement that can be made with such an instrument by addition simply of a detector is that of total fluorescence flux. But since total fluorescence flux from a sample represents the combined effects of many different things (sample preparation, sample optical properties, physical and chemical quenching of the fluorophore, competition of absorption of light and scattering by nonfluorescence components of the sample, summed contributions from other fluorescent material, etc.), this measurement, while it can be made with precision, tends to be useless. Thus, in quantitative fluorescence microscopy, you can not simply add a detector. The following sections detail some of the concerns and changes needed with respect to the illuminating light path.

## 1. SOURCE

The mercury arc burner is very bright at a few selected wavelengths, the mercury lines. It is a good illumination source for systems where the fluorophore being studied has strong absorbance at one of these lines and where the measurement being made depends on emission spectral properties. In other cases, particularly in the growing number of cases where excitation spectral properties are being monitored, the lamp used needs to have broad uniform spectral output. The relative output and spectral distribution curves of Fig. 3 show the performance of some of the possible choices. Goldman (1968) and Ploem (1971) have discussed the selection of light sources for fluorescence microscopy. One thing not shown by these data is the dependence of spectral output on environmental and electrical factors. For example, tungsten filament based lamps have a spectral output dependent on the current flowing in the filament. Figure 4 shows a portion of spectra measured for a typical filament lamp burner as a function of the operating voltage (and current). This effect is responsible for the week to week variation in measurement calibration curves discussed by Bright *et al.* (1987a) and can be a source of considerable variation if such lamps are used without rigorous procedures to set their input current before each use.

Arc lamps, on the other hand, have more stable spectral output, but suffer from arc wander which contributes noise to the fluorescence signal they excite. It is the authors' opinion that the best general purpose source for fluorescence measurements is a Xenon arc lamp supplied by a particularly stable power supply with very precise current regulation. Using power supplies of the design of DeSa (1970), we have experienced extremely

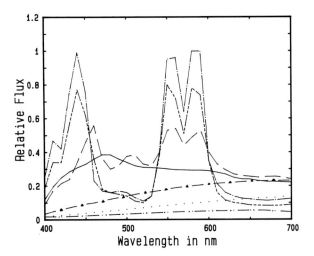

FIG. 3. Lamp emission spectra. Shown are the relative spectral outputs of a variety of microscope illuminators taken with a 10-nm band pass. Data shown are for 12 V quartz halide illuminators (30 W, dashed line with two dots; 60 W, dotted line; and 100 W, dashed line with triangles), high pressure mercury arc lamps (50 W, long and two short dashes; and 100 W, dashed line with dot), xenon arc (75 W, solid line) and the new tin halide lamp (dashed line). Data were obtained from Carl Zeiss, Inc.

stable, long lasting performance. Most commercial microscope power supplies are not very stable or well regulated. Other alternatives for stabilizing arc lamps have been discussed in the literature (Green *et al.,* 1968; Breeze and Ke, 1972; Katzir and Rosmann, 1973; Oldham *et al.,* 1987).

Arc lamps can also be the source of electronic problems. Radio frequency noise from the arc can be picked up by electrophysiological equipment. In addition, since the circuits that are used to start arc lamps often send high voltage transients through ground loops, these starting transients can cause failure of digital electronic components that are improperly grounded or isolated. A standard procedure of this laboratory is to turn off all digital electronic components prior to starting an arc lamp.

## 2. ILLUMINATOR GEOMETRY

When wavelength selection components are moved to the auxiliary location (Aux) in the illumination light path and toward the detector in the imaging light path, many of the components of the epiilluminator cube (epicube in Fig. 1) must be removed. The auxiliary excitation filter may interfere with spectral selections by a monochromator as would the barrier filter on the imaging light path. If wavelength selection is always made

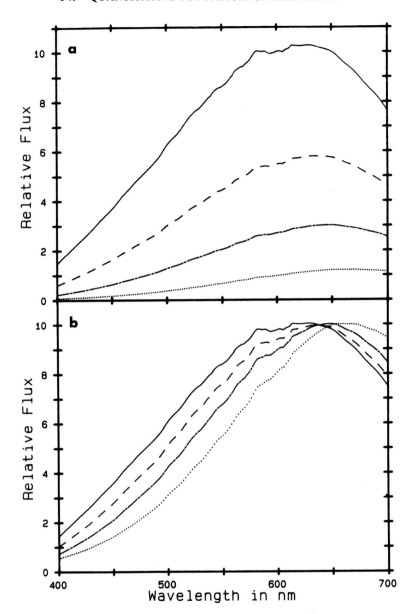

FIG. 4. Spectral and flux changes of a 100 W quartz halide lamp with operating voltage. (a) Unnormalized data showing the relative output from the lamp operated at 6, 8, 10, and 12 V. (b) Normalized data showing the spectral changes at 6, 8, 10, and 12 V (right to left). Using an optical pyrometer (Leeds and Northrup, Philadelphia, Pennsylvania) these curves are associated with color temperatures of 2090°, 2250°, 2420°, 2550°C, respectively.

elsewhere, the dichroic mirror may be replaced by a beam splitter and excess, unnecessary lenses removed from the light paths.

### 3.   POSITIONING APERTURES AND MONOCHROMATORS

One approach to the multiple wavelength selection necessary for meaningful quantitative fluorescence would be to automate interchange of dichroic filters. Multiple position filter sliders are made by several microscope manufacturers. Leitz uses a rotating carousel in some cases. When the measurement is based on very different wavelengths of two fluorophores such as the pH measurement technique based on the ratio of fluorescence intensities of two dyes originally described by Geisow (1984), this approach is certainly feasible. However, we are often interested in defining the shape of the spectrum of a single fluorophore or of closely related fluorescent species. In this case fluorescence measurements at multiple closely spaced wavelengths become necessary and the excitation wavelength selection components must be removed from the beam splitter compartment. By replacing the dichroic filter with a beam splitter and filtering the light at the source, considerable flexibility is obtained. The simplest geometry for such an illuminator system involves a lamp, collimator, and wheel containing multiple interference filters (Fig. 5). Several laboratories currently use this approach (see Bright *et al.,* 1987a,b, for references). The collimator allows optimum use of the filters so that the light beams striking them are all normal to their surface assuring accuracy of wavelength position selection and minimum bandwidth. If such filters are placed in diverging or converging light rays, their performance is degraded with a broadening of the band pass.

A more flexible alternative to discrete interference filters is the use of graded or variable spectrum interference filters such as those manufactured by Ealing Corporation (South Natick, Massachusetts) and Optical Coatings Laboratories (Santa Rosa, California). These devices allow continuously variable selection of wavelengths with relatively broad bandwidths over most of the visible wavelength range and can be mounted as in Fig. 5 in the same type of wheel as discrete filters. The data imposed on the image of Fig. 6 indicate how the center band position of illumination varies over the sample when such a device was used in the illumination light path of a research spectrofluorometer (Wampler, 1986).

Finally, the wavelength may be selected by a prism or grating monochromator. Optimum performance requires a more complex optical arrangement where the optics at the entrance to the monochromator match the source lamp's output to its collection angle. The output optics then focus this light or collimate it in order to couple it to the microscope optics.

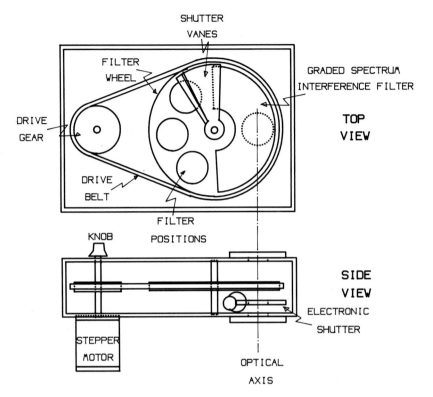

FIG. 5. Filter wheel monochromator. The wheel, mounted in a light tight aluminum housing with electronic shutter, has positions for three discrete 1-in. filters and a graded spectrum interference filter sector. Shutter vanes may be placed to block any portion of the wheel.

## B. Overview of the Imaging Light Path in Fluorescence Microscopy

### 1. SPECIMEN AS A PART OF THE OPTICAL COMPONENTS

In fluorescence and, in fact, in all microscopy, the specimen being examined is an integral component of the optical system. Refraction and reflection within the sample alters the signal being measured before it even reaches the objective lens. The image of the two capillaries in Fig. 7 illustrates the point. One capillary is filled with fluorescein dye, the other with water containing a bubble of air. Both are surrounded by water and covered with a cover slip. The image obtained shows the reflections on the air bubble – water surface from the fluorescence in the other capillary. Thus it is obvious that refraction and reflection give false signals where there is no fluorescence.

Theoretical studies and measurements on model systems by Kerker *et al.* (1979) indicate that optical effects cannot only change the light flux from an area of a specimen, but that localized fluorescence in subcellular particles may be affected spectrally. In addition, Benson and Knopp (1984) have shown that signals can be significantly contaminated by light from areas both above and below the plane of focus, and that high NA lenses can collect a large percentage of their total fluorescence signal from out-of-focus emitters.

Together, these combined effects make correlation of a measured fluorescence signal with a specific chemical or environmental factor within a cell difficult. They make simple fluorometric measurements without wave-

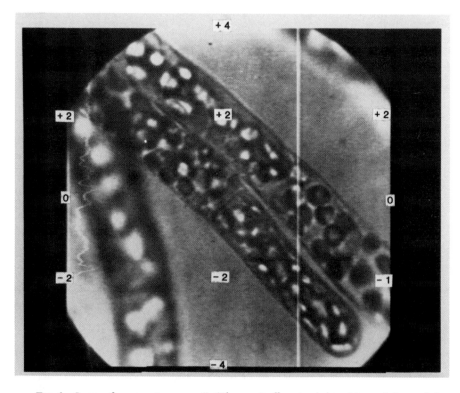

FIG. 6.    Image of moss protonema cells *(Physcomitrella patens)* viewed through the graded spectrum interference filter showing the wavelength deviation from the center band position over the measurement field in nanometers.

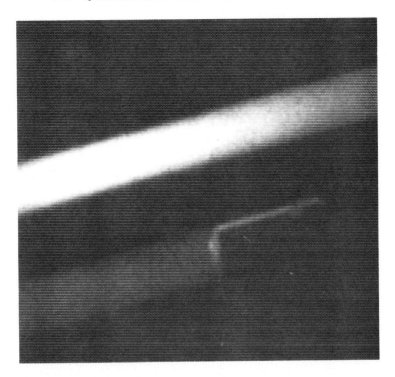

FIG. 7. Image of two capillaries showing reflective and refractive effects seen from the lower capillary (filled with water and an air bubble) from the fluorescence of the upper capillary filled with fluorescein solution.

length sensitivity fairly useless and have precipitated the change in focus of the field toward microspectrofluorometric analysis. The negative factor of high NA in collecting large portions of out-of-focus emission reinforces the arguments for detector sensitivity above and supports alternative approaches to spatial resolution in fluorescence such as laser excitation scanning and confocal microscopy when imaging specimens are much thicker than the depth of focus of the optics employed (Brakenhoff *et al.*, and Agard and Sedat, Volume 30, this series).

## 2. THE OBJECTIVE

The choice of objective lens is critical for quantitative fluorescence microscopy. Each objective used must have high transmission efficiency and contribute no fluorescence from its internal parts. In most applications there are two additional criteria, high light collection efficiency and good resolution. Both of these needs seem to translate into objective lenses with high NA. However, in fluorescence measurements things are not always as they seem.

Numerical aperture is a measure of the effective collection angle of the lens. A lens looking through air can capture a portion of the light limited by the refraction angle (Snell's law). In reference to the central point in the specimen, this angle is defined by the product of the refractive index of the intervening media (air, $n = 1$) and the half angle subtended by the lens itself, i.e.,

$$NA = n \sin \Phi \tag{1}$$

Since $\sin \Phi$ has its largest value (1.0) at 90°, with air as the medium the maximum NA is 1.0. In order to obtain larger collection angles, the lens must be submerged in a medium of larger refractive index. Thus, the highest NA lenses are oil immersion objectives. A 1.4 NA objective viewing a sample through oil immersion ($n = 1.5$) intercepts light through a plane half collection angle of 69° and a solid angle of 4 steradians. Since there is $4\pi$ steradians in the closed sphere, this means that if light emission from the specimen were isotropic, 32% of the emitted light would be within this collection cone. Partial reflection of some of this light at the lens surface would reduce the actual collection efficiency further.

In normal, bright field microscopy, high NA also means good resolution since the resolution of a microscope is limited in proportion to the reciprocal of the NA. However, one negative consequence of the use of high NA lenses is the close working distances. In order for the plane angle to be large, the lens must focus on a small nearby point. This, of course, is due to the fact that as any lens of a given physical size moves further from the focal point, the plane collection angle decreases and so, therefore, would the NA by Eq. (1).

A second consequence of high NA objectives is a shallow depth of focus, roughly proportional to the inverse square of the NA (in units of microns) (see also Taylor and Salmon, Chapter 13, this volume). Thus, a 1.4 NA lens has a depth of focus of approximately 0.5 $\mu$m. As discussed above and indicated by the study of Benson and Knopp (1984), this narrow depth of field does not, however, exclude collection of large amounts of out-of-focus fluorescence when emitters are present within the volume illuminated by the illumination cone. This can cause loss of contrast and considerable degradation in the resolution of fluorescence microscopy when studying thick samples. High NA lenses are also not recommended for photometry if the luminous field diaphragm is to be varied in size in order to isolate different size particles (Goldman, 1968). Thus, in some cases it may be wise to trade detector sensitivity for lower NA in order to gain photometric flexibility or resolution in a fluorescence image.

Highly corrected lenses may also not be necessary in fluorescence microscopy where the imaging light path generally involves filtration to a relatively narrow range of wavelengths. A planapochromatic objective may contain a dozen or so lens elements. Even with the best antireflection coatings, the transmission loss due to reflections will be of the order of 15% with additional losses from the absorption of the glues and coatings.

## 3. Optics between Objective and Eyepiece

The returning fluorescence signal in a conventional epifluorescence microscope must pass the beam splitter and any intermediated optics used by the manufacturer to control the internal light path. Regardless of the efficiency of the antireflection coatings and the size of these optical elements, some light losses will occur. Typically, broad-band multilayer antireflection coatings reduce on axis (normal angle of incidence) reflection to around 0.5%. Off axis their performance degrades again with a typical value of say 1% at 45° incidence. Ten such surfaces in a light path would then reduce the transmission by from 5 to 10%.

## 4. Role and Position of Ocular

The ocular is a series of lenses which are placed in the light path to collect the primary image and convert it to the virtual image. The function of the ocular is to magnify, control field of view, and correct or compensate for abberations in the objectives. Thus, in many systems, the primary image is not fully corrected and an ocular is necessary to provide the detector with a fully corrected image to quantify. Increased magnification of the primary image by the ocular also allows for higher spatial resolution in both photometric and image-based systems.

All oculars control the field of view presented to the detector. Inside each type of ocular there is an ocular field diaphragm which limits this field. The location of the diaphragm coincides with the primary image position. This same location may also be used for pin holes and apertures to reduce the area of the field presented to a detector. In addition, as mentioned above, oculars compensate for chromatic abberations in the objectives in most microscopes. A recent development is the use of a system where a color compensating tube lens corrects for chromatic abberations rather than a compensating eyepiece. This is purported to yield better corrected images and increase flexibility of the microscope.

## 5. Placing a Detector

In quantifying fluorescence images, the position of the detector becomes critical in obtaining the desired data. Placement may depend on whether the information desired is primarily spatial or temporal. While in most systems detectors are located at an image plane formed by an ocular or projection eyepiece, there are instances where it is advantageous to place the detector directly at the primary image plane. This is particularly true in situations involving very low light levels where removal of unnecessary air to glass interfaces is desirable in order to increase optical efficiency. In the light path of the microscope light rays are bundled closely at the primary image plane yielding a potentially more intense concentration of light but lower magnification. Thus, placement of the detector is dictated by whether the investigator wishes to measure only total emitted light or to spatially resolve the fluorescence signals.

Placing the detector at the virtual image plane offers advantages in increased spatial resolution. Typically with vidicon-based image digitizing systems, images are digitized at either $512 \times 512$ or $1024 \times 1024$ matrices limiting the number of points of information available for quantification. By increasing the magnification to the detector, detail can be measured with a greater precision, i.e., a single picture element (pixel) represents a smaller area. The disadvantage of increasing the magnification to the detector is that there is usually a significant light loss associated with the introduction of complex lens elements and their coatings (see above). In addition to light loss, the spectral properties of the antireflection coatings can inhibit the transmittance of desired wavelengths. Glare can also be introduced in the system through dust, polishing marks, and fingerprints on these surfaces. Generally, it is a good idea to limit the number of air to glass interfaces to only the essential necessary to accomplish enough spatial separation for quantification.

## II. Accessories for Quantitative Fluorescence Analysis

For quantitative measurements, two additional types of components are often added: apertures and wavelength selecting components. Apertures are added to control stray light on both illuminating and imaging light paths, and to limit the measurement area for photometry. In general, wavelength selection of both the emission and excitation beams requires additions to both light paths of filters or monochromators.

### A. Positioning of Pin Holes and Apertures

Pin holes and apertures are positioned in the light path to achieve a variety of results. Depending upon where in the light path the aperture is positioned, the effect can modify data, control glare, and reduce internal reflections which might bias data collected from the specimen. Examining the illumination light path, the most common areas for the introduction of apertures are at the field diaphragm, at the rear focal plane of the objective, and at other nonimage planes. In the imaging light path apertures are found in the rear focal plane of the objective, at the primary image, and before the detector. Each of these areas have different ramifications in quantifying a fluorescence image.

The luminous field diaphragm plane is coincident with the image of the specimen. When the full field is to be illuminated, an aperture set to just outside the field of view reduces stray light or glare in the body of the microscope. Since it is generally advantageous to only radiate the specimen area to be quantified, thus eliminating the photobleaching of adjacent areas, pin holes and apertures can be inserted to limit the field of illumination on the specimen. The illuminated field stop can either be a pin hole, slit, or any other shape which might be useful in measuring the specimen.

The next significant image plane on the illumination path is the rear focal plane of the condenser, or in the case of epifluorescence, the rear focal plane of the objective. In this image plane, apertures are introduced to limit the size of the illuminating cone angle of light passing through the condenser/ objective. This constriction of the cone angle of light limits the amount of light passing through the system by reducing the angle of light available at the specimen plane and what is imaged by the objective. Restricting this cone decreases the effective NA and increases the depth of field of the lens and may be an advantage to reduce corruption of the spatially resolved fluorescence signal by out-of-focus light as discussed above (Section I,B,2). Modulators and phase rings are introduced in this plane and can have a limiting effect on the amount of light transmittance of the system. Generally it is advantageous not to limit the aperture in this plane with such devices. In the case of modulators not only do you limit the aperture of the objective, you also introduce another air to glass interface which can contribute to a reduction in transmittance.

In the imaging light path the field diaphragm serves the same function of area selection as in the illumination light path. In the primary image plane light rays are bundled close together. Apertures in this image plane are used to restrict the field size presented to the ocular and, in photometry, to limit the measured spot in the image. The diffraction limit for unambiguous spatial localization of measurements demands an aperture diameter equivalent to about 0.5 $\mu$m at the specimen (Zimmer, 1983; Piller, 1977). Apertures here also serve the function of eliminating glare and stray light. This plane is also referred to as the ocular field diaphragm. It is used to further restrict the size of the image presented to the detector and can limit parasitic light from striking the detector and biasing the measurements. As with the field diaphragm, the ocular field diaphragm can be a pin hole, slit, or of their shapes which might conform to the area to be quantified. It is generally recommended to make your measurements directly on the optical axis thus minimizing the effects of spherical and chromatic aberrations.

## B.   Wavelength Selection and Isolation

### 1.   Fixed Wavelength Selection (Filters)

There are essentially four types of wavelength selecting filters that might be used in the microspectrofluorometer: gelatin filters, colored glass filters, interference filters, and cold and hot edge filters. The transmission characteristics of all of these filters consists of three components, reflection loss at the front surface, absorption or interference loss within the filter, and reflection loss at the back surface. Typically these filters are uncoated and the reflection losses can be as great as 5–10% even at normal incidence angle.

Gelatin filters such as the Kodak Wratten filters consist of gelatin film containing dissolved organic dyes. The film is coated with lacquer for protection. They are available with a wide range of transmission properties, broad-band pass, and long wavelength transmission (cut on) and short wavelength transmission (cut off). The gelatine base has a low melting temperature and is hygroscopic, thus these filters are not suited to harsh environments and must be protected from temperatures above about 50°C, or moisture. Intense light in the absorption band of a gelatin filter can cause local heating and melting. In addition, prolonged exposure to UV light can bleach the dyes and change the filter's characteristics.

Colored glass filters such as the Corning, Hoya, and Schott glass series also offer a wide range of transmission characteristics. The Hoya and Schott glass filters are constructed of optically ground and polished optical quality glass. As with the gelatin filters, since their operation depends on absorption of the unwanted light, local heating can occur due to this absorption. In the extreme case this heating can cause a colored glass filter to crack.

Interference filters use thin film, partially reflecting layers to obtain a Fabry-Perot interference effect which selectively enhances transmission of some wavelengths and greatly diminishes transmittance at others through destructive interference. They are typically available in a series of band-pass filters with fixed half bandwidths. One side of these filters is usually highly reflective to the off band wavelengths and if it faces the source, internal heating is reduced. If it does not face the sources, stray light effects are often seen.

The center band wavelength of interference filters is dependent on the angle of incidence. Near the normal ($< 10°$) this is seldom a concern, but at severe angles it can cause a significant shift in the filtered band position (typically 10% lower wavelength at 45°). Similarly and for the same reason, interference filters placed in uncollimated light will have a broader bandwidth than indicated by their specifications.

Edge interference filters make ideal cut on and cut off filters for fluorescence studies. They tend to be designed with much sharper transition from reflection to transmission than a corresponding colored glass cut on filter. Their use as wavelength selective beam splitters in most fluorescence microscopes simplifies the optics needed dramatically.

## 2. Variable Wavelength Selection (Monochromators)

The choice of monochromator for fluorescence microscopy is dictated primarily by efficiency. High spectral resolution is generally not required, but the stray light characteristics of a system should be considered. Many early systems used prisms to select wavelengths (Rousseau, 1957; Olson, 1960; Pearse and Rost, 1968; Rost, 1973; Ploem et al., 1974). Today, grating monochromators are much more available. They have high efficiency and simple triangular transmission curves which tend to be independent in shape regardless of the selected wavelength. However, since gratings disperse light in several overlapping orders, there will be contamination of the selected light with light from other orders. For an instrument operating in the first grating order, set so that the detector "sees" wavelength $\lambda$, the second order stray light is from the spectral region $\lambda/2$. This stray light problem is obviously more important if the monochromator is on the emission side of the spectrofluorometer than if it is on the excitation side.

Grating monochromators are not equally efficient at all wavelengths. Their peak efficiency is determined by the angle and shape of the grating's grooves. The peak efficiency wavelength is called the blaze wavelength. With ruled diffraction gratings efficiency falls off fairly quickly on either side of the blaze wavelength (down to 50% at $\frac{2}{3}\lambda$ and at $2\lambda$). Holographically recorded gratings (Hayat and Pieuchard) have the advantage that the efficiency curve is less wavelength dependent, but their peak efficiency in the visible region of the spectrum is generally less than that of a similar ruled grating.

With imaging detector systems the constraints for wavelength selection on the emission side are more restrictive since the image quality must be maintained. With low light video detectors one solution has involved using a simple variable spectrum interference filter. This type of monochromator has a relatively broad bandwidth, but maintains good image quality. With biological and organic fluorophores good quality fluorescence spectra can be measured with this device (Wampler, 1986). Monochromators employing circular graded interference filters can easily be incorporated into a microscope (Fig. 5) and, via computer control, allow rapid, repetitive selection or wavelengths and spectral scanning (Rich and Wampler, 1981; Wampler, 1986).

Another alternative to rapid scanning in a photometric type of system was described by Olson (1960). In this case a prism monochromator was used dispersing the image of the photometric aperture. A spiral slit was then used at the plane of the dispersed image to select wavelengths seen by a photomultiplier tube. This slit (in a wheel) could be turned rapidly to continuously scan spectral data.

Optical multichannel analyzers (OMA) can also be used in conjunction with a spectrograph for photometric measurements (Jotz et al., 1976). With newer, more sensitive OMAs such as those manufactured by Princeton Applied Research (New Jersey), rapid and sensitive spectral analysis can be performed on the microscopic spot delimited by an aperture in the specimen image plane.

Finally, an important new approach has just been announced where high frequency modulation of a birefringent crystal mounted on a piezoelectric transducer is used to set up a standing wave pattern which, in turn, diffracts a narrow wavelength band of light (Kurtz et al., 1987). The so-called acousto–optical tuned filter has been said to give 1 nm spectral resolution with rapid wavelength selection or scanning. These devices are being commercialized by several groups including AOTF Technology, Perceptics, and Westinghouse and should become available in the near future.

## III.   Detectors for Quantitative Fluorescence Microscopy

In quantitative measurements in fluorescence microscopy, there are essentially two classes of detectors used, photomultiplier tubes and imaging devices of a variety of types. Photomultiplier tubes are the detector of choice for precise measurements provided no spatial information is desired or that it can be obtained by scanning some other component of the system (laser excitation beam scanning or mechanical stage scanning). Image detectors are used where spatial detail is needed and it must be obtained rapidly.

## A.   Photomultiplier Tubes

The advantage of the photomultiplier tube (PMT) is due to its excellent amplification characteristics. Several types of PMTs have been developed, each with its own distinctive characteristics (see Boutot *et al.,* 1983; and Candy, 1985, for recent reviews). Each design employs the photocathode film as the quantum detector which emits photoelectrons in response to absorption of light. The emitted photoelectrons are drawn from the vicinity of the photocathode and accelerated toward a nearby secondary emitter by a positive voltage gradient. In conventional PM tubes the secondary emitter is called a dynode. Each dynode in the amplification chain is maintained at a more positive potential than the one preceding it, and each electron impact accelerated by this potential produces several secondary electrons which in turn strike the next dynode producing more. The resulting electron gain is commonly $>10^6$ electrons and with the best modern tubes can exceed $10^8$.

The commercial photometers for microscopes generally use side window tubes. Their main advantage in such applications is that they are compact and inexpensive. While they don't get much respect in the physics and engineering literature, a direct comparison of specifications with end window tubes of the same manufacturers is often favorable. However, most custom-designed instrument systems use end window tubes and, with the push to higher sensitivity and photocounting electronics, these tubes offer the most range of flexibility and extended performance.

In addition to the conventional end window construction with a series of discrete dynodes for electron multiplication, a newer technology has emerged using a continuous channel electron multiplier (see Boutot *et al.,* 1983). The channel electron multiplier is a tube lined with a semiconductor coating. A gradient of potential is maintained from one end of this tube to the other. Individual photoelectrons enter the tube, strike the semiconductor coating, cause emission of secondary electrons, etc. Recently, miniaturization of these channels has led to development of a variety of types of compact arrays of channels called microchannel plates (MCPs). MCPs are used in very fast, high gain PMTs and as compact image intensifiers for low light imaging. Commercial MCP PM tubes which use two channel plates in sequence achieve electron gains of $4 \times 10^5$ with very fast pulse characteristics which are particularly useful for fluorescence lifetime measurements. Whether DC or photon counting electronics are used, the typical output signal of a PMT is amplified to such an extent that it can be amplified and processed with little addition of noise.

The performance characteristics of PMTs of most importance to this discussion are the quantum efficiency, the overall gain of the device, and the dark current. High gain tubes allow the use of low cost, easy to build amplifiers with amplifier noise making a negligible contribution to the total noise in the signal. Quantum efficiency (Q) measures how many photoelectrons are produced per photon of light. Typically, quantum efficiency is very wavelength dependent and peaks at less than 30% in the blue region of the spectrum. The spectral sensitivity curve of a PMT is determined by the wavelength dependent quantum efficiency of the photocathode material and how that material is deposited on the light collecting surface. Examining the published quantum efficiency spectra from several manufacturers, one sees that there are few, if any, tubes with quantum efficiency above 10% in the spectral region above 600 nm. In addition, without special and more expensive window materials (quartz, fused silica, etc.) most common PMTs have quantum efficiencies less than 10% for all wavelengths less than 300 nm.

At first, it might seem that the best PMT quantum efficiency spectrum would be the flattest, highest spectrum extending from UV to the red part of the visible spectrum. This type of photocathode would allow spectral studies on a whole variety of fluorophores. However, there are tradeoffs for this flexibility. First is cost, since extended response tubes (both ends of the spectrum) are more expensive. Second, broad response means broad sensitivity to stray light and, typically, stray light is the final limit in fluorescence analysis. In addition, since the work function of red sensitive tubes is lower, they give more noise in the signal even in the absence of stray light. In these tubes thermal events eject electrons spontaneously. Thus, for optimum performance red sensitive tubes are usually cooled with the cooling system adding to both the cost and complexity of the overall system. If the measurement goal is to define subtle spectral shape differences or to look at the fluorescence from a wide variety of fluorophores, then an extended S-20 photocathode on a UV windowed tube may be best. However if the spectral range of emission can be anticipated and only limited spectral information is required, a better choice might be a tube with more limited response.

The dark current of a PMT is primarily a function of the extent of red spectral response as mentioned above and the photocathode area. However, since thermal electrons can also originate from dynodes, the area of secondary emitter surface is also important. In photon counting, this latter contribution is less important since the pulse height from these electrons is lower and can be discriminated against.

## B.   Imaging Detectors

Imaging detectors, on the other hand, are generally not high gain devices. In addition, because of the response time requirement for readout of the spatial information with reasonable repetition rates, they are also generally much noisier than the conventional PMT–amplifier combination in reading out the same signal from a pixel. Aikens et al., Chapter 16, and Spring and Lowy, Chapter 15 (this volume), detail the characteristics of both vidicon and solid state image detectors. A wide variety of technologies have been used in designing image detectors. Progress in this area is monitored by the contributions to the *Symposia on Photo-Electronic Image Devices* published by Academic Press as part of their "Advances in Electronics and Electron Physics" series. The most recent volumes in this series (Morgan, 1985a,b) cover advances in image intensifiers, solid state detectors, and a variety of different types of camera tubes. In general, the sensitivity of solid state detectors and target vidicon tubes is insufficient for low light applications. However, intensified and cooled detectors have been developed which have single photon sensitivity (see below; Lowrance, 1979; Reynolds and Taylor, 1980; Wampler, 1985a,b).

## 1. INTENSIFIED VIDICONS

Some improvement in the sensitivity of conventional vidicons has been achieved by incorporating image intensifier stages in front of the target. The simplest such devices use a detecting photocathode and a high voltage potential to accelerate the photoelectrons before they strike a vidicon target. Electron optics are used to keep the image focused. These devices are called intensified target vidicons and there are two main types depending on target construction, SEC vidicons where the target is a secondary electron conductor (MgO, KCl, MgF, and Ag), and SIT (silicon intensified target) vidicons where the target is a silicon diode array. Commercial SEC vidicons are about ten times more sensitive than conventional vidicons while SIT cameras can be up to 500 times more sensitive. Only in the case of the SIT vidicon is single photon detection accessible with cooling and slow scan readout (Milch, 1979; Gruner *et al.*, 1978).

Solid state array detectors may also be cooled and used alone, or intensified with improvements in sensitivity, (Airey *et al.*, 1985; Aikens *et al.*, Chapter 16, this volume). Intensified position sensitive detectors have been employed as photon counting imagers (Tsuchiya *et al.*, 1985). However, unlike the other low light detectors mentioned here, the position sensitive devices are not integrating and are limited to event rates small enough to allow the subsequent circuitry to handle each event as it occurs. This limits them to very weak light signals and makes detection of dynamic changes (even camera focusing) difficult.

## 2. SINGLE PHOTON IMAGE DETECTOR SYSTEMS

Single photon sensitivity is desirable if we are going to quantitate spatially resolved fluorescence signals without constant problems with fluorophore concentration, photobleaching, and stray light. The technologies employed by most commercially available cameras generally preclude their use in real-time detection of quantum limited images due to high dark noise which limits their usable integration time for weak signals. As mentioned, cooling (in the case of SIT and ISIT vidicons, and CID and CCD cameras) has been used to reduce this dark signal to allow long-term integrations, but, of course, the added sensitivity is gained at the expense of time resolution. An ideal imaging device would exhibit the gain and dynamic range of a photomultiplier with real time resolution.

Detector systems formed by coupling video cameras to high gain image intensifiers have been used to answer this need in several fields, not just in microscopy. Perhaps the earliest and most extensive development efforts occurred in astronomy (see Meaburn, 1976). The first such systems for biological and microscopic applications were developed by Reynolds and co-workers (Reynolds and Botos, 1970; Reynolds et al., 1978; Milch, 1979; Gruner et al., 1978). Reynolds' basic system uses a very high gain ($\sim 4 \times 10^6$), low noise (from 6 to 20 dark events/cm/second) 4-stage image intensifier, coupled to either a SIT or SEC (secondary electron conductor) vidicon. The only major limitation of these systems is the physical size of the intensifier itself and size and voltage requirements (35 kV) of its supporting electronics. The IDG low light video system (Rich and Wampler, 1981) was designed to be similar in performance, but more compact. It employs a small two stage MCP intensifier requiring less sophisticated support electronics. It also achieves single photon sensitivity, but with considerably higher noise ($\sim 21,000$ dark events/cm$^2$/second). While this is indeed much higher, it still represents only 3% of the pixels in a single field image (1/60th second, $256 \times 256$ pixels) and in fluorescence applications stray light is more limiting than the dark noise of the detector. However, image processing techniques can be used to significantly reduce the effect of noise (Wampler, 1985c) and the detector system is small enough to easily mount in either a conventional or inverted microscope configuration. A similar system, developed for astronomical applications, is described by Airey et al., (1985). It employs a high gain double MCP intensifier which has excellent noise characteristics ($< 200$ dark events/cm$^2$/second). This intensifier is optically coupled to a CCD camera with $244 \times 190$ pixels.

Since these systems were developed, commercial low light level video cameras have been manufactured with sensitivities for single photon detection (see Wampler, 1985b; Spring and Lowy, Chapter 15, this volume). Some of these cameras are even more compact using fiber optic coupling between the intensifier stage and the camera tube. The most sensitive of these commercial systems known to the author are the intensifier–camera systems marketed by Photonic Microscopy, Inc. (Oak Brook, Illinois). These systems utilize an unusually high gain image intensifier coupled to a vidicon camera and include digital processing electronics.

All of these systems, commercial and custom designed, have one very important characteristic when compared to conventional video cameras alone. Most are able to form an image with illumination levels varying over at least one million-fold by changing the gain and high voltage on the intensifier. With a variable iris diaphragm to limit exposure of the detector, even wider operating ranges can be obtained. Broad dynamic range is an advantage of any system where the first stage detector is a high gain image intensifier. The gain of these systems can be translated in fluorescence microscopy into lower excitation irradiation levels and narrower wavelength selection. They allow use of lower output, broad-band light sources, and monochromators for wavelength isolation.

### ACKNOWLEDGMENTS

The work on which this manuscript is based was supported by the National Science Foundation (PCM 79-12316), the National Institutes of Health (GM28209 and GM37255), the United States Department of Agriculture (86-CRCR-1-1954), and the University of Georgia Research Foundation. The authors are indebted to Dr. Bob White for helpful discussions and advice.

### REFERENCES

Airey, R. W., Lees, D. J., Morgan, B. L., and Trynar, M. J. (1985). *Adv. Electron. Electron Phys.* **64A**, 49–59.

Arndt-Jovin, D. J., Robert-Nicoud, M., Kaufman, S. J., and Jovin, T. M. (1985). *Science* **230**, 247–255.

Barrows, G. H., Sisken, J. E., Allegra, J. C., and Grasch, S. D. (1984). *J. Histochem. Cytochem.* **32**, 741–746.

Benson, D. M., and Knopp, J. A. (1984). *Photochem. Photobiol.* **39**, 495–502.

Benson, D. M., Bryan, J., Plant, A. L., Gotto, A. M., and Smith, L. C. (1985). *J. Cell Biol.* **100**, 1309–1323.

Boutot, J. P., Nussli, J., and Vallat, D. (1983). *Adv. Electron. Electron Phys.* **60**, 223–305.

Breeze, R. H., and Ke, B., (1972). *Rev. Sci. Instrum.* **43**, 821–823.

Bright, G. R., Fisher, G. W., Rogowska, J., and Taylor, D. L.. (1987a). *J. Cell Biol.* **104**, 1019–1033.

Bright, G. R., Rogowska, J., Fisher, G. W., and Taylor, D. L. (1987b). *Bio Tech.* **5**, 556–562.

Brumberg, E. M. (1959). *Biophysics* **4**, 97–104.

Candy, B. H. (1985). *Rev. Sci. Instrum.* **56**, 183–193.

Geisow, M. J. (1984). *Exp. Cell Res.* **150**, 29–35.

DeSa, R. J. (1970). *Anal. Biochem.* **35**, 293–303.

Goldman, M. (1968). "Fluorescence Antibody Methods." Academic Press, New York.

Green, M., Breeze, R. H., and Ke, B. (1968). *Rev. Sci. Instrum.* **39**, 411–412.

Gruner, S. M., Milch, J. R., and Reynolds, G. T. (1978). *IEEE Trans. Nucl. Sci.* **NS–25**, 562–565.

Hayat, G. S., and Pieuchard, G., eds. "Handbook of Diffraction Gratings, Ruled and Holographic." Jobin Yvon Optical Systems, Division of J.-Y. Diffraction Gratings, Inc., Metuchen, N. J. (undated).

Hirschberg, J. G., Wouters, A. W., Kohn, E., Kohen, C., Thorell, B., Eisenberg, B., Salmon, J. M., and Ploem, H. S. (1979). *In* "Multichannel Image Detectors" (Yair Talmi, ed.), Vol. 1, pp. 262–283. American Chemical Society, Washington, D.C.

Inoue, S. (1986). "Video Microscopy." Plenum, New York.

Jotz, M. M., Gill, J. E., and Davis, D. T. (1976). *J. Histochem. Cytochem.* **24**, 91–99.

Katzir, A., and Rosmann, M. (1973). *Rev. Sci. Instrum.* **45**, 453.

Kerker, M., Chew, H., MNulty, P. J., Kratohvil, J. P., Cooke, D. D., Sculley, M., and Lee, M.-P. (1979). *J. Histochem. Cytochem.* **27**, 250–263.

Kohen, E., and Kohen, C. (1977). *Exp. Cell Res.* **107**, 261–268.

Kohen, E., Thorell, B., Hirschberg, J. G., Wouters, A. W., Kohen, C., Bartick, P., Salmon, J. M., Viallet, P., Schachlschabel, D. O., Rabinowitch, A., Mintz, D., Meda, P., Westerhoff, H., Nestor, J., and Ploem, J. S. (1981). *In* "Modern Fluorescence Spectroscopy" (E. L. Wehry, ed.), pp. 295–339. Plenum, New York.

Kurtz, I., Dwelle, R., and Katzka, P. (1987). *Rev. Sci. Instru.* **58**, 1996–2003.

Lowrance, J. L. (1979). *Adv. Electron. Electron Phys.* **52**, 421–429.

Lynch, T. F. (1979). *In* "Applications of Electronic Imaging Systems" (R. E. Franseen and K. K. Schroder, eds.), pp. 36–41. Society of Photo-Optical Instrumentation Engineers (SPIE 143), Bellingham, Washington.

Meaburn, J. (1976). "Detection and Spectrometry of Faint Light." Reidel, Dordrecht.

Milch, J. (1979). *IEEE Trans. Nucl. Sci.* **NS-26**, 338–345.

Morgan, B. L., ed. (1985a). Photo-electronic image devices. *Adv. Electron. Electron Phys.* **64A**.

Morgan, B. L., ed. (1985b). Photo-electronic image devices. *Adv. Electron. Electron Phys.* **64B**.

Oldham, P. B., Patonay, G., and Warner, I. M. (1987). *Anal. Instrum.* **16**, 263–274.

Olson, R. A. (1960). *Rev. Sci. Instrum.* **31**, 844–849.

Pearse, A. G. E., and Rost, F. W. D. (1968). *J. Micros.* **89**, 321–328.

Piller, H. (1977). "Microscope Photometry." Springer-Verlag, New York.

Ploem, J. S. (1971). *Ann. N.Y. Acad. Sci.* **177**, 414–428.

Ploem, J. S., and Tanke, H. J. (1987). "Introduction to Fluorescence Microscopy." Oxford Univ. Press, Oxford.

Ploem, J. S., de Sterke, J. A., Bonnet, J., and Wasmund, H. (1974). *J. Histochem. Cytochem.* **22**, 668–677.

Reynolds, G. T. (1964). *IEEE Trans. Nucl. Sci.* **NS 11**, 147–151.

Renyolds, G. T. (1968). *Adv. Opt. Electron Microsc.* **2**, 1–40.

Reynolds, G. T. (1972). *Rev. Biophys.* **5**, 295–347.

Reynolds, G. T., and Botos, P. (1970). *Biol. Bull.* **139**, 432–433.

Reynolds, G. T., and Taylor, D. L. (1980). *Bioscience* **9**, 586–592.

Reynolds, G. T., Milch, J., and Gruner, S. (1978). *Rev. Sci. Instrum.* **49**, 1241–1249.

Rich, E. S., and Wampler, J. E. (1981). *Clin. Chem.* **27**, 1558–1568.

Rost, F. W. D. (1973). *In* "Fluorescence Techniques in Cell Biology" (A. A. Thaer, and M. Sernetz, eds.), pp. 57–63. Springer-Verlag, New York.

Rousseau, M. (1957). *Bull. Microsc. Appl.* **7**, 92–94.

Talmi, Y., ed. (1983). "Multichannel Image Detectors," Vol. 2. American Chemical Society, Washington, D.C.

Tanasugarn, L., McNeil, P., Reynolds, G. T., and Taylor, D. L. (1984). *J. Cell Biol.* **98**, 717–724.

Thaer, A. A. (1966). *In* "Introduction to Quantitative Cytochemistry." (G. L. Wied, ed.), pp. 409–426. Academic Press, New York.

Tsien, R. Y., and Poenie, M. (1986). *TIBS* **11**, 450–455.

Tsuchiya, Y., Inuzuka, E., Kurono, T., and Hosoda, M. (1985). *Adv. Electron. Electron Phys.* **64A**, 21–31.

Wampler, J. E. (1985a). *In* "Bioluminescence and Chemiluminescence: Instruments and Applications" (K. van Dyke, ed.), Vol II, pp. 123–145. CRC Press, Boca Raton, Florida.

Wampler, J. E. (1985b). *In* "Chemiluminescence and Bioluminescence Today" (J. Burr, ed.), pp. 1–44. Dekker, New York.

Wampler, J. E. (1985c). *Comput. Vision, Graphics Image Process.* **32,** 208–220.

Wampler, J. E. (1986). *In* "Applications of Fluorescence in the Biomedical Sciences" (D. L. Taylor, A. S. Waggoner, R. F. Murphy, F. Lanni, and R. R. Birge, eds.), pp. 301–319. Liss, New York.

Zimmer, H. G. (1983). *In* "Micromethods in Molecular Biology" (V. Neuhoff, ed.), pp. 297–328. Springer, New York.

# Chapter 15

# *Characteristics of Low Light Level Television Cameras*

KENNETH R. SPRING AND R. JOEL LOWY

*Laboratory of Kidney and Electrolyte Metabolism*
*National Heart, Lung, and Blood Institute*
*National Institutes of Health*
*Bethesda, Maryland 20892*

269

# I. Introduction

When fluorescent probes are illuminated in the presence of oxygen, bleaching and photodynamic damage result (Foskett, 1985; Giloh and Sedat, 1982; Plant *et al.*, 1985). The extent of these adverse consequences is a function of the number of fluorescent molecules excited by the illuminating beam. Two approaches have been used to reduce bleaching: (1) removal of oxygen; and (2) reduction of the intensity, or duration of the illumination. Although the addition of antioxidants to the mounting medium for fixed cells effectively reduces bleaching (Giloh and Sedat, 1982), this approach is unacceptable for studies on living preparations. Investigators are left only with reduction of excitation light intensity as a practical solution to the problem. Since the extent of bleaching is proportional to the integral of the exciting light flux (Foskett, 1985; Plant *et al.*, 1985), the investigator must balance illumination intensity and duration to acquire the desired image with a minimum of damage to the preparation. With very low intensity excitation light the number of photons emitted by the fluorochrome are so limited that the resultant image is of poor quality. In this chapter we will describe the general approach to acquisition of low light level video images and the characteristics of the devices used to obtain such images.

# II. Requirements for a Low Light Level Video Camera

## A. Sensitivity

The photon fluxes corresponding to various illuminance levels are shown in Table I. It is apparent that at illuminances of $1 \times 10^{-8}$ foot candles (fc) or less the photon flux is so low that usable video information cannot be obtained in a single video frame (33 msec). The images obtained will only prove useful if extensively averaged or integrated. Long periods of image integration of a preparation illuminated with very low excitation intensities are often used in an attempt to minimize bleaching and still obtain acceptable image information. Such an approach suffers both from photodynamic effects during the relatively long exposure period as well as decreased temporal resolution. It is better, in our opinion, to increase the excitation light intensity and shorten the exposure and thus the integration time such that the duration–intensity product and signal-to-noise ratio are unchanged. This approach enables improved temporal and spatial resolution with sufficient measurement speed to accommodate the rapid changes which may occur in living cells.

TABLE I

MICROSCOPE IMAGE APPEARANCE AND BRIGHTNESS

| Appearance of image | Illuminance[a] | | Irradiance[b] (nW/cm$^2$) | Photon flux[b,c] | |
|---|---|---|---|---|---|
| | fc | lx | | #/second | #/frame |
| Bright field | $10^{-2}$ | $10^{-1}$ | 16.6 | $41.5 \times 10^9$ | $13.3 \times 10^8$ |
| Detection limit of eye | $10^{-5}$ | $10^{-4}$ | 0.017 | $41.5 \times 10^6$ | $13.3 \times 10^5$ |
| Detection limit of best image tube | $5 \times 10^{-8}$ | $5 \times 10^{-7}$ | $0.85 \times 10^{-4}$ | $2.0 \times 10^5$ | $0.65 \times 10^4$ |
| Practical lower limit | $10^{-8}$ | $10^{-7}$ | $0.17 \times 10^{-4}$ | $0.4 \times 10^5$ | $0.13 \times 10^4$ |

[a] fc denotes foot candles and lx equals lux; 1 lux is equivalent to 0.093 foot candles.
[b] At a wavelength of 500 nm.
[c] In number of events per second or per video frame (33 msec).

There are two approaches to the production of a video camera with sufficient sensitivity for low light level work: employment of a device with high quantum efficiency such as a charged coupled device (CCD) (see Aikens et al., Chapter 16, this volume) or use of a photoemissive device such as an image intensifier. Both approaches have been used with success for video imaging of living preparations (Arndt-Jovin et al., 1985; Bright and Taylor, 1986; Bright et al., 1987; Wick, 1985). In both cases the quality of the resultant image is largely determined by the noise characteristics of the detector and associated electronics, and sensitivity is limited by the signal to noise of the camera.

## B. Signal-to-Noise Characteristics

The noise of the ideal low light level video camera is solely due to that associated with emission from the photocathode. In practice the noise due to subsequent amplifier stages usually dominates the noise of the device (Csorba, 1985; Electro-Optics Handbook, 1974). Noise arising at the photosensor is typically temperature dependent and can be reduced by cooling of the detector. Cooled CCD detectors have been used with great success in astronomy (Timothy, 1983) and biology (Arndt-Jovin et al., 1985; see Aikens et al., Chapter 16, this volume). In these systems, cooling to liquid nitrogen temperatures virtually eliminates noise arising from the photosensitive surface; noise is then primarily associated with readout of the stored charge (Aikens et al., Chapter 16, this volume; Arndt-Jovin et al., 1985; Timothy, 1983). Although cooling of photomultiplier tubes is used routinely to reduce detector noise (dark current), lowering the temperature of image tubes is not a common practice.

A signal-to-noise ratio of 3 is typically taken as the detection limit for imaging systems (Csorba, 1985; Electro-Optics Handbook, 1974; Inoue, 1986). As with photomultiplier tubes, signal to noise may be increased by image averaging with the resultant reduction in temporal resolution. The tradeoff between spatial and temporal resolution is of prime consideration in the selection and use of low light level video systems. It is generally not sufficient for the biologist to merely ascertain the presence of a fluorescent signal; we wish to know considerably more. A signal to noise of 3 is too low to make measurements with confidence either of the intensity of the fluorescent signal or of its spatial distribution. At low photon fluxes the only way to increase the information content of the video image is to accumulate images. Since detector noise is random, image averaging increases the signal-to-noise ratio by approximately the square root of the number of frames (Bright et al., 1987; Inoue, 1986). Image averaging may be done by means of a digital image processing system (Bright and Taylor, 1986; Bright et al., 1987; Inoue, 1986; Wick, 1985) or by integration of the light input on the face of the photodetector. Such integration is crucial for the successful employment of CCD-based cameras (Aikens et al., Chapter 16, this volume; Arndt-Jovin et al., 1985) and may also be used on image tube systems (so called "gated tubes"). In gated devices, light is allowed to fall on the face of the detector in the absence of any scanning beam. The signal will be integrated on the tube photosensor (up to the charge storage capacity of the photocathode) and may then be read out when the tube is gated to the "on" state.

Proper determination of the signal-to-noise ratio of a image tube is a complex undertaking requiring specialized test equipment generally not available in the laboratories of biologists. An approximation of the performance of a camera may be made by measurement of the signal amplitude of a high contrast resolution test target slide on the microscope stage. An example of such a target (the USAF 1951 resolution test slide) and the resultant intensity profile across the field are shown in Fig. 1. The amplitude of the signal minus the dark current divided by the square root of the noise amplitude gives a measure of the signal to noise of the camera (Bright

---

Fig. 1.  Image of a USAF 1951 resolution test target for the determination of signal to noise of a video camera. The signal is taken as the amplitude of the white bar in 8 bit resolution digitizer gray levels, where black is 0 and white is 255. The noise is calculated as the square root of the amplitude of the signal over the black bar. The dark current of this camera was equal to the level of the video signal over the black bar. At the bottom of the screen is shown the amplitude of the signal along a single scan line; the location of the selected line is shown by the horizontal line in the middle of the field. The signal to noise determined from this figure was 14.

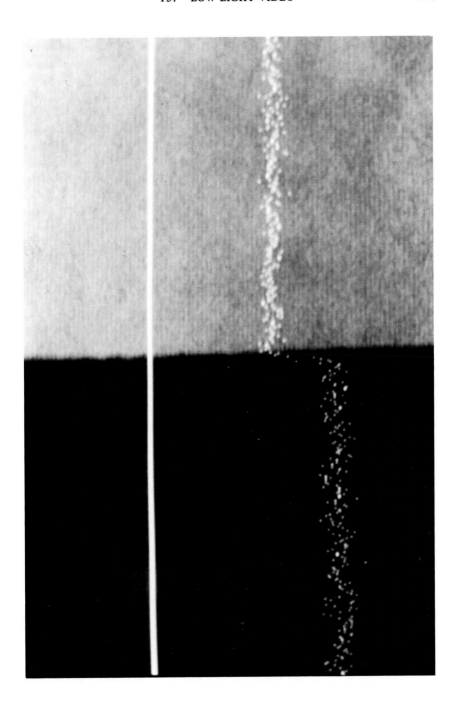

and Taylor, 1986; Inoue, 1986; Electro-Optics Handbook, 1974). This procedure may be carried out at any desired illuminance at wavelengths selected by the investigator to match experimental conditions.

Another significant source of noise in low light level images is related to the quantal nature of photons. At $1 \times 10^{-8}$ fc illuminance of the photocathode, about 1300 photons impinge on a 1-cm$^2$ photosensor in a video frame. This corresponds (Table I) to about 40,000 photons arriving at the face plate per second. For a photocathode with a 20% quantum efficiency about 8000 photons are detected and amplified. If the investigator is using an imaging system with a resolution of $512 \times 512$ pixels, each pixel will display a signal due to a photon once every 32.8 seconds. If the investigator were to confine his view to a single pixel, it would show a true signal about once in every 1000 video frames. Since the detector also exhibits noise, the likelihood of the observed signal being real must be weighed against the possibility of it being due solely to random noise. The solution to this problem has generally been to sacrifice spatial and temporal resolution for the resultant improvement in signal detection ability (Bright *et al.*, 1987). Quantal noise cannot be eliminated by improvements in detector operating characteristics, although detector noise can be reduced to insignificance in photon-counting imaging systems (Csorba, 1985; Tsuchiya *et al.*, 1985; Wick, 1985).

## C.  Spectral Sensitivity

Most imaging detectors exhibit one of the spectral curves shown in Fig. 2; silicon photosensors, such as the CCD, have different spectral properties depending on the design of the device (Aikens *et al.*, Chapter 16, this volume). The best photocathodes for most present day image tube applications in video microscopy are the S-20 or multialkali. Quantum efficiency of these photosensors is in the 20–30% range; the S-25 photocathode is about 10 to 15% quantum efficient. Silicon photosensors exhibit extremely high quantum efficiencies of 60–80%, but it is difficult to take advantage of this characteristic without cooling the detector (Aikens *et al.*, Chapter 16, this volume; Timothy, 1983). The extended red sensitivity of some photocathodes (S-25 and silicon) can be used in many microscope applications (Waggoner *et al.*, Volume 30, this series). However, the red and infrared emissions from the light source must be reduced to prevent high background signals from reducing contrast. Many image intensifiers have fiber optic input windows which do not transmit at wavelengths below 350 nm, limiting their use at these wavelengths (Csorba, 1985; Electro-Optics Handbook, 1974).

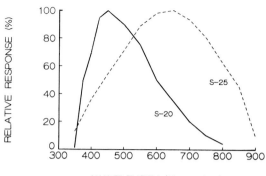

FIG. 2. Typical spectral sensitivity curves for two photocathodes commonly used in low light level cameras. The S-20 and related photocathodes (multialkali) are used in SIT, ISIT, specialized image tubes, and some image intensifier tubes. The S-25 photocathode is used in image intensifiers designed for night vision and red-sensitive applications.

# D. Spatial Resolution

Although a chapter of this series is devoted to determination of the modulation transfer function of imaging systems (Young, Volume 30, this series), it is of some value to review the determinants of spatial resolution at low light intensities. Figure 3 shows spatial resolution, expressed in TV lines per video image, as a function of the input light intensity at the face plate of

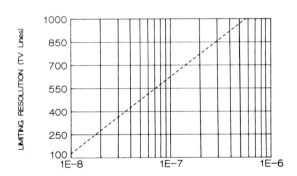

FIG. 3. The theoretical relationship is shown between detector limiting resolution and face plate illuminance in foot candles (fc). The resolution is calculated for the number of detected photons and is therefore independent of the quantum efficiency of the detector (Electro-Optics Handbook, 1974).

the detector. At the lowest illuminance, $1 \times 10^{-8}$ fc, resolution is limited to about 100 TV lines/image regardless of the operating characteristics of the detector (Electro-Optics Handbook, 1974). The spatial resolution of an imaging device is usually expressed as the 3% modulation point obtained from images of a test target with 100% contrast. Resolution expressed in this fashion is really the limiting value obtained at relatively high light intensities. Images obtained at lower light levels exhibit less resolution; ideally, averaging of such low light level images should result in the same resolution as obtained at high light intensities. Figure 4 shows an example of the limiting resolution of an intensified system (upper left), a low light level image of the same target (upper right), an 8 frame average (lower left), and a 16 frame average (lower right). It is our opinion that the best practical

FIG. 4.   Resolution test target at relatively high light levels (A), low illuminance (B), after averaging 8 frames (C) or 16 frames (D) of the low illuminance image. Input resolution (in TV lines) calculated for each panel are as follows: A, 685; B, 429; C, 479; D, 537. Reprinted with permission of Video Scope International, Ltd.

approach to determination of resolution of a microscope imaging system is utilization of a test target such as that shown in Fig. 4. A negative version of the target may be used to simulate fluorescent signals of the desired wavelength. The microscope constitutes an ideal optical bench for determination of the performance characteristics of a television camera. Intensity, wavelength, image size, contrast, and test target resolution are all readily controlled by the investigator. There is no need, in our opinion, to invest in sophisticated video camera testing equipment when a microscope is at hand.

## E.  Temporal Resolution

The speed of response of low light level cameras depends on photoconductive and capacitative effects at the photocathode (Electro-Optics Handbook, 1974). The response speed of video cameras is expressed as the lag of the image which is the percentage of the signal remaining in the next field (Csorba, 1985; Electro-Optics Handbook, 1974; Inoue, 1986). Since it is the object of the investigator to use the camera as a two-dimensional photometer, it is necessary to ascertain the time at which the signal obtained is a valid indicator of the face plate illuminance. In most detectors, lag is an exponential process whose time constant may be used to predict the magnitude of the response after the desired interval. For example, a 50-msec time constant camera responds to 67% of the final value in 50 msec, 89% in 100 msec, and 96% in 150 msec after the change in signal occurs.

When the time constant of the detector response is known, the investigator may choose a sampling speed which meets the experimental time and measurement accuracy constraints. Lag is increased in most image tubes when the signal current is reduced as in low light level microscopy. This occurs because the scanning electron beam does not fully discharge the charge accumulated on the camera tube target. The manufacturer's lag curves for image tubes must be viewed with attention to the corresponding illuminance when choosing a detector for low light level video (Electro-Optics Handbook, 1974; Inoue, 1986).

It is relatively easy for the investigator to determine the response speed of a video camera under the illumination conditions appropriate for the experiment. The field should be illuminated at the intensity level desired and the intensity suddenly changed to a lower or higher value. The intensity change should be of sufficient magnitude for accurate quantitation (50% is ideal) and fast compared to the response speed of the camera. The introduction or removal of a neutral density filter of the appropriate transmission into the optical path is all that is required. This may be done, for example, by the rotation of a filter wheel (Wampler, 1986) or by the change in voltage controlling a modulator (Spring and Smith, 1987). The resultant video images may be recorded on tape or disk and analyzed later for the intensity as a function of time. If the recording and analysis equipment is not available, a storage oscilloscope record of the video signal amplitude may be used to estimate the time constant of the camera.

The investigator must make a realistic appraisal of the response speed requirements of the experiment. While it may seem desirable to have the fastest system possible, the cost both in dollars and in system complexity necessitates compromise. Lag in image tube cameras is such that no more than 15 frames/second can be captured with fidelity (90–95% response). As mentioned above, at lower light intensities this rate diminishes considerably to no more than 4 frames/second (Electro-Optics Handbook, 1974; Inoue, 1986). The most rapid conventional systems involve a fast-response image intensifier coupled to a CCD camera operated at video rates (Arndt-Jovin et al., 1985; Spring and Smith, 1987). These devices are virtually free of lag and may be gated for the very fast (microsecond) acquisition of single images. Their continuous recording and display rate is still limited to 30 frames/second. Solid state cameras are now commercially available with framing rates of 400–1000 frames/second. These devices could be combined with high speed intensifiers to produce a true 1–2 msec response system; such a system would require about 30 times higher face plate illuminance to produce images of similar quality to those obtained at normal framing rates.

## F.   Gain Uniformity

Image tubes always exhibit some spatial nonuniformity in the photo-cathode sensitivity; solid state detectors (CCD) are generally more uniform in their responsivity. Additional nonuniformities in the video signal result from camera electronics involved in scanning the electron beam. Such "shading" is more substantial in low light level cameras because of distortion introduced by the image intensification stage (Bright and Taylor, 1986; Bright et al., 1987; Csorba, 1985; Spring and Smith, 1987); elaborate gain correction maps may be required to achieve photometric accuracy over the entire image (Bright et al., 1987; Williams et al., 1985). Ratio imaging, discussed in another chapter, has been used extensively to compensate for such nonuniformities (Bright et al., Volume 30, this series). Figure 5 shows how ratio imaging corrects for an intentionally introduced nonuniformity in illumination in an otherwise uniform-response low light level video camera. The ability of the investigator to discriminate small changes in signal amplitude diminishes, even with ratio imaging, if the extent of

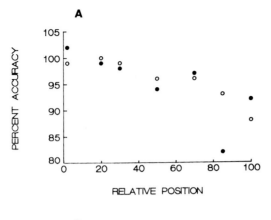

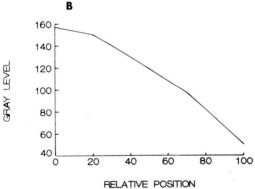

FIG. 5. Percentage accuracy of the measurement of a 10% perturbation in the intensity of the input illuminance as a function of the relative position of the measuring box. A relative position of 0 corresponds to the center of the image and 100 corresponds to the right edge of the field. The intentionally produced gradient in input illuminance, B, shows that input level at the edge of the field dropped to less than half of the value at the center. A shows the accuracy of the intensity ratio measurement for a 10% change in input illuminance superimposed on the gradient. The filled circles show the ratios obtained from 8 frame integrations of the images at the two intensities. The open circles show the ratios obtained from 256 frame integrations of the same input. The accuracy of the measured ratio falls from 100% at the center of the field to about 90% at the edge of the field where the gradient in gray level is most pronounced.

shading becomes too large (Bright *et al.*, 1987, Volume 30, this series). Certain types of image intensifier tubes (see below) are free of gain nonuniformities and may be used in conjunction with tube or solid state detectors to produce a camera free of shading (Csorba, 1985; Spring and Smith, 1987; Timothy, 1983).

## G.  Gain Linearity

When a low light level camera is used as a photometer it is assumed that the camera output is a linear function of the face plate illuminance (Bright and Taylor, 1986; Inoue, 1986). In video camera nomenclature, the tube is said to have a $\gamma$ of 1. Graphs of image tube signal current versus face plate illuminance are readily available (Electro-Optics Handbook, 1974; Inoue, 1986) and show a linear relationship over a wide range of input light intensities. These graphs do not always represent the operating characteristics which prevail during quantitative imaging. If the gain of the camera is fixed, as is the case in most photometric applications, the usable range of camera operation is greatly reduced. Figure 6 shows a typical curve for a newvicon tube camera in which the linear range of operation is limited to a 10- to 20-fold variation in input intensity. The investigator is presented with the problem of adjusting the specimen illumination, or degree of staining, so that the resultant fluorescence intensity falls within the linear operating range of the camera (Bright and Taylor, 1986; Bright et al., 1987). This constraint can severely limit the utility of some image-tube cameras and has led to the employment of CCD based systems because of their larger dynamic range (Aikens et al., Chapter 16, this volume; Timothy, 1983).

When image tube camera gain is allowed to change, the linear operating range can be greatly extended (Electro-Optics Handbook, 1974; Inoue, 1986). Difficulties in gain stability and reproducibility have limited this approach in image tube low light level cameras (Bright and Taylor, 1986; Bright et al., 1987). Image intensifiers exhibit linear input–output relations

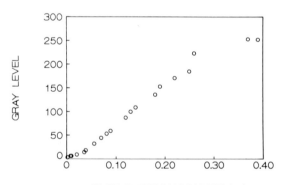

FACE  PLATE  ILLUMINANCE  (lux)

Fig. 6.  Output gray level of a newvicon tube camera is shown as a function of face plate illuminance in lux (where 1 lux = 0.093 fc). The linear operating range of this fixed gain camera is limited to the region from 20 to 220 gray levels. At lower intensities, camera noise obscures the changes in input illuminance; at higher illuminances, camera output saturates and no longer corresponds to the input.

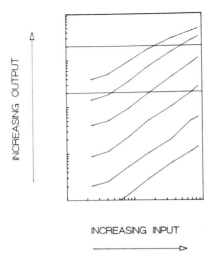

INCREASING OUTPUT

INCREASING INPUT

Fig. 7. A family of gain curves for an image intensifier showing the effect of alterations in intensifier gain setting; gain is increased by raising the voltage across the microchannel plate of the image intensifier. The bottom curve is at minimum intensifier gain and the top curve is at maximum intensifier gain. The linear relationship between output and input signals shows that gain is constant at each setting. At maximum gain (top curve), some nonlinearity appears in the relationship as the limit of microchannel plate current is reached. The horizontal lines near the top of the figure show the approximate output intensity range which corresponds to the sensitivity of a typical newvicon tube camera used as a second stage. The curves show that the usable operating range of the intensifier could be extended if a more sensitive camera were used as the second stage.

over a wide operating range (Csorba, 1985; Spring and Smith, 1987). The gain of these devices can be rapidly and reproducibly changed to extend the dynamic range of the imaging system (Csorba, 1985; Spring and Smith, 1987). Figure 7 shows the relationship between input illuminance and output brightness for an image intensifier with external gain control. The linear operating range of such a device is largely determined by the sensitivity of the video camera to which it is coupled rather than by the characteristics of the intensifier (Spring and Smith, 1987).

## H.   Geometric Distortion

The extent to which the output image truly represents the geometry of the original specimen varies among different low light level cameras. The distortion of some cameras is obvious to the observer upon imaging a test target (Bright and Taylor, 1986; Bright et al., 1987). Correction of geometric distortion in the image processor is a complex and computationally intensive task. The lowest distortion systems are based on solid state detectors with or without image intensification (Aikens *et al.,* Chapter 16, this volume; Spring and Smith, 1987). It is our experience that geometric distortion of 3% or more is readily detectable from observation of the image of a grid or linear target.

# III.   Criteria for Camera Selection and Evaluation

## A.   Choosing the Right Camera for Your Application

In the section which follows, we will review the characteristics of the low light level cameras that are currently available for use in quantitative light microscopy. Before the investigator can make an informed decision about the device of choice for his experimental application, the relevant parameters should be defined. They are as follows:

1. *Sensitivity.* Must the images be obtained at or near photon limited conditions?
2. *Resolution.* Do you need to see much detail or is an area average sufficient?
3. *Spectral Sensitivity.* No photocathode is optimum for all wavelengths. What is required for the fluorochromes in use?
4. *Speed.* How frequently must images be obtained? Include the time required for filter changes, camera lag, and image storage in this estimate.

## B.   Evaluating Camera Performance

The camera should be attached to a microscope equipped for bright field illumination of a resolution test target slide. The camera output should be displayed on an oscilloscope screen as well as on a monitor. The oscilloscope is an essential component of the test apparatus; it enables the determination of the amplitude of the video output of the camera. Because oscilloscopes are high impedance devices, the camera video output cannot be solely connected to it but must be connected to a proper termination (75 $\Omega$). Check camera output with the shutter closed for sync pulse amplitude and shape. Open the shutter and project an input of sufficient intensity to yield a full video output signal. The output signal amplitude should conform to the video standard, generally 1-V amplitude peak to peak. The intensity, wavelength, and area of the test target visualized may now all be adjusted to suit the experimental requirements. Intensity is most accurately controlled with crossed polarizers although stacks of neutral density filters may also be employed. Wavelength selection should be accomplished by use of interference filters; care should be taken to ensure adequate suppression of infrared from the lamp. Evaluation of camera sensitivity, spatial resolution, signal to noise, temporal response, and geometric distortion may now be readily accomplished as outlined in previous sections. Shading or gain nonuniformity can only be assessed if the illumination beam is uniform, thus diffusion disks and careful lamp adjustment may be required. Adjustment of camera pedestal and gain should be done with the aid of the oscilloscope to ensure that the output obtained is within the operating range of recording or processing equipment. The automatic features of many television monitors can make poor quality video outputs look very good; the recorded images may be far inferior. Ideally, all monitors should be calibrated by the use of a video test signal generator. Both brightness and contrast knobs should be marked with reference points for the black and white levels which correspond to video black level and maximum video

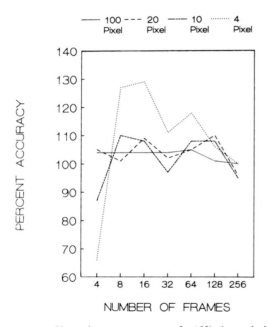

FIG. 8. The accuracy of intensity measurements of a 10% change in input brightness is shown as a function of the number of video frames integrated as well as of the size of the sampling window. The largest sample window is a box 100 pixels on each side and the smallest window is 4 pixels on each side. All of the pixel intensity values are averaged within the window to produce a mean gray level. As window size decreases, more frames must be integrated to achieve the same measurement accuracy as is realized with the larger areas. At the maximum integration (256 frames), all window sizes give equally accurate estimates of the intensity change. At the minimum integration (4 frames), only the 20 × 20 and 100 × 100 pixel windows accurately depict the intensity change.

signal. In the absence of a test signal generator the image processor can be used to generate a gray scale of known amplitude for calibration of the monitors. Only after the camera has been evaluated in this fashion can a realistic appraisal be made of its performance compared to other detectors. Fluorescence images, except those of a standardized test object, are notoriously difficult to employ in the assessment of video camera performance.

## C.  The Tradeoffs: Space versus Time

As outlined in the previous sections, low light level imaging involves a balance between spatial and temporal information. The accuracy of an illuminance measurement improves as the sample area and sampling time increase. Figures 8 and 9 give examples of the tradeoff between spatial and temporal resolution. This type of evaluation should be performed on every

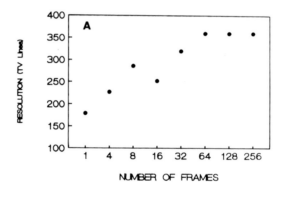

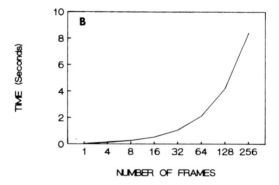

FIG. 9.  Horizontal resolution in TV lines of an intensified camera is shown as a function of the number of frames integrated (A). (B) Time required for the integration. The improvement in spatial resolution must be balanced against the diminution in temporal resolution.

imaging system to determine the validity of the measurements obtained. In the tests shown in the figures, a reduction of illuminance of 10% was produced and the performance of the system evaluated. Our requirement for acceptable performance was that this 10% perturbation should be detectable with 90% or better precision. We examined the diminution in spatial or temporal resolution required to achieve our goal. Increasing the size of the sample window improved measurement precision without sacrificing speed, but decreased spatial resolution (Fig. 8). Increasing the sampling time improved measurement precision without loss of spatial resolution, but with significant loss of temporal resolution (Fig. 9). An evaluation of this type is required to determine the measurement accuracy of an imaging system.

## IV. Performance Characteristics of Existing Low Light Level Cameras

Detailed specifications and performance graphs of tube and camera characteristics are readily available from the manufacturers (Electro-Optics Handbook, 1974) and in a recent book (Inoue, 1986). It is not our intention to reproduce such information but instead to give an overview of the salient features of each class of detectors.

### A. Silicon Intensified Target Cameras (SIT)

These image tube cameras are useful for moderately low light intensity inputs (to $10^{-5}$ fc), the rule of thumb being that the SIT cannot see what is not visible to the dark-adapted eye. SIT cameras have a very wide dynamic range in the automatic gain configuration, high limiting resolution (700 TV lines), low lag (25 msec for 67% response) at high light levels ($10^{-4}$ fc), and moderate lag (50 msec) at their low light limit ($10^{-5}$ fc). The photocathode is of the multialkali family with peak sensitivity at about 440 nm. The SIT is a good choice for imaging applications where bleaching and photodynamic damage are not serious problems and where precise control of camera gain is not needed. Applications such as immunofluorescence at multiple wavelengths can take advantage of this camera type (see Waggoner *et al.*, Volume 30, this series).

### B. Intensified Silicon Target Cameras (ISIT)

This camera tube contains a SIT fiber optically coupled to an image intensifier. The intensifier is of the "first generation" type (Csorba, 1985;

Electro-Optics Handbook, 1974) and therefore exhibits both gain nonuniformity and considerable geometric distortion (Bright and Taylor, 1986; Bright et al., 1987). ISIT cameras are capable of photon-limited imaging at light levels as low as $5 \times 10^{-8}$ fc. They have good limiting resolution (600 TV lines), low lag (25 msec) at $10^{-5}$ fc, and moderate lag (40 msec) at $10^{-6}$ fc. The multialkali photocathode of the image intensifier has a peak sensitivity near 480 nm. ISIT cameras have poor signal to noise at low light levels and are best utilized in conjunction with image processors (Bright and Taylor, 1986; Bright et al., 1987). In the automatic gain mode ISIT cameras have an extraordinary operating range, however, gain reproducibility and stability are poor under manual control (Bright and Taylor, 1986; Bright et al., 1987). The ISIT has been the low light level camera most frequently used in the fluorescence microscopy of thin, living cells and a relatively wide range of wavelengths.

## C.  Other Image-Tube Cameras

A number of specialized image tubes have been developed but never popularized. These include the Isocon, secondary electron conduction (SEC), and electron-bombarded-silicon diode-array target (EBSICON) tubes. These tubes are also available fiber optically coupled to image intensifiers. While these image tubes display many desirable features, their high cost and limited availability have virtually precluded their use in light microscopy. They do not represent practical solutions for the biologist to the problems of low light level imaging.

## D.  Image Intensifiers

Image intensifiers have been used by biologists for many years in conjunction with film and video cameras. These devices and their operating characteristics have been described in detail in a number of recent publications (Csorba, 1985; Reynolds and Taylor, 1980; Spring and Smith, 1987; Wick, 1985). In brief, they amplify the light falling on their photocathode and produce an output image on a phosphor screen. Modern intensifiers (so-called second- or third-generation devices) employ microchannel plates as electron multipliers and do not introduce geometric distortion or gain inhomogeneities into the output image. They are linear devices with a very large dynamic range in the automatic gain mode; gain may be readily and reproducibly altered under manual control. Image intensifiers are available which approach the theoretical limit for photon detection and resolution (Csorba, 1985; Spring and Smith, 1987). The response speed of these devices is usually much faster ($\mu$sec) than any biological requirements and is primarily determined by the properties of the intensifier output phosphor. Spatial resolution of image intensifiers can be very high, exceeding the requirements of typical image processors (Spring and Smith, 1987). A 50-fold range of intensities is faithfully reproduced within a given scene by these devices. Most low light level cameras employ an image intensifier as the first stage. Modular intensifiers coupled by relay lens optics to the detector of choice have been used for quantitative microscopic applications (Reynolds and Taylor, 1980; Spring and Smith, 1987). Many image intensifiers were designed for night vision applications and employ red-sensitive photocathodes and the infrared emissions of some lamps must be considered.

## E. Solid-State Cameras (CCD)

The characteristics of cooled solid-state detector cameras are described in detail in another chapter (Aikens et al., Chapter 16, this volume). These devices exhibit high resolution, extraordinary intrascene dynamic range (several thousand-fold), very high quantum efficiency, and the capability of on-chip integration of the light input. Their principle disadvantages are associated with the requirement for slow readout (to reduce readout noise) and the relatively high noise floor (Aikens et al., Chapter 16, this volume; Timothy, 1983). The result is a device which is best utilized as a slow-scan image integrator rather than a real time low light level video camera. Since solid-state detectors exhibit no lag, blooming, geometric distortion, or shading, they are far superior to image tubes. More experimentation will be required to fully evaluate the role of cooled CCD cameras for imaging fluorescence signals from living cells. Furthermore, there is room for improving the performance of these devices designed for microscopic applications. The combination of an image intensifier and solid-state detector results in a fast, sensitive camera without many of the problems of the image tube systems (Csorba, 1985; Spring and Smith, 1987). The principle disadvantage of such systems is the limited resolution and sensitivity of the solid-state detectors presently available. However, technical advances are expected in the near future (see Aikens et al., Chapter 16, this volume).

## F. Position-Sensitive Detectors

When the photon flux is very low, an image intensifier assembly may be used to detect and amplify individual photons. The resultant electron stream impacts onto a surface equipped with coordinate readout electronics. Such a photon-counting imaging system has been developed for light microscopy (Tsuchiya et al., 1985; Wick, 1985). The speed of the camera is limited by the coordinate readout electronics to about 10,000 counts/second; complete images require many seconds or minutes to accumulate. As many as five microchannel plates may be combined in the image intensifier to increase the amplitude of the electrical pulse derived from the photoelectron and ensure discrimination of the signal from the noise of the system. These devices achieve the ultimate in detector sensitivity and dynamic range. Their poor temporal resolution limits their applicability in biology.

# V.  Future Developments

It is our opinion that future improvements in low light level imaging will come not only from better cameras but also from improvements in the fluorescence microscope. The confocal laser microscope is one such development which produces fluorescence images of high resolution and contrast with shallow depth of field (Brakenhoff, Volume 30, this series). The present limitation to the use of such microscopes for the study of living cells lies in the mechanics of specimen scanning. The excitation beam in a scanning microscope resides at any point in the image for a brief time. If the microscopic field is to be digitized at a resolution of $512 \times 512$ pixels and at video rates, something that is not possible now, the scanning laser beam can reside at each point for only about 0.1 $\mu$sec. In the conventional fluorescence microscope the image is integrated on the face of the camera for 33 msec before being read out by the scanning beam. Since photodynamic damage and bleaching are proportional to the duration–intensity product of the excitation light, the intensity of the exciting beam in the scanning microscope could be increased 330,000-fold over that in the conventional microscope without any adverse effects on the preparation. However, in the scanning microscope only the region being detected is illuminated, and a conventional video camera is not an appropriate detector. Therefore, new classes of video-rate detectors will have to be developed to properly utilize the confocal scanning microscope.

It is clear that the rate of development of low light level image acquisition systems is now very high and will remain so for the next few years. The selection of the camera or other detectors will depend on the requirements of the experimental situation (i.e., how many wavelengths needed, spatial resolution, temporal resolution, etc.). One central requirement will continue to be the availability of a powerful imaging system including digitizer, image processor, image analysis programs, image display capabilities, and image archiving. It is predicted that multiple types of detectors will be required to perform the myriad of experiments possible in the future.

## REFERENCES

Arndt-Jovin, D. J., Robert-Nicoud, M., Kaufman, S. J., and Jovin, T. M. (1985). *Science* **230**, 247–256.
Bright, G. R., and Taylor, D. L. (1986). *In* "Applications of Fluorescence in the Biomedical Sciences" (D. L. Taylor, F. Lanni., A. S. Waggoner, R. F. Murphy, and R. R. Birge, eds.), pp. 257–288. Liss, New York.
Bright, G. R., Fisher, G. W., Rogowska, J., and Taylor, D. L. (1987). *J. Cell Biol.* **104**, 1019–1033.
Csorba, I. P. (1985). "Image Tubes." W. W. Sam and Co.
Electro-Optics Handbook (1974). RCA Co., Lancaster, Pennsylvania.
Foskett, J. K. (1985). *Am. J. Physiol.* **249**, C56–C62.
Giloh, H., and Sedat, J. W. (1982). *Science* **217**, 1252–1255.
Inoue, S. (1986). "Video Microscopy." Plenum, New York.
Plant, A. L., Benson, D. M., and Smith, L. C. (1985). *J. Cell Biol.* **100**, 1295–1308.
Reynolds, G. T., and Taylor, D. L. (1980). *Bioscience* **30**, 586–592.
Spring, K. R., and Smith, P. D. (1987). *J. Microsc.* **147**, 265–278.
Timothy, J. G. (1983). *Publ. Astron. Soc. Pac.* **95**, 810–834.
Tsuchiya, Y., Inuzuka, E., Kurono, T., and Hosada, M. (1985). *Adv. Electron. Electron Phys.* **64**, 21–31.
Wampler, J. E. (1986). *In* "Applications of Fluorescence in the Biomedical Sciences" (D. L. Taylor, F. Lanni, A. S. Waggoner, R. F. Murphy, and R. R. Birge, eds.), pp. 301–319. Liss, New York.
Wick, R. A. (1985). *Photon. Spectra* **May,** 133–136.
Williams, D. A., Fogarty, K. E., Tsien, R. Y., and Fay, F. S. (1985). *Nature (London)* **318**, 558–561.

# Chapter 16

## Solid-State Imagers for Microscopy

### R. S. AIKENS

*Photometrics Ltd.*
*Tucson, Arizona 85745*


### D. A. AGARD AND J. W. SEDAT

*Department of Biochemistry & Biophysics and*
*Howard Hughes Medical Institute*
*San Francisco, California 94143*

METHODS IN CELL BIOLOGY, VOL. 29

# I.  The History of Solid-State Imagers

Silicon solid-state devices were introduced into the electronics industry early in 1960. The properties of silicon, when exposed to electromagnetic radiation in the visible spectrum, were predicted, but not exploited in the early phase of device development. Silicon was first used to make discrete devices, transistors and diodes which, because of their high cost, were used principally in research, aerospace, and industrial applications. Light-sensitive phototransistors and photodiodes were developed, but were considered an interesting by-product of the electronics industry during that period.

Small-scale integration allowed designers to assemble many transistors into basic logic elements, a development which had major implications in the digital electronics industry. The power consumption, size, and cost of computers were dramatically reduced through the use of integrated circuits. Large-scale integrated circuits (LSI) and very large integrated circuits (VLSI) followed, resulting in digital systems which were used to control everything from giant industrial installations to sewing machines. Costs dropped to the point where a hand-held calculator could be manufactured for less than a dollar.

During this amazing evolution of silicon-based electronic devices, advances in photoelectronics were made, but at a much slower pace. Light-sensitive silicon sensors were integrated into linear arrays and, finally, into two-dimensional area arrays. These devices were called solid-state imagers, focal plane arrays, or, simply, imagers. A few companies began to market solid-state imagers for research and industrial applications. These devices were expensive, and their performance was marginal. Most of the device research in focal plane arrays was carried on by a few people, primarily in government-funded institutions.

Practical and affordable solid-state sensors and cameras appeared late in the 1970s. In the mid-1980s, the interest in silicon-based solid-state imagers rose dramatically, due to the potential consumer markets for television cameras. However, devices designed specifically for use in television cameras, while inexpensive, were not of the quality required for quantitative imaging and radiometric measurements.

The need for higher quality imaging sensors was recognized by several companies, and imaging devices specifically designed for use in nontelevision environments were produced. The photometric acuity, sensitivity, and resolution of these new imagers exceeded that previously seen. Special slow scan, high-performance cameras were designed in order to take advantage of the superior performance these imagers could offer. These cameras were made possible by bringing together several diverse technologies involving high-performance signal processors, solid-state coolers, precision digitizers, and high speed digital controllers.

Several types of silicon array imaging devices were introduced over the years during the development of the solid-state electronics industry. The charge injection device (CID), photodiode array (PDA), and the charge coupled device (CCD) were three generic devices which proved most successful and are in use today. The CCD has received the most attention in recent times because of its superior detection capability and performance characteristics and is now the imager of choice for use in quantitative imaging systems. The discussion which follows covers the theory of CCD operation, CCD performance characteristics, and the utility of CCD technology in optical microscopy.

## II. Theory of CCD Imagers

### A. Properties of Silicon

Because most CCD imagers are made from silicon, it is useful to examine the properties of this element and the way in which it interacts with light. Silicon can be prepared in large boules to very high purity in its crystalline form. Figure 1 presents a two-dimensional view of a three-dimensional silicon crystal lattice to illustrate several effects fundamental to the understanding of imager performance.

In the crystalline form, each atom of silicon is covalently bonded to its neighbor. Energy greater than the gap energy, about 1.1 V, is required to break a bond and create an electron hole pair. Incident electromagnetic radiation in the form of photons (quanta) of wavelength shorter than 1 $\mu$m can break the bonds and generate electron hole pairs. Under constant illumination, an equilibrium is set up between the rates of incoming flux and charge recombination.

Figure 1 illustrates how bonds can be broken by incident photons. The wavelength of incoming light and the photon absorption depth are directly

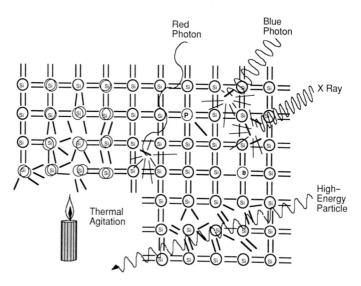

FIG. 1. Silicon lattice in various states of excitation.

related; the shorter the wavelength, the shorter the penetration depth into the silicon. Silicon becomes transparent at approximately 1100 nm and is essentially opaque to light at wavelengths shorter than 400 nm.

High-energy particles, X-rays, and cosmic rays can break many thousands of bonds; excessive exposure can cause damage to the crystal lattice. Bonds can also be broken by thermal agitation. At room temperature, approximately 50 bonds/second/$\mu m^3$ are broken and recombine on a continuous basis. The rate of electron hole pair generation due to thermal energy is highly temperature dependent and can be reduced arbitrarily through cooling.

## B. Potential Well Concept

In order to measure the electronic charge produced by incident photons, it is necessary to provide a means for collecting this charge. Figure 2 illustrates the potential well concept. A thin layer of silicon dioxide is grown on a section of silicon and a conductive gate structure is applied over the oxide. It is possible to apply a positive electrical potential to the gate, thus creating a depletion region where free electrons generated by the incoming photons can be stored. It is important to note that electrons freed by thermal agitation and by high-energy particles are indistinguishable from those generated by photon interaction. Hence, dark electrons can have an adverse effect on the detection limits for photon-induced charge. A poten-

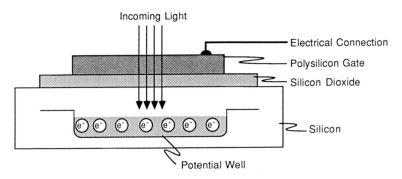

FIG. 2. Simple photodetector.

tial well will continue to collect any available electronic charge until that well is filled.

Practical potential well capacities range upward to a million electrons and depletion depths range from a few micrometers up to tens of micrometers in specially prepared silicon. The potential well illustrated in Fig. 2 is produced using a transparent silicon conductive gate. Because silicon is opaque to photons of short wavelengths, light shorter than 400 nm cannot enter the silicon and produce any charge for collection in the well.

## C. CCD Principles

It is possible to process a single crystal of silicon and to slice it into very thin sheets up to several square centimeters in area. Typically, silicon imaging devices are processed from wafers that are 10 cm in diameter and about 500 $\mu$m in thickness. A matrix arrangement of oxide and gate structures can be fabricated so that many thousands of potential wells are established across a large area of silicon. This gate structure can be arranged with multiple phases so that potential wells can be propagated through a silicon sheet, as illustrated in Fig. 3. In the illustration, the gates are arranged in triads. The application of an appropriate sequence of potentials causes the potential well to propagate from left to right. Any charge which has been collected, no matter what the source, is carried along in the well. The separation of individual charge packets is maintained during this process and no mixing between wells occurs. Charge packets can be transferred hundreds of times without loss of charge.

While Fig. 3 and the explanation given above oversimplify the process, they serve to illustrate the principle of charge transfer, a concept essential to understanding charge-coupled devices. In reality, the charge is not carried along the surface at the silicon – silicon dioxide interface, but is contained in

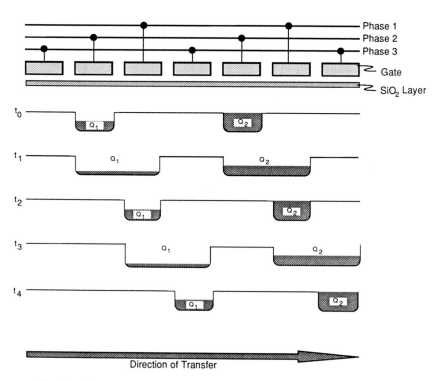

FIG. 3.   Charge transfer in a three-phase CCD using the potential well concept.

a buried channel just below the surface. This buried channel is free of interference from interface states and so assures effective charge transfer. Although the gates are illustrated in a single plane, they actually overlap, thereby ensuring properly shaped drift field that is required for efficient charge transfer.

## D.   Classical CCD Imager Implementations

It is possible to extend the one-dimensional charge concept illustrated in Fig. 3 to two dimensions. Figure 4 shows a stylized two-dimensional CCD imager. The device illustrated is composed of an array of potential wells arranged in columns. The geometry of the buried channel establishes the columns and prohibits charge from migrating between columns. Along the columns, the charge is contained in individual photosites by potentials that are applied to a multiphase gate structure, as discussed in the previous section. By this means, it is possible to support a two-dimensional array of independent potential wells. The device illustrated is square and has

262,144 independent sites, each capable of storing photon-induced electronic charge. In the illustration, each element of the array is square, with 20 $\mu$m on a side. This two-dimensional array of potential wells is called the *parallel register.*

An image that is focused on the parallel register produces a pattern of charge in proportion to the total integrated flux incident on each photosite. The CCD array can integrate and collect charge over a prescribed period of time with the total charge collected at a photosite being proportional to the product of flux level and the integration time.

The *serial register,* shown at the top of Fig. 4, is itself a one-dimensional CCD and plays an important role during CCD readout.

Figure 5 illustrates the CCD readout sequence. A programmed sequence of changing gate potentials causes all charge packets stored in the parallel register to be shifted in parallel one row toward the serial register. The charge stored in the top row is shifted from the parallel register into the serial register. Once in the serial register, charge packets are individually shifted toward the output amplifier. The output amplifier produces a measurable signal that is proportional to the quantity of the charge in each charge packet. The analogy of CCD action to that of a bucket brigade has often been used to explain CCD operation (Kristian and Blouke, 1982; Janesick and Blouke, 1987).

After the serial register is emptied of charge, a second row of charge packets is shifted into that register from the parallel register for transport to

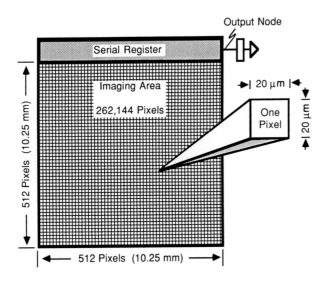

FIG. 4. Typical 512 × 512 CCD.

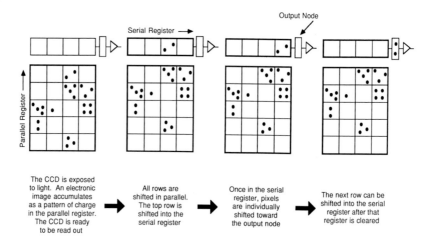

FIG. 5.    CCD readout.

the output amplifier. As charge is shifted out of the parallel register, empty rows are left at the bottom of that register until it is completely emptied of charge. A new exposure can then be made and the process can be repeated.

All CCDs depend on the efficient transport of charge from the photosite to the on-chip output amplifier. Because the charge from wells located far from the output amplifier must undergo many hundreds of transfers, the charge transfer efficiency (CTE) is of concern. A scientific-grade CCD exhibits a CTE of 0.99999, where 1.0 is perfect. CTE is of special concern at low charge levels where a small loss of charge can cause significant degradation of the resulting image.

## E.    CCD Architectures

Three types of CCDs are used for quantitative electronic imaging. Figure 6 illustrates basic CCD structures that are in current use.

The *full frame* CCD has the most simple architecture. This classical CCD has a single parallel register used for photon collection, charge integration, and charge transport. Incident radiation must be blocked during the read-out process in order to prevent "smearing" of the image. An electronic shutter is typically used to block incident radiation.

The *frame transfer* CCD has a parallel register that is really composed of two CCDs that are arranged in tandem. The CCD register that is next to the serial register, the storage array, is covered with an opaque mask and provides temporary storage for collected charge for readout. The other CCD register, the image array, is identical in size to the storage array and is used

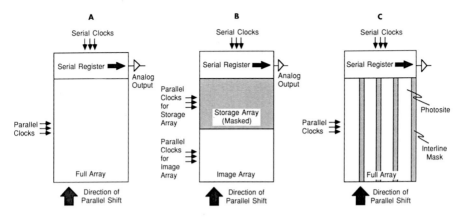

FIG. 6. CCD architectures. (A) Full frame CCD, (B) frame transfer CCD, (C) interline transfer CCD.

for imaging. After the image array is exposed to light, the entire resulting electronic image is rapidly shifted in the parallel register from the image array to the storage array for readout. This shift typically takes one millisecond. While the masked storage array is read, the image array may integrate charge for the next image. A frame transfer CCD can operate continuously without a shutter at television frame rates. If the mask on the storage array is removed, the entire imager can be used with a shutter as a full frame imager.

The *interline transfer* CCD has a parallel register that has been subdivided so that the opaque storage register fits between columns of pixels. The electronic image accumulates in the exposed area of the parallel register, as it does in the frame transfer CCD. At readout, the entire image is shifted under the interline mask. The CCD shift register lies under the interline masks. Readout proceeds in a fashion similar to the full frame CCD. Interline transfer CCDs exhibit poor sensitivity to photons, because a large portion of each photosite is covered by the opaque strip mask. Also, the mask acts like a venetian blind, which leads to loss of important image information.

## F. Thinned CCDs

Light normally enters the CCD through gates of the parallel register. These gates are made of very thin polysilicon, which is reasonably transparent at long wavelengths, but becomes opaque at wavelengths shorter than 400 nm. Thus, at short wavelengths, gate structure attenuates incoming light.

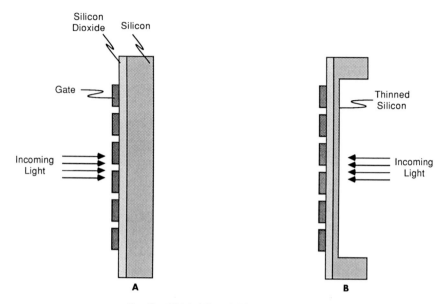

FIG. 7.    Thick (A) and thinned (B) CCDs.

It is possible, using acid etching techniques, to uniformly thin a CCD to a thickness of approximately 10 $\mu$m and focus an image on the backside of the CCD register, where there is no gate structure. Thinned CCDs exhibit high sensitivity to light from the soft X-ray to the near-infrared regions of the spectrum. Figure 7 illustrates the structures of thick and thinned CCDs. Thinning is applicable only to full frame and frame transfer devices.

## III.    CCD Camera Implementations

### A.    CCD Camera Applications

CCD camera implementations have appeared in many forms, each driven by the application. In order to achieve the highest possible sensitivity, astronomers use cryogens to cool the CCD, thus eliminating the unwanted charge normally produced by thermal generation at room temperature. This unwanted charge is called *dark current,* because it is produced in the absence of light. On the other hand, high energy physicists have employed CCDs in ultrahigh speed cameras to observe transient phenomena where dark current is not relevant. The demands of conventional video applications are far less stringent and low-cost CCD television cameras are available from a dozen or more manufacturers. A myriad of CCD camera systems which emphasize a particular application have been developed, but in the research environment, two types of electronic imaging systems are in current use.

## B. Video Cameras

Videography, using conventional television system components for electronic imaging, is employed in countless research and industrial applications with acceptable, but often compromised, results.

Video cameras are often used because they are inexpensive, readily available in off-the-shelf components, portable, and have reasonable frame rates. A large amount of effort has gone into adapting video cameras to solve imaging problems in microscopy. The utility of intensified vidicons and CCDs operated at video rates has been demonstrated, and for some applications these systems are still the best choice (see Spring and Lowy, Chapter 15, this volume). Video frame digitizers, frame grabbers, and image processing systems are all available for the user who wishes to implement a traditional electronic imaging system. Compared to a slow-scan cooled CCD camera, video cameras have a low intrascene dynamic range, high noise level, and poor long wavelength sensitivity. They have little flexibility in readout modes and are not photometric.

Video cameras provided with CCD imagers do not take full advantage of the capabilities of the CCD. Conventional video systems operate at 60 fields/second, then are interlaced to produce images at 30 frames/second. Because video systems are designed to comply with television formats and bandwidths, severe limits are placed on the performance of the CCD imager.

## C. Slow Scan Cameras

The CCD is a multielement optical detector with high dynamic range and excellent linearity over four orders of magnitude. In order to take advantage of the high performance a CCD has to offer, special slow scan cameras have been designed to operate at a significantly lower speed than conventional video cameras. Slow scan cameras are sometimes called "still imaging cameras" because, unlike video cameras which operate continuously at 30 frames/second, the slow scan camera takes electronic snapshot images. Slow scan cameras support a variety of CCDs and operating modes.

The electronic photograph, or *frame,* readout from the CCD in the slow scan camera can be digitized and stored in a computer memory. CCD readout is slower than that in a video camera and may take from one decisecond up to several seconds, as each pixel is digitized with up to 14 or even 16 bit precision. The benefits of slow scan readout are ultralow noise, maximum CCD performance, and photometric precision in the image data. Cooling the CCD to reduce dark current to negligible levels allows exposure times of seconds or even hours.

Slow scan cameras are provided with a shutter or optical gate. A conventional shutter can be used to acquire exposures as brief as a few milliseconds or as long as an hour; a microchannel plate image intensifier can be used to gate exposures of a few nanoseconds.

Because slow scan cameras do not have to conform to the restrictions imposed by television standards, many operational modes are available to the user. Slow scan cameras can be operated with upgraded lower bandwidth electronics to allow ultralow noise performance with dynamic range as high as 50,000 : 1. Contemporary slow scan cameras exhibit noise levels as low as 10 electrons. The CCD can be used as a full frame imager or can be programmed to read out subarrays anywhere within the CCD area. Resolution is extended to millions of pixels instead of the 300,000 available in the best video systems.

Slow scan camera systems produce large quantities of data. A 2000 × 2000 pixel CCD with a dynamic range of 20,000 : 1 requires 8 megabytes of storage for each image. The user of modern electronic imaging systems must decide which image data are significant and, through data reduction algorithms, store only those data that have value or else be overwhelmed with mountains of information.

The disadvantages of slow scan cameras are few. Because slow scan cameras do not exhibit high time resolution, it is not possible to obtain a high-speed sequence of images. The precision readout electronics do not allow frame rates of more than four or five frames/second. Therefore, practical focusing can best be performed using a video-rate low light level camera adjusted to parfocality with the cooled CCD. Also, the cost of a slow scan CCD camera can be high. The need for a cooling system, precision digitizers, and high-performance electronics places the price for this class of instrument significantly higher than conventional video equipment. The price performance tradeoff of slow scan versus video cameras must be evaluated in the context of each application. A variety of high-performance slow scan CCD cameras suitable for quantitative microscopy are available from several manufacturers.

## IV.  CCD Performance

### A.  Resolution

The resolution of a CCD is determined by the geometry of the specific device being used. A rectangular resolution element, often called a picture element or *pixel,* varies in size from a few micrometers up to 30 $\mu$m. Pixels are always rectangular in shape. Nonsquare pixel geometries lead to difficulties in data reduction due to asymmetrical spatial image sampling. In full frame imagers, there is no dead space between pixels. Charge generated by photons striking the CCD between pixels diffuses to the nearest potential well. Because some charge packets in the parallel register must undergo thousands of transfers, *charge transfer efficiency* (CTE) plays a role in resolution. Scientific-grade imagers can exhibit a CTE as high as 0.999998, and resolution is not degraded when a CCD is correctly adjusted.

CCD resolution often is expressed as the modulation strength in an output signal that is produced by a black and white bar pattern imaged on the device, and is called the *modulation transfer function* or MTF. The bar pattern used to measure MTF has a sinusoidal contrast function and varies in spatial frequency. It should be noted that imaging detail that has a spatial frequency close to CCD pixel sampling spatial frequency gives rise to beat frequencies within the image. The highest spatial frequency appearing in an image from a microscope is set by the diffraction limit of the microscope optics. The CCD spatial sampling frequency must be at least twice the highest spatial frequency appearing in the microscope image in order to avoid aliasing, moiré, or beat frequencies. A scientific-grade CCD with square discrete photosites and a fixed geometry exhibits nearly ideal behavior in terms of sampling theory.

Often, CCD resolution is made to comply with standard television formats that are not optimized for quantitative imaging applications. Recently, nontelevision devices with square formats and resolution consistent with demands of optical microscopy have become available. Specifically, CCDs with formats of $512 \times 512$ pixels and $1024 \times 1024$ pixels have appeared on the market. Pixel size, combined with the total number of pixels available, determines the overall physical size of a CCD. In general, CCDs have imaging areas of $1-2$ cm$^2$. Unlike electron beam-scanned image sensors, solid-state imagers do not exhibit any geometrical distortion. This is because the individual pixels are precisely located by the structure of the device.

## B. Linearity

Light incident on the CCD creates electronic charge. This is an intrinsically linear process, and charge so produced can be detected and measured at the output amplifier. With suitable electronics, the output signal can be amplified, processed, and digitized to yield a digital representation of the image which is focused on the CCD. Figure 8 shows the ideal relationship between incoming radiation and output signal for an imaging system. Ideally, there should be a linear relationship between the light level and the

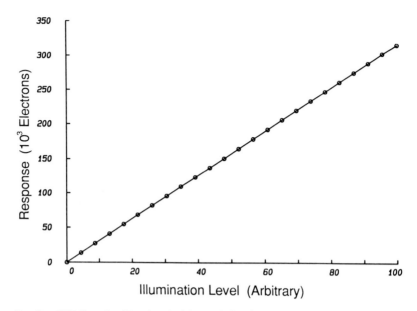

Fig. 8. CCD linearity. [Reprinted with permission from Epperson, P. M., Sweedler, J. V., Denton, M. B., and Sims, G. R. Electro-optical characterization of the Tektronix TK512M-011 charge-coupled device. *Opt. Eng.* **26**, (1987).]

digital number representing that level. Scientific-grade CCDs exhibit a linear signal transfer relationship within a few hundredths of a percent. Because CCDs are used in many different types of cameras, the camera design often determines the overall linearity of an imaging system.

Video cameras do not usually have good linearity, because high-speed electronics often do not exhibit linear behavior. The video camera is therefore considered to be a qualitative imaging device. Sometimes video cameras are purposefully made nonlinear to match a photographic response or to optimize the video display due to nonlinearities in viewing apparatus.

High-performance cameras have been designed to exploit the excellent linearity characteristic of the CCD. These cameras employ precision analog electronics and digitizers with high dynamic range in order not to degrade the intrinsic linearity of the CCD. This high performance is achieved by cooling the CCD and operating the readout electronics at significantly slower rates than those used in a video camera.

Linearity is assumed when taking ratios between images, doing shading corrections, and using linear transforms to do image analysis. Nonlinearity in an imaging system introduces very serious errors into image processing results and can lead to erroneous interpretation of data.

## C. Signal-to-Noise Considerations

Noise is composed of undesirable signal components that arise from various sources within an electronic imaging system. The figure of merit that dictates the ultimate performance of a system is the signal-to-noise ratio (SNR).

Preamplifier noise, dark current, and photon noise are the three primary CCD noise sources. Figure 9 is a graph depicting the square root relationship between the signal and photon shot noise. The dotted horizontal line is the preamplifier noise floor.

*Photon noise,* or *photon shot noise,* is a fundamental property of the quantum nature of light. The total number of photons emitted by a "steady" source over any time interval varies according to a Poisson distribution. Likewise, the photoelectrons collected by a CCD exhibit the same Poisson distribution. This distribution dictates a square root relationship between signal and noise. Simply stated, the noise in electron units is equal to the square root of the signal. Photon shot noise is unavoidable and is always present in imaging systems. When the level of photon shot noise exceeds system noise, the CCD data are said to be photon shot noise limited.

*Preamplifier noise* sets the detection limits of a CCD. This noise is generated by the on-chip output amplifier and can be reduced to a few electrons with careful choice of operating conditions. Under low light conditions, where preamplifier noise exceeds photon shot noise, the CCD data are said to be preamplifier noise limited. Once the signal level is high

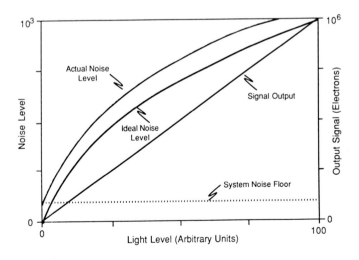

FIG. 9. Solid-state detector performance.

enough so that the photon statistics are the dominant source of noise, preamplifier noise is not relevant.

In a given situation, the available light level determines the integration time required to arrive at an acceptable SNR. Acceptable SNRs vary with the situation; the tradeoffs between light level and integration time must be considered for each circumstance.

Two examples give some insight into the tradeoffs required to maximize the SNR. Because the electron is the basic signal unit of a CCD, noise and signal quantities are expressed in electron units.

Solar astronomy is a typical high light level CCD application. For this application, it is important to detect small fluctuations in intensity over the area of the sun as a function of time. Because the light source is very bright, a slow scan CCD camera always operated under photon shot noise limited conditions. Assuming that the well capacity of the CCD is 500,000 electrons and that the wells are filled, the photon shot noise is 707 electrons. This means that the maximum SNR for solar images is 707:1. Under these conditions, a solar intensity fluctuation of 0.28% can be detected with 95% confidence.

Low light level conditions, such as those encountered in fluorescence microscopy, present different problems. For this application, the photon flux is typically low. Furthermore, it is desirable to keep the sample exposure-to-excitation ratio low to avoid bleaching. Under these conditions, CCD sensitivity and readout noise are extremely important. Assuming that the CCD has a readout noise of 10 electrons, the image data are preamplifier noise limited when the number of photoelectrons in a pixel is less than 100. For signals above 100 electrons, the data are again shot noise limited.

Equation (1) gives the total noise, NT, for a CCD as a function of preamplifier noise, NP, and the number of electrons, NE, collected in a potential well at a pixel.

$$NT = \sqrt{NP^2 + NE} \tag{1}$$

Equation (2) gives the SNR for a CCD, expressed in terms of incoming light flux, $I$, photons/second per pixel. In Equation (2), the quantum efficiency, QE, and the integration time, $T$, have been included

$$SNR = \frac{I \cdot QE \cdot T}{\sqrt{I \cdot QE \cdot T + (NP)^2}} \tag{2}$$

## D.  Dark Current Noise

The presence of dark current is an additional concern in low light level applications. It is important to ensure that photon shot noise from dark current does not exceed preamplifier or photon shot noise from the signal even when long integration times are used. CCDs can be chilled to approximately $-60°C$ by use of thermoelectric cooling. At this temperature, dark current is typically reduced to 1 electron/pixel/second. If the preamplifier noise is 10 electrons, then an integrated dark charge of 100 electrons adversely affects performance due to the dark current photon shot noise component. Clearly, exposures lasting over 100 seconds are not preamplifier noise limited, but, rather, are dark current noise limited. For extremely long exposures, liquid nitrogen is used to cool the CCD to $-120°C$. At lower temperatures, CCD performance may be degraded due to poor CTE.

## E. Sensitivity

Sensitivity is a measure of the minimum detectable signal a CCD can produce. The meaning of minimum detectable signal is somewhat arbitrary, and there is no accepted standard defining sensitivity. Of more significance is the signal-to-noise ratio which was presented in Section IV,C.

The sensitivity of a CCD imager to light is determined by two factors. The first is the quantum efficiency, which is a measure of the sensor's efficiency to generate electronic charge from incident photons. As discussed in Section II,A, electron hole pairs are produced by photons in the region from 200 to 1000 nm. Within the visible spectrum, the photon to electron conversion factor is less than unity and it varies as a function of wavelength. Figure 10 gives the quantum efficiency for both thick and thinned CCDs, clearly demonstrating the higher efficiency of the thinned device. In general, CCDs with high quantum efficiency have superior sensitivity.

The second factor determining the sensitivity is system noise level. At high light levels, system noise is overshadowed by photon statistics, and sensitivity is determined by the device quantum efficiency. At low light levels, preamplifier noise begins to dominate and sensitivity is degraded to the point where signals get lost in the preamplifier noise. In both cases, maximum sensitivity is achieved when the quantum efficiency is maximized. For low light level applications, it is necessary to reduce system noise to the lowest possible level. Noise levels of 10 electrons are attainable in high-quality CCD cameras. The relationship between noise and preamplifier noise is discussed in Section IV,C.

If the quantum efficiency at a specific wavelength is known, then the SNR for a given photon flux can easily by computed. There are many confusing definitions of sensitivity in as many different units as there are definitions. The reduction of light level to photons/second/pixel allows simple SNR calculations in fundamental units which are universally understood.

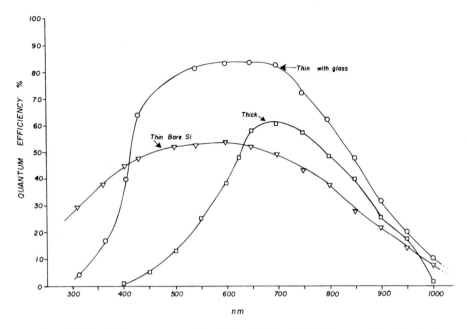

FIG. 10. Spectral response curves for thick and thinned CCD detectors. [Reprinted with permission from Denton, M. B., Lewis, H. A., and Sims, G. R. *In* "Multichannel Image Detectors" (Y. Talmi, ed.), Vol. II, pp. 133–154. Copyright (1983) American Chemical Society.]

## F.  Dynamic Range

The dynamic range of an imager is simply defined as the ratio of the device saturation charge to the system noise level, expressed in electrons. System noise level may vary between a few electrons in a high-performance slow scan cooled camera, up to many thousands of electrons in an inexpensive video camera. Saturation charge is limited by the well capacity of the CCD and varies with device architecture and pixel size. Small pixels have small potential wells, and, hence, low saturation charge. The saturation charge level scales more or less linearly with pixel area and is completely independent of system noise level. A rough rule which applies is that the saturation charge in electrons is about 1000 times the pixel area in square micrometers.

An imager with a $20 \times 20$ $\mu$m pixel has a 400 $\mu$m$^2$ area and a saturation level of about 400,000 electrons. The same imager may have a noise level from 10 electrons up to 10,000 electrons, depending on the camera configuration used. If the noise level is in fact 10 electrons, then the dynamic range is the ratio of 400,000 : 10, or 40,000 : 1. If can be seen by this example that system noise level often is the limiting factor in systems with low dynamic range.

## V. Quantitative Imaging with CCDs

## A. Shading Corrections

When striving for the highest performance, careful calibrations on bias levels and shading corrections must be made. A bias frame, taken in the absence of light, represents the system DC level plus any small structure exhibited by the CCD in the dark. Usually, a bias frame is stored and subtracted from an image to yield a zero reference.

A shading correction frame, or *flat field,* is required to make corrections for variations in system responsivity. Because the CCD has excellent linearity, only one flat field must be taken to correct an image anywhere within the available dynamic range at a specific wavelength. Shading varies with wavelength, and flat fields are often obtained at many wavelengths. A calibration sequence requires taking the image of interest, $I_R$, a bias frame, $I_B$, and a flat field, $I_F$. An arithmetic pixel-by-pixel computation yields a corrected image with photometric integrity, $I_C$. Equation (3) is the algebraic expression of the correction algorithm required to make a shading correction. Figure 11 illustrates this principle.

$$I_C = (I_R - I_B)/(I_F - I_B) \qquad (3)$$

After the correction is made, each pixel may be considered to be an independent linear photometer. This correction must be done with great care if the image is to be used for quantitative photometry. In microscopy applications, a shading correction must include variations due to the detector responsivity, variations in the illumination pattern from the light source, spatial attenuation due to dust particles, and vignetting due to optics.

A fluorescent shading correction is more difficult to obtain than a normal, white-light correction. A uniformly fluorescent slide may be used to provide the light for the shading correction. However, the light from the shading correction slide must pass through the optics in exactly the same way as the light from the specimen to be corrected. For this to occur, light for the shading correction image must come from exactly the same plane as light from the specimen. The microscope's focus cannot be changed to achieve that condition because the light path would change, possibly introducing spurious shading. The optimal solution to this problem is to prepare

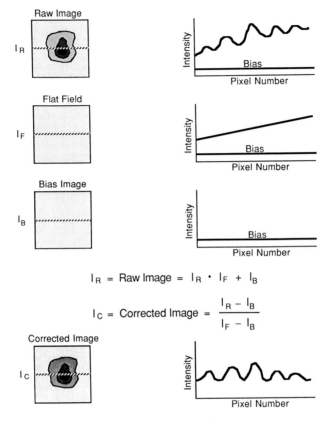

$$I_R = \text{Raw Image} = I_R \cdot I_F + I_B$$

$$I_C = \text{Corrected Image} = \frac{I_R - I_B}{I_F - I_B}$$

FIG. 11.   Image correction.

standard slides of the same pathlength as the specimen under investigation. Improperly applied shading corrections result in serious degradation of the final processed image. For an in-depth discussion of an approach for accurate decalibration of fluorescent microscope CCD images, see Hiraoka *et al.* (1987).

## B.   Exposure Times

Each application dictates the acceptable SNR. The SNR increases with total integrated light and, thus, ultimately determines the required exposure time. In video camera systems, frames can be summed at 30 frames/second in an image processor to increase the SNR, while in slow scan cameras the exposures are made directly on the CCD through the use of a shutter. In general, slow scan cameras outperform video cameras when exposure times are greater than tenth of a second.

In the observation of high-speed phenomena, the exposure time is dictated by the required time resolution, and the user has to accept the resulting SNR. The quality of images taken with a quantitative imaging system is always limited by the SNR. Every means should be employed to use available photons as effectively as possible when making a measurement.

## C. Chromatic Aberration

Microscope manufacturers go to great lengths to make their optics as efficient as possible, while preserving image quality. Most solid-state imagers respond from the ultraviolet to the near-infrared. Care must be taken to be sure that images are correctly focused. This is especially important for wavelengths which are outside the normal range of a microscope that has been color-corrected. New red probes which emit in the 700- to 900-nm region present new problems, since their emission is beyond the range of visibility. CCDs perform at their best with these near-infrared sources, but focusing must be done totally by electronic imaging methods, because the light produced by these sources is invisible to the human eye (Waggoner *et al.*, Volume 30, this series).

## D. Mechanical Interface Considerations

Most microscopes have one or more photographic or video ports on which to attach cameras. Solid-state video cameras have a universal C mount, an industry-standard, threaded aperture through which light passes. Because video cameras operate at 30 frames/second, there is no need for a shutter. The image may be viewed continuously. Video camera adapters are available, and the mechanical interface is straightforward.

Slow-scan solid-state cameras usually have a larger input port and are equipped with a 35-mm camera lens mounts. Again, because many different camera adapters are available, the mechanical interface is simple. A shutter is usually provided between the output of the microscope and the camera input window. Because slow-scan solid-state cameras employ a cooled sensor, the camera head is larger and heavier than a video camera.

It is desirable to install a second shutter between the excitation source and the excitation input on the microscope so that the excitation can be computer-controlled. An exposure of a fluorescent specimen may be controlled by the length of time the excitation source is opened.

## E.   Optical Matching Considerations

The CCD in a solid-state camera has equally spaced pixels, which set the limit of resolution. It is necessary to match the CCD resolution to the required spatial resolution in the image plane of the microscope. The magnification must be chosen so that at least two CCD pixels cover a desired resolution element in the image plane.

# VI.   Future Trends in Solid-State Detector Development

## A.   Television-Related Advances

Most of the progress made over the last 20 years can be attributed to the needs of television. The devices that were developed for television conformed to very strict format and size specifications and they were optimized for operation at video rates and broadcast bandwidths.

New, high definition television-like systems are now in the design stages, pushing CCD imaging technology to new heights, especially in resolution. These new imagers will exhibit movie-like quality with a million or more pixels. The cost of these devices, while high initially, will probably drop to a few hundred dollars as volume increases. The analogy of solid-state memory is applicable in this case. The price performance ratio for memory devices has dropped several orders of magnitude in the last 10 years. The utility of these devices for scientific imaging will be limited, due to their relatively poor performance.

## B.   Research- and Industry-Related Advances

The need for scientific-grade CCDs has been demonstrated, and device manufacturers are responding to that need. CCDs are in production with resolution in excess of 1 million pixels and with excellent photometric properties. CCDs with square formats, as opposed to the rectangular format of television devices, are not available in a $512 \times 512$ configuration. Kodak recently introduced a $1300 \times 1000$ pixel CCD which is well matched to microscopy, due to its small pixel size. Other manufacturers have in development CCDs with over a million pixels which are to appear in 1988. CCD technology has evolved to the point where little improvement can be made in device performance. Any further advances will be in the area of increased format size, higher resolution, and lower cost.

# VII.  Summary

Solid-state imagers offer a variety of options to the modern optical microscopist. Photodiode arrays, charge injection devices, and charge-coupled devices are the three basic types of solid-state imagers available for research imaging. The charge-coupled device (CCD) is the solid-state imager of choice, because of its superior characteristics and its widespread acceptance in the research environment. The performance characteristics of the CCD are well documented and understood, having been quantified by many experimenters, especially in the physical sciences. CCDs exhibit dynamic ranges up to 50,000:1, very high quantum efficiency, and ultra-low noise. The camera system in which a device is used, however, dictates the overall performance which can be achieved. The CCD imaging system as it applies to cell biology is discussed in detail in Hiraoka *et al.* (1987).

A video camera operating at 30 frames/second does not provide the resolution, low noise, dynamic range, and linearity of a slow scan, cooled camera operating at 1 frame/second. Conversely, a slow scan camera does not offer the user the facility to resolve rapidly changing events in real time. The analogy between an 8-mm movie camera (video camera) and a 35-mm snapshot camera (slow scan cooled camera) is a useful one to emphasize the different character of these two electronic imaging systems.

These two basic camera systems both employ CCD imagers, and each has very different, but complementary, characteristics. No one camera system can address the wide variety of imaging problems which face the modern microscopist. The user of this new generation of instrumentation must decide which system best fits the problem at hand.

## REFERENCES

Denton, M. B., Lewis, H. A., and Sims, G. R. (1983). *In* "Multichannel Image Detectors" (Y. Talmi, ed.), Vol. II, pp. 133–154. American Chemical Society, Washington, D.C.
Epperson, P. M., Sweedler, J. V., Denton, M. B., and Sims, G. R. *Opt. Eng.* **26,** 715–724.
Hiraoka, Y., Sedat, J. W, and Agard, D. A. (1987). *Science* **238,** 36–41.
Janesick, J., and Blouke, M. (1987). *Sky Telescope* **74,** 238–242.
Kristian, J., and Blouke, M. (1982). *Sci. Am.* **247,** 67–74.

# Index

# CONTENTS OF RECENT VOLUMES

# Volume 25

*The Cytoskeleton*

Part B.    *Biological Systems and* in Vitro *Models*

# Volume 26

*Prenatal Diagnosis: Cell Biological Approaches*

## Volume 27

*Echinoderm Gametes and Embryos*

## Volume 28

Dictyostelium Discoideum: Molecular Approaches
to Cell Biology